Terrorism, Security, and Computation

Series Editor
V.S. Subrahmanian

More information about this series at http://www.springer.com/series/11955

Chemical, Biological, Radiological, Nuclear and explosive
Department of Industrial Engineering and School of Medicine and Surgery

Maurizio Martellini • Andrea Malizia
Editors

Cyber and Chemical, Biological, Radiological, Nuclear, Explosives Challenges

Threats and Counter Efforts

 Springer

Editors
Maurizio Martellini
University of Insubria and Landau Network
 Fondazione Volta
Como, Italy

Andrea Malizia
Department of Biomedicine and Prevention
University of Rome Tor Vergata
Rome, Italy

ISSN 2197-8778   ISSN 2197-8786 (electronic)
Terrorism, Security, and Computation
ISBN 978-3-319-87236-0   ISBN 978-3-319-62108-1   (eBook)
DOI 10.1007/978-3-319-62108-1

Printed on acid-free paper

This Springer imprint is published by Springer Nature
The registered company is Springer International Publishing AG
The registered company address is: Gewerbestrasse 11, 6330 Cham, Switzerland

Contents

Editorial Board

Introduction

The international security landscape is under stress, because of a worldwide new security concept and of the increasing threats of non-state actors, including terrorist groups, as well as of the difficulty to design long-term counter measures and security initiatives. In particular, conventional terrorist attacks could increase the general instability, but we cannot exclude nonconventional, asymmetric, hybrid attacks by non-state actors or states through proxy actors. Among the nonconventional attacks, the governmental agencies, think tanks, and academies should consider the persisting proliferation of chemical, biological, radiological, nuclear, and explosive (CBRNe) assets and the related cyber (Cy) systems involved. It is essential to analyze the evolution of the threats in order to enforce the safety, security, and CBRNeCy risk management. In general, to achieve this goal, a multidisciplinary approach is needed, a multilayer strategy is demanded, and different users should be involved, spanning from the academy to the NGOs/think tanks and to the governmental agencies.

Not only could the CBRNeCy threats directly impact on several critical infrastructures, but they have a wider impact; therefore, a large spectrum of challenges should be considered. These are related to global security issues, like the reduction of fossil energy resources, the massive exploitation of potable water resources, and, in general, catastrophic events related to climate change. The control of energy and water resources might be pursued, in an asymmetric hybrid warfare scenario, through CBRNeCy events. On the other side, major environmental destructive events might be triggered by criminal or unintentional actions such as the Bhopal chemical accident. Moreover, major nuclear/radiological events, like the Fukushima Daiichi one, that are the consequence of a tsunami or an earthquake, can also be the result of a deliberate attack against the safety and security systems of a nuclear power plant. From the academic point of view, the risk management of these major CBRNeCy events, considering their low probability and their high destructive potential, falls under the definition of "black swan" events that require a further boost in the preparedness, prevention, mitigation, and response phases, with respect to conventional events.

An additional CBRNeCy threat is represented by the growing diffusion and availability of scientific knowledge and expertise in this field that represents the "human dimension of proliferation." The mitigation of this risk could be achieved only through "intangible" measures, like education, training, and proliferation awareness raising. In a theoretical social science framework, this contributes to create a "CBRNeCy taboo," "CBRNeCy norms," and "CBRNeCy codes of conduct."

In every industrialized country there are multiple entities (governmental agencies, ministries, universities, think tanks, NGOs, etc.) with specialized teams in very specific fields, but the complexity of CBRNeCy events requires professionals that not only have specific know-how but are also able to look at this phenomenon with a comprehensive approach. Furthermore, an enhanced coordination among these entities is paramount.

This monograph will deal extensively with the security and safety of CBRNeCy assets and management, as well as with the strengthening of the security and safety culture, and will show which risks may emerge and how to face them through an enhanced risk management approach.

This monograph should be the first one tackling the CBRNeCy threats, their risk mitigation measures, and the relevance of raising proliferation awareness and education/training reinforcing CBRNeCy security and safety. It will also present international instruments and legislations/regulations to deal with them as, for instance, UNSC Resolution 1540 of 2004. In general, it should be desirable to transform the CBRNeCy security, in a holistic sense, into a new international mechanism to be placed side by side with the traditional arms control international treaties such as the NPT, BTWC, and CWC.

More importantly, with respect to other "technical manuals," this monograph will address a multitude of stakeholders with different professional backgrounds and will have a multidisciplinary nature as a consequence of the need to consider crosscutting areas like the convergence of biology and chemistry, the development of edging technologies, and, in the cyber domain, the impelling risks of the use of malwares against critical subsystems of CBRNe facilities as, for instance, against the supervisory control and data acquisition (SCADA) subsystems of fertilizer industries or refineries.

In conclusion, facing CBRNeCy threats cannot be achieved only by the aggregation of independent competences. The purpose of this monograph is to introduce a new key concept concerning the holistic and comprehensive approach to CBRNeCy. The editors and the authors, with this monograph, want to demonstrate how an integrated and cooperative scientific and technical CBRNeCy approach can evolve into a new comprehensive discipline.

A Reflection on the Future of the CBRN Security Paradigm

Maurizio Martellini, Tatyana Novossiolova, and Andrea Malizia

Abstract This paper is focused on the concept of CBRN security paradigm and how this concept is affecting the international community on develop and maintain an appropriate effective measures to account for and secure such items in production, use, storage or transport; on the develop and maintain an appropriate physical protection; on the develop and maintain appropriate effective border controls and law enforcement efforts. Basically on the develop and maintain all the actions needed to reduce risks.

Keywords CBRN security paradigm • WMD • CW • BTWC

1 Introduction

For decades, issues concerning the proliferation of weapons of mass destruction (WMD) – biological, chemical, and nuclear – have been largely addressed within the framework of disarmament and arms control. In the second half of the twentieth century, key international multilateral agreements were negotiated that effectively set the legal grounds for the prohibition of entire classes of WMD (biological and

M. Martellini (✉)
University of Insubria and Landau Network Fondazione Volta, Como, Italy
e-mail: maurizio.martellini@uninsubria.it

T. Novossiolova
Landau Network Fondazione Volta, Como, Italy
e-mail: tnovossiolova@gmail.com

A. Malizia
Department of Biomedicine and Prevention, University of Rome Tor Vergata, Rome, Italy
e-mail: malizia@ing.uniroma2.it

© Springer International Publishing AG 2017
M. Martellini, A. Malizia (eds.), *Cyber and Chemical, Biological, Radiological, Nuclear, Explosives Challenges*, Terrorism, Security, and Computation,
DOI 10.1007/978-3-319-62108-1_1

chemical) and the control the spread of others (nuclear). The norms of customary international law have thus been embedded in statute law binding on all States Parties to the respective treaties – the 1968 Non-Proliferation Treaty (NPT), the 1975 Biological and Toxin Weapons Convention (BTWC), and the 1993 (1997) Chemical Weapons Convention (CWC). States are the referent object of all three treaties: it is states that are responsible for both implementing and observing the treaty provisions and that, at the same time are considered the primary source of potential threats.

With the dawn of twenty-first century marked by the tragic events of 9/11 and the subsequent Anthrax Letters Attacks, the limitations of the traditional state-centred lens through which security has been predominantly viewed and assessed have become acutely apparent. Against the backdrop of intense globalisation coupled with rapid scientific and technological advancement, the fragmented realities of post-modernity have given rise to novel security challenges which hardly recognise borders and against which established means of defence often fall short of delivering the intended objectives. Issues such as international terrorism, organised crime, illicit trafficking, and smuggling that previously have been addressed in silos can no longer be dealt with in isolation from other concerns, including the problem of development and proliferation of WMD. Likewise, given the shift in attention to the effects of events rather than their causes, the spectre of possible security risks involving WMD-related knowledge and materials has drastically expanded encompassing natural disasters (e.g. disease outbreaks, physical destruction of nuclear and/or chemical plans as a result of tsunami, hurricane, or other naturally occurring catastrophic event), accidents (e.g. infrastructural failures, laboratory leaks), and deliberate attacks (e.g. terrorist attacks, sabotage).

Within the context of a rapidly evolving security landscape, new strategies and tactics are required, in order to adequately prevent, detect, respond to, and mitigate potential risks. The multifaceted nature of novel security concerns related to WMD knowledge and materials calls for a redefinition of traditional disarmament and arms control approaches, that is designed to enhance their flexibility and adaptability and thus maximise their effectiveness and efficiency. A fundamental element of this process of redefining WMD security is the emergence of a CBRN – chemical, biological, radiological, and nuclear – security paradigm that is underpinned by a comprehensive set of measures, policies, and practices aimed at addressing risks related to CBRN knowledge and materials, regardless of whether the origins of such risks are naturally occurring events, accidents, or acts of deliberate misuse.

2 The CBRN Security Paradigm: Origins and Evolution

By design, the CBRN security paradigm is a relatively recent development which has been steadily evolving over the past two decades. In some respects its origins can be traced back to 2004 when the United Nations Security Council unanimously adopted Resolution 1540 (UNSCR 1540) on *Non-Proliferation of Weapons of Mass*

Destruction – a Resolution that was adopted under Chapter VII of the United Nations' Charter which made it legally binding on all states. Under UNSCR 1540:

all States shall take and enforce effective measures to establish domestic controls to prevent the proliferation of nuclear, chemical, or biological weapons and their means of delivery, including by establishing appropriate controls over related materials and to this end shall:

(a) Develop and maintain appropriate effective measures to account for and secure such items in production, use, storage or transport;
(b) Develop and maintain appropriate effective physical protection measures;
(c) Develop and maintain appropriate effective border controls and law enforcement efforts to detect, deter, prevent and combat, including through international cooperation when necessary, the illicit trafficking and brokering in such items in accordance with their national legal authorities and legislation and consistent with international law;
(d) Establish, develop, review and maintain appropriate effective national export and trans-shipment controls over such items, including appropriate laws and regulations to control export, transit, trans-shipment and re-export and controls on providing funds and services related to such export and trans-shipment such as financing, and transporting that would contribute to proliferation, as well as establishing end-user controls; and establishing and enforcing appropriate criminal or civil penalties for violations of such export control laws and regulations. [1]

The provisions of UNSCR 1540 have been reinforced by subsequent Resolutions, such as UNSCR 1673, UNSCR 1977, and UNSCR 2325. Yet it is worth noting that whilst UNSCR 1540 is the first international legal instrument to address all three classes of WMD and call upon all states to implement relevant measures, its scope remains largely limited to criminal activities carried out by non-state actors. States are bound by the provisions of the UNSCR 1540 to report on steps that they have taken to implement the Resolution. They are also encouraged to prepare on a voluntary basis, National Implementation Action Plans to map out their priorities for implementing the key provisions of the Resolution [2]. The Security Council Committee established pursuant to Resolution 1540 (1540 Committee) and its Group of Experts administers the collection of national reports and other documentation. The 1540 Committee further has a clearinghouse role to facilitate assistance to others for implementation of the Resolution [3].

The CBRN security paradigm has also manifested itself in the context of international multilateral agreements. Whereas the focus of the BTWC, NPT, and the CWC as noted in the previous section has been on disarmament and arms control, including destruction and reduction of existing stockpiles of weapons, in the recent years it has shifted to stakeholder engagement, capacity building, and fostering sustainable systems for oversight at institutional, national, regional, and international level.

One area in which considerable progress has been made is nuclear security. Since 2002, the International Atomic Energy Agency's (IAEA) Board of Governors has been adopting a *Nuclear Security Plan*. The primary objective of the fourth edition of the *Nuclear Security Plan, 2014–2017* is 'to contribute to global efforts to achieve effective security wherever nuclear and other radioactive material is in use, storage and/or transport, and of associated facilities by supporting States, upon request, in their efforts to meet their national responsibilities and international obligations, to reduce risks and to respond appropriately to threats.' [4] To this end, the *Plan* covers seven programme elements, including:

- Information Collation and Assessment;
- External Coordination;
- Supporting the Nuclear Security Framework Globally;
- Coordinated Research Projects;
- Assessment through Self-assessment and/or through Peer Review Missions;
- Human Resources Development;
- Risk Reduction and Security Improvement [5].

In order to enhance nuclear security capacity building, the International Network for Nuclear Security Training and Support Centres (NSSC Network) was set up in 2012 in Vienna by representatives of 30 IAEA Member States. The NSSC Network is a collaborative network of security training and support centres which seeks to:

- 'Promote a high level of nuclear security training and support services as a cornerstone in the development of sustainable national, regional and global nuclear security training and support centres;
- Facilitate cooperation and assistance activities (including technical and scientific), to optimize the use of available resources, and to leverage those resources to meet specific needs.'

The Network comprises three Working Groups. These are: Working Group A on Coordination and Collaboration; Working Group B on Best Practices; and Working Group C on Information Management and other Emerging Issues. Membership is open to all IAEA Member States, observers to the IAEA and other relevant stakeholders involved, or planning to be involved, in the provision of training and/or technical and scientific support in the area of nuclear security [6].

A similar trend has been observed in the area of chemical security. In a Statement delivered on 20 February 2012, the Director General of the Organisation for the Prohibition of Chemical Weapons noted that:

> Despite the existence and progressive strengthening of clear norms against chemical weapons, criminal or terrorist use of either chemical weapons or the use of toxic chemicals as chemical weapons remains a concern. Especially within the contemporary international security environment, there is a real threat of non-state entities acquiring and using dangerous weapons [7].

Besides the risk of non-state actors acquiring chemical weapons, the Statement also recognised that:

> The obligations of the Convention extend through national laws to all citizens of a country including the individual scientist and engineer. Many chemists, academics, scientists, engineers, technicians, however, have little or no exposure during their training and professional life to the ethical norms and regulatory requirements of the CWC. At the same time, advances in the life sciences are creating enormous opportunities. While their potential for benefit is undisputed, these could also be prone to abuse. Education and awareness-raising about the norms and principles enshrined in the CWC are therefore becoming increasingly important.

At the Third Review Conference of the CWC in 2013, States Parties underscored their:

Determination to maintain the Convention's role as a bulwark against chemical weapons; to that end to promote, inter alia, outreach, capacity building, education and public diplomacy; [Emphasis as original; see para.9.15]

and

acknowledged the role of education, outreach and awareness-raising as a relevant activity for the national implementation of the Convention, including awareness among academia and relevant scientific communities of the provisions of the Convention, the domestic laws and regulations relevant to the Convention. Accordingly, the Third Review Conference welcomed the establishment of the SAB temporary working group on education and outreach [8].

The Eight Review Conference of the Biological and Toxin Weapons Convention (BTWC) held in November 2016 when considering the national implementation of the Convention reinforced the language on the value of education and awareness-raising adopted by its predecessor:

13. The Conference notes the value of national implementation measures, as appropriate, in accordance with the constitutional process of each State Party, to: […]

(a) encourage the consideration of development of appropriate arrangements to promote awareness among relevant professionals in the private and public sectors and throughout relevant scientific and administrative activities and;
(b) promote amongst those working in the biological sciences awareness of the obligations of States Parties under the Convention, as well as relevant national legislation and guidelines;
(c) promote the development of training and education programmes for those granted access to biological agents and toxins relevant to the Convention and for those with the knowledge or capacity to modify such agents and toxins;
(d) encourage the promotion of a culture of responsibility amongst relevant national professionals and the voluntary development, adoption and promulgation of codes of conduct. [9]

With regard to Article VII of the BTWC which pertains to the provision of assistance in case of an alleged use of biological and toxin weapons, the Conference further recognised

'capacity building at the national and international levels as the most immediate imperative for enhancing and strengthening the capacity of the States Parties to promptly and effectively detect and respond to the alleged use or threat of use of biological weapons.'

More specifically, the Conference drew attention to the

'the need for States Parties to work nationally, and jointly, as appropriate, to improve, in accordance with their respective circumstances, national laws and regulations, their own disease surveillance and detection capacities for identifying and confirming the cause of outbreaks and cooperating, upon request, to build the capacity of other States Parties. The Conference notes that the International Health Regulations (2005) are important for building capacity to prevent, protect against, control and respond to the international spread of disease; such aims are compatible with the objectives of the Convention' [10].

The concept of CBRN security has been considered within the framework of various international ad-hoc initiatives. A case in point is the Global Partnership against the Spread of Weapons of Mass Destruction which was established during the G8 (currently G7) Summit held in 2002 in Kananaskis, Canada. In 2009 the

Global Partnership Working Group (GPWG) issued a document titled 'Recommendations for a Coordinated Approach in the Field of Global Weapons of Mass Destruction Knowledge Proliferation and Scientist Engagement' which drew attention to the fact that:

> 2. The proliferation of WMD expertise, or any sensitive knowledge in the chemical, biological, radiological, and nuclear (CBRN) areas, remains a serious concern. Preventing the illicit use of such knowledge is one of the most difficult non-proliferation challenges to address, as we are dealing with scientists, engineers and technicians who, in some cases (those doing biological research, for instance), may not consider their expertise and current activities as potentially vulnerable to misuse by others for whom their "proliferation-critical" knowledge could represent a route to developing a WMD capability. They should be made aware that their legitimate work could have dual-use applications and be diverted for malicious purposes.

The document went on to underscore that

> 4. Closer attention is now needed to engaging scientists and raising awareness and responsibility among them, to prevent their knowledge in legitimate scientific disciplines to be diverted for unintended malicious purposes, and to strengthen frameworks within which to prevent the spread of sensitive information and to promote collaborations to advance common non-proliferation objectives.
> [and that]
> 5. Chemical, biological, radiological and nuclear research and applications are receiving growing attention in this perspective. Education and training are becoming increasingly important, notably in areas where the knowledge and expertise are rapidly advancing [11].

The CBRN security paradigm has further been endorsed within the context of collective security, something evident in the activities of the North Atlantic Treaty Organisation (NATO). Starting in 2006, NATO launched its system of Centres of Excellence (COE), international military organisations that train and educate leaders and specialists from NATO Member and Partner Countries [12]. There are 24 NATO COEs covering a wide variety of areas such as civil-military operations, cyber defence, military medicine, energy security, naval mine warfare, defence against terrorism, cold weather operations, and counter-IED. The scope of work of NATO COEs includes but is not limited to:

- Assisting in doctrine development;
- Identification of lessons learned;
- Improvement of interoperability and capabilities;
- Testing and validating concepts through experimentation.

The NATO Joint CBRN Defence COE became officially operational in 2007 [13]. It is a NATO military body and a multi-national organisation featuring the following Member Countries: Austria, Czech Republic, France, Germany, Greece, Hungary, Italy, Poland, Romania, Slovakia, Slovenia, the United Kingdom, and the USA. The NATO Joint CBRN Defence COE offers recognised experience and expertise in such areas as:

- NATO Transformation Process;
- Operational Support by providing a CBRN Defence advice; and
- Support of CBRN Defence Education, Training and Exercises [14].

Another milestone in the evolution of the CBRN security paradigm was the launch in 2010 of the EU CBRN Centres of Excellence Initiative. Initially the legal basis for the EU CBRN Centres of Excellence Initiative was Regulation (EC) No 1717/2006 of 15 November 2006 establishing an Instrument for Stability (IfS) which was superseded in 2014 by the Instrument contributing to Stability and Peace (IcSP), the latter currently being managed by the European Commission's Directorate-General for International Cooperation and Development (DG DEVCO). The primary aim of the EU CBRN Centres of Excellence Initiative is to address the need to strengthen the institutional capacity of Partner Countries to mitigate CBRN risks through, *inter alia,* enhancing local ownership, fostering local expertise, and promoting long-term sustainability [15]. To this end, the Initiative is structured in a way that avoids a traditional top-down approach: it is centred on a worldwide network of local experts and collaborating partners.

The EU CBRN Centres of Excellence Initiative is currently present in more than 55 Partner Countries grouped around eight EU CBRN Centres of Excellence Regional Secretariats, located mainly in Africa, Asia, and the Middle East. At the national level, each Partner Country appoints a National Focal Point for the EU CBRN Centres of Excellence Initiative. The National Focal Point is responsible for supporting the creation of an inter-ministerial CBRN National Team, comprising relevant representatives from ministries, national agencies and institutions representing relevant communities involved in CBRN risk mitigation, e.g. police and law enforcement, defence, the judiciary, government officials dealing with science, technology, industry, and trade, civil protection and emergency services, universities and research centres, public laboratories, intelligence services, and diplomats.

The EU CBRN Centres of Excellence Initiative seeks to facilitate regional cooperation among Partner Countries, in order to enhance their CBRN risk mitigation capabilities. To this end, a specific cycle of activities has been defined, starting from the Partner Countries' needs assessment at the local level (bottom-up approach) to the definition of project objectives, the selection of implementers, the actual project implementation, the monitoring of activities, and evaluation of project outcomes and impacts overall.

3   CBRN Security: Next Steps

CBRN security is a *new organising principle* of the international multilateral relations that deal with international security in a holistic approach on CBRN knowledge and materials, and a possible mechanism for doing so it is through the so-called "*soft law*".

Given the multifaceted security challenges arising from the intersection between complex globalising dynamics and the rapid pace of advancement of science and technology, manifested in the diffusion of knowledge and materials outside their traditional domains, it is vital to keep the momentum of dialogue and interaction among the communities dealing with biological security, chemical security, and radiological/nuclear security.

A possible end goal of such an enhanced interaction could be a common methodology for addressing CBRN risks. From a functional point of view, the CBRN domains are the skeleton of the CBRN security concept but they are not exclusive. In perspective, CBRN security needs to be interpreted in a holistic way by including safety issues, possible future challenges, such as the problem of Improvised Explosive Devices (IEDs) chemical precursors, cyberattacks against CBRN critical infrastructures, and the proliferation of sensitive tacit knowledge. The EU CBRN Centres of Excellence Initiative, the EU Community of Users on Secure, Safe, and Resilient Societies [16] and the IAEA NSSC-Network, among others, could serve as reference models in the elaboration of a CBRN risk mitigation methodology.

One observation that merits specific attention is the fact that at present the CBRN security initiatives remain largely fragmented, not harmonized, limited in geographical scope, and focusing on a different range of stakeholders. Addressing these and other related obstacles requires a further adjustment of the CBRN security paradigm through, *inter alia,* institutionalised dialogue and focused deliberation and hence, the transformation of the CBRN security concept and practices into a new "*formal Institution*".

Indeed, CBRN security might be institutionalised at a lower level with respect to the international treaties dealing with arms control and disarmament, which do not cover CBRN risk mitigation. The reference framework for doing so could be the so-called counter proliferation initiatives, such as the 2003 Proliferation Security Initiative [17]. Another possible reference framework model for a formal CBRN security Institution could be the establishment of an *ad hoc* "*CBRN UN-Governmental Group of Experts (CBRN UN-GGE)*", similarly to the UN-GGE launched for the global Information and Telecommunication security [18], or the development of an arrangement tailored on the so-called Intergovernmental Panel on Climate Change (IPCC) [19].

The CBRN UN-GGE could be tasked with the development of codes-of-conduct, guidelines, principles and standards on CBRN risk mitigation. It could also serve as a "*clearance platform*" for discussion and data sharing, including the exchange of best practices and lessons learned, as well as the international harmonization of the national laws/regulations and national governance systems on CBRN security.

The CBRN UN-GGE could also be responsible for the administration of a database with relevant information, in order to facilitate multi-stakeholder engagement. As kick-off, a simple move could be to set a common digital agenda for all CBRN initiatives (e.g. workshops, fact finding missions, professional association gatherings, etc.), which could play an essential role in informing the formulation of an envisaged "*CBRN Global Action Plan*". Each State might provide national assistance and technical cooperation, as appropriate, to the CBRN UN-GGE, and formulate, if not already in place, a National Action Plan (NAP) that is compatible and consistent with the measurable objectives of the CBRN UN-GGE.

States' adherence to the CBRN UN-GGE needs to be un-discriminatory, voluntary and not restricted to the States Parties to international disarmament and arms control agreements, such as the NPT, CWC, and BTWC. A close collaboration with those and other existing international or multilateral instruments needs to be explored and actively pursued.

Furthermore, a CBRN UN-GGE could pave the way to adopt a *new UNSC Resolution* (including and generalizing the UNSCR 1540) demanding the enforcement of laws and regulations at national levels against the deliberate and criminal uses of CBRN expertise, materials and technologies.

References

1. United Nations, Security Council, *Resolution 1540*, S/RES/1540 (2004), 28 April 2004. Available at http://www.un.org/en/ga/search/view_doc.asp?symbol=S/RES/1540%20(2004) (accessed 3 April 2017).
2. For further information see 1540 Committee: Security Council Committee Established Pursuant to Resolution 1540 (2004), *General Information* available at http://www.un.org/en/sc/1540/national-implementation/general-information.shtml (accessed 4 April 2017).
3. For further information see 1540 Committee: Security Council Committee Established Pursuant to Resolution 1540 (2004), *Assistance*, available at http://www.un.org/en/sc/1540/assistance/general-information.shtml (accessed 4 April 2017).
4. International Atomic Energy Agency, *IAEA Nuclear Security Plan for 2014–2017,* available at http://www-ns.iaea.org/security/nuclear-security-plan.asp?s=4 (accessed 4 April 2017).
5. International Atomic Energy Agency, *IAEA Nuclear Security Plan for 2014–2017,* available at http://www-ns.iaea.org/security/nuclear-security-plan.asp?s=4 (accessed 4 April 2017).
6. Further information about the International Network for Nuclear Security Training and Support Centres. 2017. – NSSC Network is available at http://www-ns.iaea.org/security/nssc-network.asp?s=9&l=76 (accessed 4 April 2017).
7. Address by Ambassador Ahmet Uzumcu, Director General of the Organisation for the Prohibition of Chemical Weapons (OPCW), *Perspectives in the Context of the Third Review Conference of the Chemical Weapons Convention,* IUPAC Workshop, 'Trends in Science and Technology Relevant to the Chemical Weapons Convention (CWC)', 20 February 2012, Spiez, Switzerland. Available at https://www.opcw.org/fileadmin/OPCW/ODG/uzumcu/IUPAC_DG_Statement_Feb_2012.pdf (accessed 4 April 2017).
8. Organisation for the Prohibition of Chemical Weapons, *Report of the Third Special Session of the Conference of the States Parties to Review the Operation of the Chemical Weapons Convention*, RC-3/3, 19 April 2013, The Hague. Available at https://www.opcw.org/fileadmin/OPCW/CSP/RC-3/en/rc303__e_.pdf (accessed 1 April 2017).
9. United Nations, The Eighth Review Conference of the States Parties to the Convention on the Prohibition of the Development, Production and Stockpiling of Bacteriological (Biological) and Toxin Weapons and on their Destruction, Geneva, 7–25 November 2016, *Final Document*, BWC/CONF.VIII/4. Available at http://www.unog.ch/__80256ee600585943.nsf/(httpPages)/57a6e253edfb1111c1257f39003ca243?OpenDocument&ExpandSection=3#_Section3 (accessed 4 April 2017).
10. United Nations, The Eighth Review Conference of the States Parties to the Convention on the Prohibition of the Development, Production and Stockpiling of Bacteriological (Biological) and Toxin Weapons and on their Destruction, Geneva, 7–25 November 2016, *Final Document*, BWC/CONF.VIII/4. Available at http://www.unog.ch/__80256ee600585943.nsf/(httpPages)/57a6e253edfb1111c1257f39003ca243?OpenDocument&ExpandSection=3#_Section3 (accessed 4 April 2017).
11. G8, *Recommendations for a Coordinated Approach in the Field of Global Weapons of Mass Destruction Knowledge Proliferation and Scientist Engagement*, 2009. See http://www.g8.utoronto.ca/summit/2011deauville/2011-gpassessment-en.html#engagement (accessed 4 April 2017). Full text of the Recommendations is available at http://www.mofa.go.jp/policy/economy/summit/2009/report_gp-a2.pdf (accessed 4 April 2017).

12. For further information see North Atlantic Treaty Organisation, *Centres of Excellence,* available at http://www.nato.int/cps/en/natohq/topics_68372.htm (accessed 4 April 2017).
13. For further information see JCBRN Defence COE, *History of JCBRN Defence COE,* available at http://www.jcbrncoe.cz/index.php/history (accessed 4 April 2017).
14. For further information see JCBRN Defence COE, *JCBRN Defence COE Mission and Tasks,* available at http://www.jcbrncoe.cz/index.php/organization-65/mission-64 (accessed 4 April 2017).
15. Further information on the EU CBRN Centres of Excellence is available at http://www.cbrn-coe.eu/ (accessed 4 April 2017).
16. European Commission, Directorate-General for Migration and Home Affairs (DG HOME), *A Community of Users on Secure, Safe, and Resilient Societies (CoU): Mapping EU Policies and FP7 Research for Enhancing Partnerships in H2020,* available at https://www.cbrn-networkofexcellence.org/filter-results-3/publication (accessed 4 April 2017).
17. Further information on the Proliferation Security Initiative is available at http://www.psi-online.info/Vertretung/psi/en/01-about-psi/0-about-us.html (accessed 4 April 2017).
18. For further information see United Nations Office for Disarmament Affairs, *Developments in the Field of Information and Telecommunications in the Context of International Security,* available at https://www.un.org/disarmament/topics/informationsecurity/ (accessed 4 April 2017)
19. For further information on the Intergovernmental Panel on Climate Change, see http://www.ipcc.ch/organization/organization.shtml (accessed 4 April 2017).

Selected Issues of Cyber Security Practices in CBRNeCy Critical Infrastructure

Stanislav Abaimov and Maurizio Martellini

Abstract The article highlights the strong relevance and crucial importance of cyber security defence and response capacities in CBRNeCy assets and management, including in ICS and SCADA systems. Based on the overview of the recent cyber security publications and available information on global cybercrime, it reviews types of cyber and cyber related physical attacks on CBRN Industrial Control Systems; classifies attack types and defence techniques by network layer of attack; analyses security testing approaches based on knowledge of the targeted system, and evaluates types of due protection. The proper combination of existing physical security measures and cyber security testing exercises is considered, by the authors, as one of the most efficient ways to ensure sufficient protection against increasing global cyber threats to CBRNeCy infrastructures. The paper deals also with the best security practises, and contains enumeration of the globally recognized testing techniques and methodologies required to design effective multi-disciplinary security measures, thus providing a substantial ground for their practical implementation in the areas of concern.

Abbreviations

APT	Advanced Persistent Threat
BYOD	"Bring your own device"
CBRNe	Chemical, Biological, Radioactive, Nuclear and Explosives
CBRNeCy	Chemical, Biological, Radioactive, Nuclear, Explosives and Cyber

S. Abaimov (✉)
University of Rome Tor Vergata, Rome, Italy
e-mail: stanislav.abaimov@uniroma2.it

M. Martellini
University of Insubria and Landau Network Fondazione Volta, Como, Italy
e-mail: maurizio.martellini@uninsubria.it

© Springer International Publishing AG 2017
M. Martellini, A. Malizia (eds.), *Cyber and Chemical, Biological, Radiological, Nuclear, Explosives Challenges*, Terrorism, Security, and Computation, DOI 10.1007/978-3-319-62108-1_2

11

DoS	Denial of Service
DDoS	Distributed Denial of Service
DMZ	Demilitarised Zone
ICS	Industrial Control System (or Systems)
IEEE	Institute of Electrical and Electronics
PLC	Programmable Logic Controller
RFID	Radio-frequency identification
SCADA	Supervisory Control and Data Acquisition
SIEM	Security Information and Event Management
UN	United Nations
US CERT	United States Computer Emergency Readiness Team

1 Introduction

In the age of global communication, sophisticated technologies and widely available cyber tools, industrial, corporate and political espionage merged with cybercrime has become an issue of significant global concern. The information security has been on the UN agenda since 1998, when the Russian Federation first introduced a draft resolution in the First Committee of the UN General Assembly. It was adopted without a vote (A/RES/53/70).[1] The UN has raised it high on the international agenda, calling cyber security as one of the pillars for maintenance of international peace and stability and stressing the need for a universal cyber security legal framework, global cyber diplomacy and internet governance.[2]

In the 2015 *Report of the Group of Governmental Experts on Development in the Field of Information and Telecommunications* the UN Secretary-General notes: "Few technologies have been as powerful as information and communications technologies in reshaping economies, societies and international relations. Cyberspace touches every aspect of our lives. The benefits are enormous, but these do not come without risk. Making cyberspace stable and secure can only be achieved through international cooperation, and the foundation of this cooperation must be international law and the principles of the UN Charter."[3]

Due protection, early warning and effective response are especially crucial in chemical, biological, radioactive, nuclear and explosives (CBRNe) facilities, whose damage not only entails country-level process disruptions, but also endangers human existence globally.

---

[1] http://www.un.org/ga/search/view_doc.asp?symbol=A/RES/53/70

[2] Report of the Group of Governmental Experts on Developments in the Field of Information and Telecommunications in the Context of International Security, 68th General Assembly, A/68/98, June 2013, pp. 8–11.

[3] http://www.un.org/ga/search/view_doc.asp?symbol=A/70/174

Ensuring the serene existence of humanity and advancing technological progress by using a wide range of materials and agents, the above-mentioned facilities have one area in common: automation and control systems, which coordinate the whole process. Invention of computers promoted their successful use in manufacturing and eventually in the management itself, thus enhancing the quality and speed of the production cycle, but escalating danger at the same time.

Both computer and industrial control systems (ICS) have evolved over the decades. Having emerged initially for different needs and in different centuries, they merged together in the later decades and acquired very similar architectures. From the time when, in 1936, the first principle of the modern computer was proposed by Alan Turing in his paper "On Computable Numbers", the computer has materialized, initially as a calculation machine, and made a quantum leap in its functions and use.

The world's first stored-program computer was built at the Victoria University of Manchester and ran its first program on 21 June 1948. Although considered "small and primitive", it was the first working machine to contain all elements essential to a modern electronic computer. In April 1951, the newly developed LEO I computer became operational and ran the world's first regular routine office computer job. In 1949, the first integrated circuit was invented; its mass use and the following invention of the microprocessor in 1970, led to a fast popularization of personal and industrial computers. The modern computer architecture, based on the microprocessor technology, is widely used in ICS and industrial workstations, office and personal computers, smartphones and embedded devices.

It is also worth mentioning that the automatic feedback control systems have been known and used for more than 2000 years. Some of the earliest examples are water clocks described by Vitruvius and attributed to Ktesibios (circa 270 B.C.). About 300 years later, Heron of Alexandria in his works "Automata" and "Pneumatica" described a range of mechanisms which employed a variety of feedback mechanisms. The term *feedback* was introduced in the 1920s by radio engineers to describe parasitic, positive feeding back of the signal from the output of an amplifier to the input circuit. This feedback mechanism is the basic principle in any Control System [1].

Early minicomputers were used in the control of industrial processes since the beginning of the 1960s. Thus, the IBM 1800 was an early computer that had input/output hardware to gather process signals in a plant for conversion from field contact levels (for digital points) and analogical signals to the digital domain.

In 1950, the Sperry Rand Corporation built UNIVAC I, the first commercial data processing machine. The machine tools began to be automated in the 1950s with Numerical Control (NC) using punched paper tape. This lately evolved into Computerized Numerical Control (CNC). The first industrial control computer system was built in 1959 at the Texaco Port Arthur, Texas, refinery with an RW-300 of the Ramo-Wooldridge Company [18].

Prior to the 1950s, the predominant control systems were analogical-based or were simply "on/off" controls due to switch or relay positions [8]. The first reported use of digital control systems (DCS) took place in 1956, and was placed into operation in 1959 at the Port Arthur (Texas) refinery and in 1960 at the Monsanto ammonia plant in Luling, Louisiana. These systems were supervisory in nature and the individual loops were controlled by conventional electrical, pneumatic or hydraulic controllers, but monitored by a computer.

It was in 1959 when the researchers initiated to design a digital computer that could fully control the industrial controls process. In the late 1960s, specialized process control computers arrived on the scene offering direct digital control, so that the computer architecture could implement a discrete form of a control algorithm [20]. However, research and technological advancements of these systems were expensive and they were superseded by the cheaper microcomputers of the early 1970s.

(IEEE Communications Surveys & Tutorials [13])

Supervisory controls and data acquisition (SCADA) history is rooted in distribution applications, such as power, natural gas, and water pipelines, where there is a need to gather remote data through potentially unreliable or intermittent low-bandwidth/high-latency links. SCADA systems use open-loop control with sites that are geographically dispersed. A SCADA system uses Remote terminal/telemetry units (RTUs), to send supervisory data back to a control center. Most RTU systems have some limited capacity to handle local controls while the master station is not available. However, over the years RTU systems have grown more and more capable of independently handling local controls.

Programmable logic controller (PLC) evolved to replace racks of relays in ladder form. The latter were not sufficiently reliable, were difficult to rewire and to diagnose. PLC control tends to be used in very regular, high-speed binary controls. Originally, PLC equipment did not have remote control racks, and many could not perform more than rudimentary analog controls. Only physical access could compromise the security. With the introduction of electronic, and later computer architecture, the remote access created new attack vectors and patterns.

Distributed Control Systems (DCS) generally refer to the particular functional distributed control system design that exist in industrial process plants (including CBRNe agents). The DCS concept came about from a need to gather feedback data and control the systems on a large scale in real time. It is common for loop controls to extend all the way to the top level controllers in a DCS, as everything works in real time. These systems evolved from a need to extend control systems beyond just a small cell of control units.

> The definitions of different typed of control and information processing modules are blurring as time goes on (IEEE [12]). The technical limits that drove the designs of these various systems are no longer as much of an issue. PLC platforms can now perform as a small DCS, being sufficiently reliable for SCADA systems to manage closed loop control over long distances.

Advancing technologies have merged computers and ICS as one. Remote access to controlled equipment, and even to a controlled facility, has become a standard for the majority of industries, and the issues of cyber security have become crucial. The series of critical infrastructure disruptions, caused by cyber attacks, alerted defence forces and sparked the cyber security scrutiny.

Significant interest in potential cyber-related disaster events started to emerge in the mid-1990s (e.g. the US Security in Cyber-Space (GAO 1996); Winn Schwartau "Information Warfare: Chaos on the Information Superhighway" (Schwartau 1994). In 1991, Jim Bidzos, a security industry pioneer, originated the much-repeated phrase: "Digital Pearl Harbor". Another peak of concern was in 1998 and 1999 over fears of the Y2K bug [6]. It was born on the assumption that the older versions of computer systems were not programmed to cope with date presentation in the upcoming millennium and would fail to function in a designated manner [23]. This concern has not lost its relevance nowadays.

One of the earliest publicly announced events related to the CBRN infrastructure vulnerability to cyber attacks occurred in January 2002. The malware successfully breached the perimeter network defences at Ohio's Davis-Besse nuclear power plant (tough the employees claimed the network was protected by a firewall), infiltrated a private computer network and disabled a safety monitoring system for nearly 5 h.[4]

In October 2006, the attackers gained access to computer systems at a Harrisburg water treatment plant in the USA. The ICS network was accessed after an employee's laptop computer was compromised via the Internet, and then used as an entry point to install a malware that was capable of affecting the plant's water treatment operations.

In October 2008, the derailment of the tram in the city of Lodz injured 12 people. The attacker used the repurposed television remote control to change track points through Infrared sensor. He was also suspected of having been involved in several similar incidents. The problems with the signaling system on Lodz's tram network were detected when a driver was attempting to steer his vehicle, the rear wagon of the train derailed and collided with another passing tram.[5]

The 2010 event in Iran confirmed that information technology could be used not only to trigger remote CBRN attacks,[6] but could be also perceived as a direct threat to physical CBRN ICS equipment. Stuxnet was the first malware to infiltrate and cause physical disruption in multiple ICSs in a CBRN facility (the uranium enrichment

---

[4] http://www.securityfocus.com/news/6767
[5] https://www.schneier.com/blog/archives/2008/01/hacking_the_pol.html
[6] https://www.cia.gov/library/reports/general-reports-1/terrorist_cbrn/terrorist_CBRN.htm

plant) and multiple other facilities over 2 years with similar equipment.[7] This malware was a wake-up call, which united cyber security community and CBRN defence experts by the same goal of protecting the planet [24].

In 2011, the Trojan "Poison Ivy" was used to collect intellectual property from 29 international chemical companies. It was one of the largest industrial espionage attempts in history, raising the awareness of cyber security specialists in the topic of cyber security in critical infrastructure.

In 2014, the Malware Shamoon wiped 30,000 workstations in Saudi Aramco's corporate network, raising concern over cyber attacks that can bypass firewalls and intrusion detection systems to physically affect technology networks in a large scale.[8]

In 2014, the 13 different types of malware disguised as ICS/SCADA software updates (e.g. Siemens Simatic WinCC, GE Cimplicity, and Advantech) were detected in the spear-phishing emails. After a due forensic investigation, the malware was identified as the re-purposed banking Trojan, aiming to collect private information and credentials.[9] This event confirmed the capabilities of ICT malware to be used against industrial networks.

In December 2015, the Denial of Service in a power plant and multiple substations in Ukraine triggered a power outage. In February 2016, it was acknowledged that BlackEnergy3 malware was used for the cyber attack.[10]

The Verizon data breach digest [21] describes several attacks investigated by the company, including one aimed at the systems of an unnamed water utility referred to by Verizon as the Kemuri Water Company.

In October 2016, the Domain Name System provider Dyn was targeted by a DDoS attack, whose systems support major websites and online services. The investigation is conducted by the US Homeland Security. The attackers have not been identified yet. J. McAfee claims, "The massive cyber attacks that temporarily disabled websites including Twitter, Reddit and *The New York Times* offline may be a precursor to a "cyber atomic bomb"". Several security experts believe the attacks are part of tests designed to probe for vulnerabilities ahead of a much larger attack.[11]

The events listed above indicate a constant evolution of attack capabilities of threat actors. There is no single solution to secure critical infrastructures against cyberattacks, and hence several layers of defence should be set [4]. Based on the approaches to physical and operational security and safety, this article explores comprehensive cyber security applications and strategies related to ICSs, implemented in critical infrastructure that uses CBRNe agents and technologies.

---

[7] http://www.computerworld.com/s/article/9226469/Iran_confirms_cyberattacks_against_oil_facilities

[8] http://www.darkreading.com/attacks-breaches/banking-trojans-disguised-as-ics-scada-software-infecting-plants/d/d-id/1318542

[9] http://www.darkreading.com/attacks-breaches/banking-trojans-disguised-as-ics-scada-software-infecting-plants/d/d-id/1318542

[10] http://www.ibtimes.com/us-confirms-blackenergy-malware-used-ukrainian-power-plant-hack-2263008

[11] http://europe.newsweek.com/dyn-north-korea-bureau-121-ddos-hackers-internet-attacks-513098?rm=eu

1.1 Goals and Objectives

This article aims to raise awareness and assist cyber security professionals and security management specialists in taking informed decisions to ensure due cyber protection and defence in CBRNeCy critical infrastructure, as well as to provide effective response and recovery in cyber security related events.

The goal of the research is to highlight the relevance of cyber security in CBRNeCy defence, identify network and protocol layers vulnerable to cyber-attacks and review the best security practises.

To meet the above mentioned goal and in order to provide a panoramic overview of the problem to the target audience, the following objectives were put forward:

- Set definitions and notions
- Review the background of the issue
- Review types of operational areas for computer systems
- Review types of cyber related CBRNeCy attacks
- Review network layers vulnerable to cyber-attacks
- Review types of security testing and security exercises
- Review types of cyber protection in CBRNe infrastructure
- Indicate advanced security testing techniques for effective security designs

Methodology Both quantitative and qualitative approaches were used for the collection and analysis of the available information. Quantitative literature review covered over 30 reports on ICS and SCADA cyber security, published by major security and industrial companies. Among the latter, several documents were related to cyber security and ICS implementations in critical and CBRN infrastructure, thus allowing to conduct a qualitative approach and provide grounded analysis of the existing situation.[12]

As ICS cyber security is a comparatively new area of scientific research, there is still no consistency in its terms and definitions. For the purpose of clarity, selected definitions of the key terms have been used in the article.

Following the definition of the Cornell University of Law, the term "information security" is used to define activity for protecting information and information systems from unauthorized access, use, disclosure, disruption, modification, or destruction in order to provide:

(i)   integrity, which means guarding against improper information modification or destruction, and includes ensuring information nonrepudiation and authenticity;

(ii) confidentiality, which means preserving authorized restrictions on access and disclosure, including means for protecting personal privacy and proprietary information; and

(iii) availability, which means ensuring timely and reliable access to and use of information [5].

---

[12]The publications by the following authors were the most relevant and considerably contributed to the present research: C. Baylon, I. Brown, R. Brunt, Fernandez, M. Martellini, K. Wilhoit. The same refers to the publications of the following organizations: Industrial Control Systems Cyber Emergency Response Team (ICS-CERT), Institute of Electrical and Electronics (IEEE), Cornell University of Law.

The term "cyber security" is defined as the protection of information systems from theft or damage to the hardware, the software, and to the information on them, as well as from disruption or misdirection of the services they provide identified [7].

As per Paske E. [17], "control system" is a general term that encompasses several types of control systems used in industrial production, including supervisory control and data acquisition (SCADA) systems, distributed control systems (DCS), and other smaller control system configurations such as programmable logic controllers (PLC) often found in the industrial sectors and critical infrastructures [17].

The definition of the term "CBRN defence" follows the above author, Paske E. [17] and encompasses protective measures taken in situations in which chemical, biological, radiological or nuclear warfare (including terrorism) hazards may be present. CBRN defence consists of CBRN passive protection, contamination avoidance and CBRN mitigation [17].

As per Baylon C. [3] critical infrastructure is the production, storage, logistics of CBRN materials and devices, as well as reconnaissance and disaster response [3]. In the present paper, CBRNe Infrastructure is a critical infrastructure that produces, handles, transports, recycles, decontaminates or otherwise incorporates CBRNe agents.

CBRNe cyber security is defined as the practice of security in computer systems, industrial control systems and networks in the critical industry that involves the use of CBRNe agents [14]. In the present article we introduce the term CBRNeCy, which covers Chemical, Biological, Radioactive, Nuclear, Explosive infrastructure and any other infrastructure, that incorporates the cyber management, and if damaged entails global danger.

Security Information and Event Management (SIEM) software products and services combine information security management and security event management. They provide real-time analysis of security alerts generated by network hardware and applications.

2 Main Part

As defined by the ICS-CERT [11], cyber threats to ICS refer to an individual or a group of individuals who attempt unauthorized access to a control system device and/or industrial network using a data communications pathway. Threats to control systems can come from numerous sources, including hostile governments, terrorist groups, disgruntled employees, and malicious intruders. This report introduces the following classification of attackers:

- Nation states
- Terrorists
- Industrial Espionage and Organized Crime
- Hacktivists
- Hackers
- Accidental damage or information leak

The above attackers are capable to infiltrate corporate and industrial control networks of state, municipal, military and private facilities, gather and exfiltrate valuable (confidential, classified) information, as well as intentionally decrease or disable production capacities.

For the special case of CBRNeCy assets, we should agree that any attack on a nuclear facility *or against a chemical plant is equal to a warfare attack. The best practices and norms to be pursued to avoid cyber-attacks against infrastructures should be augmented by awareness and educational activities tailored to increase prevention and preparedness. In a risk management approach, a solution might be to enforce the resilience of the Industrial Control Systems (ICSs) and their subsystems implemented in CBRNe critical infrastructures. The concept of resiliency could be in-built into the subsystems forming an ICS, like the SCADA systems, that is "security by design", so that to assure the safety of the whole CBRNe infrastructures in the case of cyber-attacks* [19].

As of today and as publicly revealed, the maximum damage caused by cyber-attacks was related to:

- CBRNe agent device rendered unstable or non-fit to purpose (Stuxnet, Iran), though theoretical analysis confirms that the attackers were capable of physically destroying the device;
- power distribution rendered offline (BlackEnergy3, Ukraine), where power supply (APS) device was reconfigured to disable the power distribution;
- communications disabled (BlackEnergy3, Ukraine).

As per the predictions by Bruce Schneier[13] and IEEE[14] [13], the possible cyber-attack capabilities in the future will relate to:

- Hybrid warfare (Cyber and any other warfare domains)
- Global Denial of Service (e.g. Internet infrastructure collapse)
- Refined delivery via email (advancements in spam and spearphishing)
- Advanced malware delivery via web applications
- Substantially increased malware sophistication and A.I.
- Attacks on critical infrastructure will increase
- "Lone-wolf" terrorism

Cyber-attacks are already a threat to the national security[15] and cyberspace is recognized by NATO as one of the warfare domains.[16] With the exponential growth of cyber technologies, the threat to embedded devices and ICS will only increase and, if not prevented, will reach the global level.

---

[13] https://www.schneier.com/blog/archives/2016/06/issues_regardin.html

[14] *IEEE 12th Symposium on Visualization for Cyber Security (VizSec 2015).*

[15] https://www.usna.edu/CyberDept/sy110/lec/crsIntro/lec.html

[16] http://www.nato.int/cps/en/natohq/topics_78170.htm

2.1   Operational Areas and Modes of Computer Systems

Management and control at any enterprise requires perfect knowledge of the operational areas of computer systems and their types. The first step in provision of standing capacity to prevent the attack or to operate in emergency situation is to identify the location and computer systems connection capacities.

All computerized management, automation and control systems can be classified by purpose and by their mobility.

In the present article, the *purpose of the computer systems* and networks in the facility defines their intrinsic value to the organization, level of security and priority of protection during the attack. *Mobility* describes the level of physical security, specific network implementations, data transfer standards, etc.

By Purpose   All computer systems and networks in the organization can be classified by purpose:

- office or corporate network
- workstations
- ICS or industrial network

The ICS network is the internal network in the facility that physically and logically links the PLCs, SCADA and other Control Units, to synchronize optimal and uninterruptable performance. The corporate network is usually connected both to the external network and the internal industrial network, to provide support and ensure service delivery. Workstations can be connected to the internal network and/or external network, usually providing limited functionality for a specific task.

By Mobility   Equipment and ICS, connected to the ICS network can be classified by mobility:

- Stationary ICS and networks
- Mobile ICS and networks

Stationary ICS and networks are physically immobile, as they do not leave the facility. The network connection is wired and/or wireless. Heightened level of physical security is provided for stationary systems, networks and facilities (e.g. perimeter walls, reinforced structures, CCTV, increased number of security personnel).

Mobile ICS may be mounted on, installed in or transported by vehicles, vessels and aircraft, while the network connection is always remote and wireless. The transport vehicles can function in mobile, stationary and deployed modes. The majority of CBRNe agents and related systems are more vulnerable during transportation, rather than in stationary mode. However, if the vehicle or vessel is in a deployed mode (mobile laboratory, research transport, standard logistics, etc.), it may be considered as a stationary system for a brief period of time (hours or days), which may give the attackers sufficient time to conduct a successful attack. For example, if the mobile laboratory is deployed, it may take hours or days for it to become mobile again, which is usually sufficient for the cyber attack to be conducted.

2.2 Cyber Related CBRNe Attacks

A cyberattack can be conducted in many ways and for many reasons. Their classification can assist in successful development of threat models and effective security policies. Cyber-attacks on CBRNe can be classified by area, industry, perceivable damage and possible attack scenario.

As cyber-attack vectors range from spear phishing and social engineering to custom designed exploits and deliberately planted routers, network layers outline the possible attack vectors on the cyber level.

The below Table 1 reflects CBRNeCY cyber-attacks by area.

Types of Cyber Related CBRNe Attacks In regards to the attack vectors on any facility as a structure, there are four generally accepted types of attacks:

1. Cyber-attack: the attackers have no physical access to systems or devices;
2. Physical attack: the attackers have no remote access to systems or devices;
3. Cyber-enabled physical attack: security system compromised to enable easy physical access for the attackers;

Table 1 Cyber attacks by area of CBRNeCy

CBRNeCy industry	Applications	Perceivable damage	Possible attack scenarios[a]
Chemical	Chemical production Logistics Storage	Destruction, Area contamination, Loss of human lives	Industrial espionage ICS controlling chemical delivery
Biological	Medical facilities Research facilities	Destruction, Area contamination, Disease outbreak, Loss of human lives	Espionage Fake medical details of public personae may result in reputational loss
Radioactive	Production Storage Logistics Decontamination	Destruction, Area contamination, Loss of human lives	Espionage Denial of service
Nuclear	Power production	Destruction, Area contamination, Planet scale disaster, Loss of human lives	Espionage, Denial of service
Explosives	Production Logistics Demolition Excavation Warfare	Destruction, Terrorism act, Area contamination, Loss of human lives	Espionage Destruction Loss of human lives Terrorism Guerrilla warfare Civil war
Cyber domain	Cyber command	Misinformation, Destruction, Area contamination, Loss of human lives	Espionage Counterespionage Psychological warfare Cyberwarfare Warfare support

[a]See the more detailed attack scenarios in Cyber-Threat Model and Scenarios

4. Physical-enabled cyber-attack: physical actions allow remove access to the previously unreachable computer system or network (e.g., rogue device is physically planted inside the facility and/or connected to the network).

In cyber-attacks the attackers have no physical access to systems or devices. Thus, Stuxnet and BlackEnergy3 malware are classified as pure cyber-attacks, as there was no direct physical contact with the target facility.

In physical attacks the attackers have no remote access to systems or devices, and they have to physically access the facility to interact with the system. A physical access to the hardware, communication wire layout and the internal wireless network of the facility may give attackers the level of access to the network sufficient to conduct a successful attack without exploitation of corporate software.

In cyber-enabled physical attacks, the security system is compromised to enable easy physical access for the attackers. The security systems may be disabled remotely for the physical attack to be conducted.

In physical-enabled cyber-attacks, the physical actions allow to remove access to the previously unreachable computer system or network. The rogue device is physically planted inside the facility and/or connected to the network, so the attackers can have the direct access to the corporate or industrial network. These techniques incorporate physical access to internal systems and networks, related to cyber security, and might be enumerated as follows:

1. Wired network physical access (wiretapping)
2. Wireless network physical access (long rage antennas, WiFi attack vectors, BYOD, etc.)
3. Insider threat (physical access by the employee)

Wiretapping is the practice of intercepting telecommunications covertly. Traditionally, wiretapping was used to monitor telephone conversations, but with the increase in Internet usage, wiretapping started to include monitoring internet communications. Previously wiretapping revolved around the electric current flowing throughout a phone line. Agents who wished to listen on someone's conversation would have to physically plug a device into the subject's phone line and interpret the electric fluctuations that represented the sounds of the conversation.[17] The same technique can be applied to the network communications if the attacker decides to connect a router or a microcomputer to a physical communication wire to intercept traffic.

As opposed to wiretapping, *physical access to wireless network* is an ability to connect to the network by physical presence in the signal proximity of the wireless routing device (Access Point or Router). The alternative way to access the internal network with wireless devices is to use techniques, such as wardriving, the act of searching for Wi-Fi wireless networks in a moving vehicle, using a portable computer or a smartphone, an external antenna and wireless attack software.

---

[17] https://cs.stanford.edu/people/eroberts/cs181/projects/ethics-of surveillance/tech_wiretapping. html

An *insider threat* is a malicious threat to an organization that comes from people within the organization, such as employees, former employees, contractors or business associates, who have detailed information concerning the organization's security practices, data and computer systems. The threat may involve fraud, theft of confidential or commercially valuable information, the theft of intellectual property, wiretapping, internal wireless access or the sabotage of computer systems.

All listed scenarios have to be expanded and customised by the cyber security experts and further used in risk profiling.

Network Architecture Layers The cyber-attack behaviour and/or pattern is adapted to the targeted network architecture layer; consequently, the defence and response measures should be adjusted to the same level or above.

Table 2 reflects the updated topology of the network architecture layers, developed by the International Organization for Standardization in 1983, including a few popular implementations.

Attacks on the layers 1, 2, 3, 4 target the routing process and mechanisms of the network infrastructure. The attack scenarios may involve Man-in-the-Middle attack, Traffic capture (sniffing), Computer System Social Engineering, etc.

Attacks on layers 5, 6 and 7 target the clear-text information (credentials, sensitive information, communications, etc.) or cyphertext, susceptible to decryption, replay or alterations. Attack scenarios may involve malware, software vulnerability exploitation, password guesting, Software Social Engineering, etc.

Every single of the seven layers is potentially vulnerable to attacks. Protection of every layer should be ensured both from the layer itself and the upper layer (e.g., HTTPS is the security upgrade to HTTP, IPSec is the security addon to IP) [16]. However, the

Table 2 Network architecture layers

Layer	OSI model	Protocols	Implementations
7	Application layer	HTTP, SMTP, SNMP, Telnet	Interacts with software applications that implement a communicating component
6	Presentation layer	MIME, XDR	Establishes context between application-layer entities
5	Session layer	NetBIOS, SOCKS, PPTP	Controls the dialogues (connections) between computers
4	Transport layer	TCP, UDP	Transferring variable-length data sequences from a source to a destination host via one or more networks
3	Network layer	IPv4. IPv6, ICMP	Transferring variable length data sequences from one node to another connected to the same network
2	Data link layer	ARP	Link between two directly connected nodes
1	Physical layer	DSL, IEEE Protocols, Bluetooth	Electrical and physical specifications of the data connection

Based on the Open Systems Interconnectivity Model, Publicly Available Standards, OSI (http://standards.iso.org/ittf/PubliclyAvailableStandards/index.html)

attack on one layer does not automatically suggest the vulnerability in the layers below, as they might be isolated and not "visible" to each other.

Special attention should be paid to the application layer, which is the nearest to the operator. The application programs function outside the OSI model, as the transmitted information is generated, calculated, or otherwise processed inside the system itself (e.g. workstation, server, PLC, etc.), rather than inside the network. The application-layer functions typically include identifying communication partners, determining resource availability, and synchronizing communication. The key difference is the application-specific implementation, which means every single application can be potentially exploited by the attackers to gain access to the whole network topology.

In addition to the above, we might add that cyber-attacks can be carried on one or multiple layers in the network architecture topology. To ensure a comprehensive level of security, cyber security specialists should conduct vulnerability assessment of all network architecture layers.

2.3   Methodology of Comprehensive Cyber Security Testing

Information security requirements include confidentiality, integrity, authentication, availability, authorization and non-repudiation. Actual security requirements are tested depending on the security requirements implemented by the system.

Security testing is the process intended to reveal flaws in the security mechanisms of information systems, that are initially designed to protect data and maintain systems functionality. It has a number of different meanings and can be completed in a number of different ways: Vulnerability Scan and Assessment, Security Assessment, Penetration Testing, Risk Analysis, etc. Each of them is significant in its own way with different purposes and outcomes. The in-depth security testing should include the following steps:

The depth of the security testing should be established, documented and co-signed by the management and the cyber security assessment team. The same document should indicate the extent to which security testing procedure imitates threat model options.

According to Common Vulnerability Scoring System (CVSS),[18] vulnerabilities are design flaws or misconfigurations that make the network (or a host on the network) susceptible to malicious attacks from local or remote users. Vulnerabilities can exist in several areas of both the corporate and industrial networks, such as in firewalls, FTP servers, Web servers or operating system. Depending on the level of the security risk, the successful exploitation of a vulnerability can vary from the disclosure of information about the host to a complete compromise of the host. The severity levels for vulnerabilities are level 1 (minimal), level 2 (medium), level 3 (serious), level 4 (critical) and level 5 (urgent).

---

[18] https://www.first.org/cvss/v2/guide

Potential Vulnerabilities include vulnerabilities that cannot be fully verified. In these cases, at least one necessary condition for the vulnerability is detected. It is recommended to always investigate these vulnerabilities further. The service can verify the existence of some potential vulnerabilities when authenticated trusted scanning is enabled. The severity levels for vulnerabilities are level 1 (minimal), level 2 (medium), level 3 (serious), level 4 (critical) and level 5 (urgent). After the vulnerabilities are confirmed and classified, the contract agreement should specify if the proof-of-concept exploit be created to illustrate the danger of the vulnerability.

Among all steps listed above, it is the penetration testing that might be implemented with the limited knowledge of the target system. Hence, the following types of penetration testing might be defined:

- white box testing – complete knowledge of the systems
- grey box testing – partial knowledge of the systems
- black box testing– no knowledge of the system or its credential

*White box testing*refers to cyber security testing with full knowledge and access to all source code and documentation on network architecture. Having full access to this information can reveal bugs and vulnerabilities more quickly than the "trial and error" method of black box testing. Additionally, the organization receives more detailed testing coverage. It provides the maximum information about the perimeter defences and attacker capabilities, as well as about the level of exposure of the client to the public.

Grey box testing means testing a system or network while having at least some knowledge of the internal architecture. This information is usually constrained to detailed design documents and architecture diagrams. It is a combination of both black and white box testing, and combines aspects of each.

Black box testing refers to testing a system without having specific knowledge of the internal workings of the system, no access to the source code, and no knowledge of the architecture. This technique is usually conducted in order to mimic the real attack scenario. Usually the most sophisticated and resource consuming.

The comprehensive cyber security analysis includes all the above seven steps listed in Table 3. Selection performance of only some of them does not provide sufficient information to design adequate security measures.

Based on the findings and outcomes, every approach should be based on a relevant threat model, designed by qualified security specialists, employed or hired by the organization.

Cyber-Threat Model and Scenarios Threat modelling is a process by which potential threats can be identified, enumerated, and prioritized from the point of view of a hypothetical attacker. The purpose of threat modelling is to provide security specialists with a systematic analysis of the attacker's profile, the most likely attack vectors, and the assets prioritised by an attacker. Threat modelling answers the following questions:

- Where are the high-value assets?
- Where the organization is most vulnerable to attacks?

Table 3 In-depth security testing steps

#	Step	Actions
1	Systems and software survey	Verify hardware configurations Check physical network layout Identify software versions
2	Vulnerability scan	Run automated vulnerability scanning tools: Port scanning software, version detection scripts Check for default credentials
3	Vulnerability assessment	Run automated vulnerability scanning tools versus detected versions of the software Validate vulnerability risk profile Remove false positives from the report
4	Security assessment	Manually Verify Step 3 findings Classify security risk level
5	Penetration testing	Simulate an attack Protocol findings
6	Security audit	Perform risk analysis Advise mitigation and/or remediation measures Report threats and outcomes
7	Security review	Review cyber security policies and procedures Apply to the internal systems and networks

- What are the most relevant threats?
- Is there an attack vector that might go unnoticed?
- Which of the network architecture layers is most vulnerable to attacks?

The attack scenario is a scenario that enumerates and describes the ways an attacker might make use of a vulnerability.[19] The attack scenarios are usually considered to be the vital part of threat modelling process.

The project PRACTICE (Preparedness and Resilience Against CBRN Terrorism using Integrated Concepts and Equipment), implemented in 2010 by the European Union, explored different CBRN crises caused by intention or accidents, to develop and enhance EU CBRN emergency preparedness planning and response. Its objective was to improve the preparedness and resilience of the EU member states and associated countries to a terrorist attack with the use of non-conventional weapons, such as CBRNe agents. According to PRACTICE research, current security management situation is characterized by a fragmented structure as regards to technology, procedures, methods and organization on national level as well as EU-level.[20] The project developed the PRACTICE Toolbox to provide knowledge framework and aid in development of security measures in critical and CBRNe specific facilities.[21]

The following CBRNe attack scenarios were developed by the Norwegian Defence Research Establishment (research institute, Norwegian Armed Forces) [9]:

---

[19] https://www.symantec.com/security_response/glossary/define.jsp?letter=a&word=attack-scenario – last accessed 2016-09-29.

[20] http://cordis.europa.eu/project/rcn/98969_en.html

[21] http://www.practice-fp7-security.eu/cms,article,2,toolbox.html

1. Chemical attack in city centre – Explosion and dispersion of sulphur mustard
2. Chemical transport accident – Train derailment causing chlorine dispersal
3. Radiological dispersal in city – Radioactive caesium spread in fire
4. Radiological attack on public transportation – Hidden radioactive source
5. Nuclear power plant accident – Release of fission products
6. Nuclear submarine accident – On board fires
7. Hoax – Unknown powder in congress centre

These scenarios are currently used to enhance the currently deployed security measures and develop response capacities on national and international levels.[22]

Some of the possible cyber-attack scenarios on CBRNe infrastructure are listed below in Table 4:

As of today, cyber-attacks can also enable or support conventional attacks. Threat actors with remote access to the facility network can:

Table 4 Cyber attack scenarios by area in CBRNe

Area	Threat	Scenario
Chemical	Denial of service and area contamination	Denial of Service in the ICS may cause the shutdown, critical malfunction and/or chemical leakage, contaminating the area inside or outside the facility
	Water supply contamination	Tampering the configurations of the water purification system, endangering consumer lives
Biological/Medical	Public figure setup	Medical records of the public figure accessed, edited with compromising values and released to the public
	Pharmaceuticals formula change	Biological research facility accessed and critical information is tampered with in order to divert or slow down the research
Radioactive	Terrorist attack	Causing the radiation outbreak with the physical or remote attack, and by accessing the ICS switching off the radiation sensors preventing it from triggering alarms, thus preventing timely response and multiplying the damage
Nuclear	Equipment corruption	The production of valuable material is intentionally slowed down
	Power blackout	ICS of a power plant is accessed to disable power distribution
	Meltdown	Critical scenario of disabling the reactor cooling system, causing the meltdown and following explosion and/or eradiation

---

[22] https://www.ffi.no/no/Aktuelle-tema/Documents/Updated%20article%20for%20the%20
Defence%20Global%20Publication.pdf

1. provide information about city traffic, mass gatherings, emergency events (to plan the attack and the target);
2. enable false alarms or disabling the positive alarms (to better control the municipal response during the attack);
3. trigger remote detonation devices or electronically controlled release valves on chemical tanks.

Extensive knowledge of the attack patterns and scenarios allows security experts to create threat models, to engineer perfect security measures which are the most effective against specific attacks and ineffective against others.

2.4 Cyber Security Management Environment

Raising awareness of all involved staff and providing basic introduction to practical cyber security ensures informed decision and compliance in the organization on all levels of corporate hierarchy.

To exercise security testing and successfully deliver knowledge and understanding to the target audience, the equipment and software environment has to be duly organized.

The deployment of the security testing environment with physical hardware is less sophisticated, but requires more maintenance. The most common ways of modern cyber security training deployment (to emulate vulnerable computer or computer network) is virtualization, which provides effective solution for logistics and time-consuming maintenance.

Hardware virtualization, or platform virtualization, refers to the creation of a virtual machine that acts as a real physical computer with an operating system. Software executed on these virtual machines is separated from the underlying hardware resources. For example, a computer that is running Microsoft Windows may host a virtual machine that looks like a computer with the Ubuntu Linux operating system, and vice versa; thus Ubuntu-based software can be run on the virtual machine.

In hardware virtualization, the host machine is the hardware and software on which the virtualization takes place, and the guest machine is the virtual machine. The definitions *host* and *guest* are used to distinguish the software that runs on the physical machine from the software that runs on the virtual machine. With the implementation of snapshots, it is possible to deploy the sophisticated networks, configured only once, avoiding repetitive configurations in the future and in case of malfunction, caused by the vulnerabilities, exploits and attack tools; to refresh or restore the target system in seconds.

The testing programmes should be adapted to the reality and ensure all possible response scenarios against multiple ways to breach the system. Information provided to the employees should be comprehensive and relevant to their primary responsibilities and might include:

- Presentation on the specific types of cyber-attacks and their detection (e.g., social engineering, spear phishing, malware, etc.).
- Simulation exercises for the non-ICT personnel to ensure compliance with cyber-security policies.
- Tutorials and trainings with clear guidance and tools provided for ICT personnel and personnel engaged in work with advanced systems.
- Tutorials and training in finding vulnerabilities and launching a relevant exploit for cyber security personnel.

The cyber security risk register should always include the assumption that some of the staff, especially the disgruntled employees, will use this knowledge to their advantage, to compromise the integrity of the facility. In spite of this fact, the knowledge on the attacks and techniques used will increase compliance of the staff through understanding of the threat, and significantly increase overall safety of the facility.

2.5   Systems Protection in CBRNeCy Infrastructure

High security status of the CBRNeCy facility requires policies and procedures to evolve alongside the technology itself. Technical means of systems security should be implemented even in isolated facilities, both for the systems and in the systems.

Physical Security of the Equipment Physical access to the electronic systems enables the attackers advanced access to the internal network and/or direct extraction of the data storage devices (USB drives, laptops, hard drives, etc.).
    Physical Security of the electronic systems inside the organization includes:

- Security personnel
- Radio-frequency identification (RFID) implementations
- Biometrics
- Closed-circuit television (CCTV) and smart cameras
- Network layout

Radio-frequency identification (RFID) uses electromagnetic fields to automatically identify and track tags attached to objects. The tags contain electronically stored information. Unlike a barcode, the tag does not need to be within the line of sight of the reader, so it may be embedded in the tracked object. For example, an RFID tag attached to an automobile during production can be used to track its progress through the assembly line; RFID-tagged pharmaceuticals can be tracked through warehouses. RFID cards are used by personnel to access restricted areas and systems: server facilities, data centres, communication and power distribution hubs, etc.
    Biometrics refers to metrics related to human characteristics. Biometrics authentication is used in computer science as a form of identification and access control. Biometric identifiers are the distinctive, measurable characteristics used to label and describe individuals, and are often categorized as physiological characteristics relating to the shape of the body, e.g. fingerprint, palm veins, face recognition, DNA,

palm print, hand geometry, iris recognition and retina, voice. Though behavioural characteristics might be also used in biometrics, such as body movements, writing patterns, typing speed.

Closed-circuit television (CCTV) is the use of video cameras to transmit a signal to a specific place, on a limited set of monitors. Positioning of the cameras and their connection to the CCTV network should be designed as per the best practices, to avoid blind spots and unauthorized access to the CCTV network.

It is not sufficient to provide security only to the physical network layout, wiring and physical protection of the hardware inside and outside the facility. Cyber-attacks are carried out remotely, thus ICS, Workstations and networks should be protected on the hardware and software level.

Cyber Security Appliances and Applications Cyber security is the protection of information systems from theft or damage to the hardware, the software, and to the information on them, as well as from disruption or misdirection of the services they provide. It includes protection against unauthorised network access, data and code injections, and due to malpractice by operators, whether intentional, accidental, or due to them being tricked into deviating from secure procedures.

The software solutions that assist in providing due cyber security measures for the corporate and industrial networks are as follows:

- Antivirus Software
- Firewalls
- Intrusion Detection and Protection Systems
- Honeypot implementations

Antivirus Software Antivirus software is a computer software used to detect, identify, prevent and remove malicious software. Traditional antivirus software relies heavily upon signatures to identify malware. As soon as it arrives to the anti-virus company, it is analyzed by malware researchers or by dynamic analysis systems. Once it is determined to be a malware, a proper signature of the file is extracted and added to the signatures database of the antivirus software, for it to be downloaded with the next update. The signature database update is only available with the network connection to the antivirus developer or with manual delivery (e.g. Disks, USB drives, etc.). Both options should be outlined in the security policies and procedures of the facility.

Although the signature-based approach can effectively contain malware outbreaks, malware authors have tried to be a step ahead of such software by writing "oligomorphic", "polymorphic" and, more recently, "metamorphic" viruses, which encrypt, encode and compress parts of themselves or otherwise modify themselves as a method of disguise, so as not to match virus signatures in the dictionary.

Anti-virus programs are not always effective against new viruses, even those that use non-signature-based methods that should detect new viruses. The reason for this is that before releasing them, the virus designers test their new viruses on the major antivirus applications, some of which are even available online for free. Thus being able to consider implementing the latest evasion techniques and modifications for the developed malware.

Firewalls Firewall is a network security system for monitoring and control over the incoming and outgoing network traffic based on predetermined security rules [2, pp. 32–33]. A firewall typically establishes a barrier between a trusted, secure internal network and another external network, such as the Internet.

Firewalls are often categorized as either network firewalls or host-based firewalls. Network firewalls are a software appliance running on general purpose hardware or hardware-based firewall computer appliances that filter traffic between two or more networks. Host-based firewalls provide a layer of software on one host that controls network traffic in and out of that single machine.

Attempting to bypass the firewall rules to access social media or open communication link from the inside of the facility network may result in the creation of an open channel for the attackers to launch the attack.

Intrusion Detection and Prevention Systems An Intrusion Detection System (IDS) is a device or software application that monitors a network or systems for malicious activity or policy violations [15]. Any suspicious activity or detected violation is typically collected centrally using a Security Information and Event Management (SIEM) system. A SIEM system combines outputs from multiple sources, and uses alarm filtering techniques to distinguish malicious activity from false alerts.

Intrusion detection and prevention systems are primarily focused on identifying possible incidents in systems and networks, logging information, reporting attempts and learning. IDS have become a necessary addition to the security infrastructure of nearly every organization.

Honeypot Implementation Honeypot is a computer security mechanism set to detect, deflect, or counter the attempts at unauthorized use of information systems. Generally, a honeypot consists of data (for example, in a network site) that appears to be a legitimate part of the site but actually isolated and monitored, and that seems to contain information or a resource of value to attackers, created to attract their attention.

Two or more honeypots on a network form a honeynet. Typically, a honeynet is used for monitoring a larger and more diverse network in which one honeypot may not be sufficient. Honeynets and honeypots are usually implemented as parts of larger network intrusion detection systems.

Currently firewalls, antivirus software and IDS have become the norm for the organizations. In highly secure environments these may not be sufficient.

Notable Mentions It is generally accepted that the cyber security relies on three pillars: physical security, network security and computer security. However, that is often not enough to provide sufficient security.

The following physically enabled cyber security policies are worth mentioning:

- Network segmentation and DMZ implementation
- BYOD policies on critical mission network
- Compliance with international recommendation (United Nations[23,24] IEEE[25])

---

[23] https://ccdcoe.org/cyber-security-strategy-documents.html

[24] https://www.un.org/disarmament/topics/informationsecurity/

[25] http://standards.ieee.org/

In computer security, a DMZ or *demilitarized zone* (sometimes referred to as a perimeter defence network) is a physical or logical subnetwork that contains and exposes an organization's external-facing services to a usually larger and untrusted network, often the Internet. The purpose of a DMZ is to add an additional architectural layer of security to an organization's local area network; an external network node only has direct access to equipment in the DMZ, rather than any other part of the network. The name is derived from the term "demilitarized zone", an area between nation states in which military operation is not permitted.

Bring your own device (BYOD) refers to the policy of permitting employees to bring personally owned devices (laptops, tablets and smart phones) to their workplace, and to use those devices to access privileged company information and applications. Bringing personal devices to the high security facility is often not permitted, as it is rightfully considered a threat to integrity of the facility.

The global nature of cyberspace and threats to its reliable functioning makes international cooperation indispensable. The ITU National Cybersecurity Strategy Guide [22] regards a coordinated international response as the only answer and possible solution. The governments may consider promulgating an international strategy for cyberspace to coordinate all activities. The nations may use a combination of international cooperation model, such as bilateral agreements[26] aimed at building capacity in all the priority areas.

Best Practices in ICS Cyber Security As per ICS-CERT [10], the best practices in cyber security for ICS are as follows:

1. Maintain an Accurate Inventory of Control System Devices and Eliminate Any Exposure of this Equipment to External Networks;
2. Implement Network Segmentation and Apply Firewall;
3. Use Secure Remote Access Method;
4. Establish Role-Based Access Controls and Implement System Logging;
5. Use Only Strong Passwords, Change Default Passwords, and Consider Other Access Controls;
6. Maintain Awareness of Vulnerabilities and Implement Necessary Patches and Updates;
7. Develop and Enforce Policies on Mobile Devices;
8. Implement an Employee Cybersecurity Training Program;
9. Involve Executives in Cybersecurity;
10. Implement Measures for Detecting Compromises and Develop a Cybersecurity Incident Response Plan .

Supplemented with the regular security assessment, those best practices enhance cyber security of the critical infrastructures and provide a solid ground for further research and development of the global cyber security.

---

[26] https://www.gov.uk/government/uploads/system/uploads/attachment_data/file/62647/CyberCommunique-Final.pdf

Word of Caution Testing software and equipment is a highly sophisticated set of tools and utilities that can be used for defence and offence. They are usually delivered inside the facility or uploaded into the network. Cyber security assessment tools can be misused by the attackers. All actions, performed by the experts, if not conducted properly, can create undesired vulnerabilities in the industrial and corporate networks.

During a live cyber security assessment procedure the following events may occur:

- Copies of password, password files and/or configuration files, stored in unencrypted text file and storage devices, may be left unattended in the vulnerable systems.
- Unattended backdoors and malware can provide easy access to the attackers.
- Disabled services may disrupt further functionality of the entire system.
- Stress-testing may cause system malfunction and partial or global shut down of the ICS.
- Improperly handled vulnerability may result in Permanent Denial of Service, rendering the equipment unrecoverable.
- Improperly handled vulnerability may remain unnoticed in the system and remain there for an extended period, causing sudden unexpected malfunctions.

In the *white box* testing, the source code of proprietary software is disclosed for the effective security assessment and shared. Security testing contractors can be targeted, as they usually have lower security requirements than the facility itself. Security policies of the facility should outline source code sharing, security procedures for the software testing and emergency response.

The organization requesting a security audit should consider the legal protection of the outcomes, which should be duly revised in draft to avoid public disclosure of sensitive information. Moreover, the auditor should legally guarantee the documented and appropriate level of responsibility, confidentiality, liability and compliance.

Safety and security during audit and cyber security assessment of any organization is the responsibility of the organization, cyber security experts and the legal experts. For the CBRNe infrastructure, it is vital to secure every step of the audit procedure and support it with the detailed documentation.

Acknowledgements One of the authors, Stanislav Abaimov, would like to express the sincere gratitude to Professor Giuseppe Bianchi for his trust, support and highly professional guidance.

References

1. Bennett, S.: A Brief History of Automatic Control. IEEE (1996)
2. Boudriga, N.: Security of mobile communications. Boca Raton. CRC Press (2010)
3. C. Baylon, R. D.: *Cyber Security at Civil Nuclear Facilities,*. Clatham House Report (2015)
4. Chatham House: Emerging Risk Report – 2016, Use of Chemical, Biological, Radiological and Nuclear Weapons by Non-State Actors. Chatham House, The Royal Institute of International Affairs (2016)

5. Cornell University of Law: *44 U.S. Code § 3542 - Definitions*. (1992) Retrieved from Cornell University of Law Web site: https://www.law.cornell.edu/uscode/text/44/3542
6. Fernandez, I.: Cybersecurity for Industrial Automation & Control Environments: Protection and Prevention Strategies in the Face of the Growing Threats. Frost & Sullivan (2013)
7. Gasser, M.: *Building a Secure Computer System*. Van Nostrand Reinhold (1988)
8. Hayden, E.: An Abbreviated History of Automation & Industrial Controls Systems and Cybersecurity. SANS Institute (2015)
9. Hege Schultz Heireng, M. E.: THE DEVELOPMENT AND USE OF CBRN SCENARIOS FOR EMERGENCY PREPAREDNESS ANALYSES. FOI (2015)
10. ICS-CERT: 10 Basic Cybersecurity Measure. US-CERT (2015)
11. ICS-CERT: Industrial Control Systems Cyber Emergency Response Team (2016) Retrieved from https://ics-cert.us-cert.gov
12. IEEE: IEEE Communications Surveys and Tutorials. IEEE (2012)
13. IEEE Communications Surveys & Tutorials: Introduction to Industrial Control Networks. IEEE (2013)
14. Martellini, M.: Deterrence and IT Protection for Critical Infrastructures. Springer (2013)
15. NIST: Guide to Intrusion Detection and Prevention Systems. NIST (2007)
16. NIST: Technical Guide to the Information Security Testing and Assessment. National Institute of Standards and Technology Special Publication (2008)
17. Paske, E. L.: Cyber Security of Industrial Control Systems, Global Conference on Cyber Space (2015)
18. Stout, T. M., & Williams, T. J.: Pioneering Work in the Field of Computer Process Control. IEEE Annals of the History of Computing (1995)
19. US Department of State: Cyber Security for Nuclear Power Plants. Washington: US Department of State (2012)
20. Vanessa Romero Segovia, A. T.: History of PLC and DCS (2012)
21. Verizon: Data Breach digest. Scenarios from the field. Verizon (2016)
22. Wamala, F.: National Cybersecurity Strategy Guide. International Telecommunication Union (2011)
23. Wilson, C.: Cyberpower and National Security (2009)
24. Wilson, C.: Cyberterrorism: Understanding, Assessment, and Response. Swansea University (2014)

NATO's Response to CBRN Events

Bernd Allert

NATO is committed to working with Allies, partners, and other international organisations to combat the proliferation of weapons of mass destruction and defend against chemical, biological, radiological and nuclear (CBRN) threats. (The Secretary General's Annual Report 2016, Brussels/Belgium 2017, p. 22.)

Jens Stoltenberg, NATO Secretary General.

Abstract At its summits, NATO places a high priority on preventing the proliferation of Weapons of Mass Destruction (WMD) and defending against Chemical, Biological, Radiological and Nuclear (CBRN) threats. Some of the summit declarations led to concrete results that are still influencing presence and future, such as NATO's Weapons of Mass Destruction Non-Proliferation Centre at NATO Headquarters in Brussels, the Combined Joint CBRN Defence Task Force (CJ-CBRND-TF), the Joint CBRN Defence Centre of Excellence (JCBRND COE) in Vyškov, the Joint CBRN Defence Capability Development Group (JCBRND-CDG) and – of course – NATO's Comprehensive, Strategic-level Policy for Preventing the Proliferation of WMD and Defending Against CBRN threats. The Framework Nations Concept (FNC) Cluster on CBRN Protection might join these success stories.

B. Allert (✉)
Lieutenant Colonel (DEU-AR) at Joint CBRN Defence Centre of Excellence,
Vyškov, Czech Republic
e-mail: allertb@jcbrncoe.cz

© Springer International Publishing AG 2017
M. Martellini, A. Malizia (eds.), *Cyber and Chemical, Biological, Radiological,
Nuclear, Explosives Challenges*, Terrorism, Security, and Computation,
DOI 10.1007/978-3-319-62108-1_3

1   Introduction

NATO summit meetings provide periodic opportunities for Heads of State and Government (HoSG) of all NATO member countries to evaluate and provide strategic direction for future Alliance activities. The Alliance hold Summit meetings very often at key moments in its evolution. They are important events of the Alliance's decision-making process. Summits introduce new policies, invite new members into the Alliance, launch major initiatives and reinforce partnerships. These summits are meetings of the North Atlantic Council (NAC) at its highest level possible – that of HoSG. Since 1949, there have been 26 NATO summits. The last one took place in Warsaw (Poland) from 8 to 9 July 2016 and Belgium will host the next one in Brussels 2017, in the new NATO Headquarters. NATO summits are always held in a NATO member country and are chaired by the NATO Secretary General.

The prevention of proliferation of Weapons of Mass Destruction (WMD) and their means of delivery, as well as the defence against Chemical, Biological, Radiological, and Nuclear (CBRN) threats have not always been on the agenda of NATO summits (Fig. 1).

However, there were some very important discussions regarding NATO's CBRN defence measures that still affect today's policies, such as the summit in Prague [3], Strasbourg/Kehl [7], and Cardiff (Caerdydd)/Wales [10]. WMD proliferation poses a direct military threat to Allies' populations, territory, and forces, and therefore continues to be a matter of serious concern for the entire Alliance. The principal non-proliferation goal of the Alliance and its members is to prevent the proliferation of WMD from occurring. Should the proliferation occur, NATO intends to reverse it primarily through diplomatic means, and not military ones.

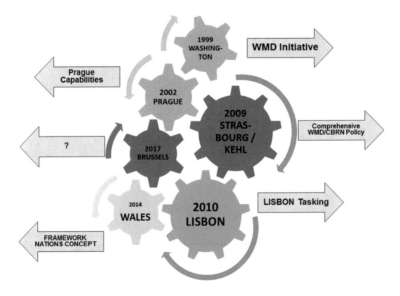

Fig. 1 NATO Summits and their results

NATO does always reiterate its full support for international non-proliferation regimes and their strengthening. The Alliance's policy of support for arms control, disarmament and non-proliferation (ADN) plays an important role in the achievement of the Alliance's security objectives. The preventing of the proliferation of Weapons of Mass Destruction and their means of delivery is included. NATO expects all member states to fully implement their arms control, disarmament, and non-proliferation commitments, and to strengthen existing international arms control and disarmament accords and multilateral non-proliferation and export control regimes. Early admission of all NATO members into all appropriate existing non-proliferation regimes plays a positive role. In this regard, at nearly every summit the Alliance:

- confirms NATO's commitment of reinforcing the Nuclear Non-Proliferation Treaty, as the cornerstone of non-proliferation and disarmament, and ensuring full compliance with it by its members as States parties to the Treaty;
- confirms the importance of the Biological and Toxin Weapons Convention (BTWC), the Chemical Weapons Convention (CWC) and the Hague Code of Conduct against the Proliferation of Ballistic Missiles;
- supports United Nations Security Council Resolution 1540 (2004) to take cooperative action to prevent non-state actors from acquiring WMD, and to end illicit trafficking of WMD and related substances;
- increases joined efforts to reduce and safeguard nuclear and radiological material;
- prevents proliferation of WMD and their means of delivery, and declares to work together to achieve these objectives.

With respect to these international treaties, the states of concerns have changed during the last decades from Libya (2004), Democratic People's Republic of Korea (2006), and Iran (2008) to Iraq and Syria (2016). NATO was concerned of Libya's, Iraq's, and Syria's chemical weapons programmes, and of North Korea's and Iran's nuclear programmes. Respective means of delivery, e.g. ballistic missiles, were and still are of interest likewise.

In addition, NATO must of course be able to deploy forces and to sustain operations in order to achieve operational objectives. This includes in particular an environment faced with CBRN threats.

Finally yet importantly, NATO summits were, are, and will be always a part of NATO's strategic communications directed to addressees inside and outside the Alliance.

2 Weapons of Mass Destruction (WMD) Initiative

At the 1999 Washington Summit [2], NATO's assessment on the proliferation of WMD and their means of delivery led to concrete results: the Weapons of Mass Destruction Initiative. However, the idea was already born 5 years earlier at the 1994 Brussels Summit [1] when NATO's HoSG constituted WMD as a threat to international security and as a matter of concern to NATO. All HoSG decided to intensify and expand NATO's political and defence efforts against proliferation, at the same time taking into account the work of other international fora and institutions. They directed to begin immediately in appropriate fora of the Alliance to develop an overall policy framework to consider how to reinforce ongoing prevention efforts as well as how to reduce the proliferation threat, and protect the Alliance against it.

Finally, at the Washington Summit, NATO launched an Initiative that has built upon achievements since the Brussels Summit to improve overall Alliance political and military efforts in this area.

The WMD Initiative should ensure a debate at NATO in order to strengthen a common understanding among Allies on all WMD related issues, and how to respond to them. Furthermore, the quality and quantity of intelligence and information sharing among Allies on proliferation issues ought to be improved. The Initiative supports the development of a public information strategy by Allies to increase awareness on proliferation issues and Allies' efforts to support non-proliferation efforts. Already existing Allied programmes which increase military readiness to operate in a WMD environment and to counter WMD threats should be enhanced. Procedures for Allies to assist one another in the protection of their civil populations against WMD threats should be improved as well. The WMD initiative integrates political and military aspects of Alliance work in responding to proliferation of WMD. With its WMD Initiative from 1999 NATO created a fundament, which is very important for NATO's response to CBRN events to date.

In addition, NATO's HoSG initiated the creation of a WMD Centre within the International Staff at NATO to support these efforts. Since then, the WMD Centre was renamed to WMD Non-Proliferation Centre, and is structurally embedded as a section in the Emerging Security Challenges (ESC) Division at NATO Headquarters. The WMD Non-Proliferation Centre comprises of personnel from NATO's International Staff and International Military Staff as well as national experts (so called "Voluntary National Contributions"). The Centre's core work is to support the Alliance itself as well as all NATO members and NATO's Partner nations in their CBRN defence efforts. The WMDC shall strengthen dialogue on WMD issues, and enhances consultations on non-proliferation efforts. The Centre assesses risks and threats to support defence efforts that improve the Alliance's preparedness to respond to the risks of WMD and their delivery systems. To summarize, the establishment of the WMD Centre has become an on-going success story of NATO's efforts to combat the proliferation of WMD and to defend against CBRN threats.

3   Prague Capabilities Commitments

The next milestone on NATO's way to protect its population, territories and forces against CBRN threats was doubtless the 2002 Prague Summit [3]. NATO's HoSG approved the Prague Capabilities Commitment (PCC) as part of the continuing

Alliance effort to improve and develop new military capabilities for modern warfare in a high threat environment. Individual member states made specific political commitments to improve their capabilities among others also in the area of CBRN defence. Inter alia, NATO took into consideration a closer cooperation and coordination with the European Union. The Alliance's efforts to improve its own capabilities through the PCC and those of the European Union through the European Capabilities Action Plan (ECAP) should be mutually reinforcing. The ECAP included also projects on CBRN defence.

The HoSG voiced to support the enhancement of the role of the WMD Centre within the International Staff to assist the work of the Alliance in countering CBRN threats.

Reducing NATO to a solely military organisation is not reflecting its significance completely. NATO is much more - it is a political and civilian organisation, too. Consequently, NATO expressed its commitment to implement fully the Civil Emergency Planning (CEP) Action Plan for the improvement of civil preparedness against possible attacks against the civilian population with Chemical, Biological or Radiological (CBR) agents. NATO intended to enhance its ability to provide support, when requested, to help national authorities to deal with the consequences of terrorist attacks, including attacks with CBRN against critical infrastructure, as described in the CEP Action Plan.

The HoSG endorsed the implementation of five Nuclear, Biological and Chemical (NBC)[1] weapons defence initiatives, which will enhance the Alliance's defence capabilities against weapons of mass destruction: a Prototype Deployable NBC Analytical Laboratory, a Prototype NBC Event Response Team, a virtual Centre of Excellence for NBC Weapons Defence, a NATO Biological and Chemical Defence Stockpile, and a Disease Surveillance system. Additionally, HoSG confirmed their nation's commitment to augment and improve their national NBC defence capabilities in order not to rely only on NATO's capabilities and capacities.

In 2003, to implement the NBC Analytical Laboratory and the NBC Event Response Team the NATO Combined Joint CBRN Defence Task Force (CJ-CBRND-TF) was established, and declared fully operational 1 year later. The Task Force consists of two main bodies: a CBRN Joint Assessment Team (JAT) made up of specialists capable of providing experience in case of CBRN events, and a multinational CBRN Defence Battalion. Other equally important components of the Battalion include Headquarters providing command and control (C2), CBRN reconnaissance, biological detection, and decontamination capabilities, as well as Deployable Analytical CBRN laboratories (Fig. 2).

Allied Command Operations (ACO) generates the Task Force on a rotational base. One NATO nation acts as a lead nation, supported by up to now 21 NATO countries. Once Ukraine contribute a decontamination platoon. The Task Force may be deployed independently, or as a part of NATO Response Force (NRF) if there is a CBRN event that affects any NATO territory or population, including deployed armed forces. The CJ-CBRN-TF deployed already to support high visibility events like the Olympic

---

[1] "Chemical, biological, radiological, nuclear" and "CBRN" replaced later the term "nuclear, biological, chemical" and the abbreviation "NBC", which are not used anymore.

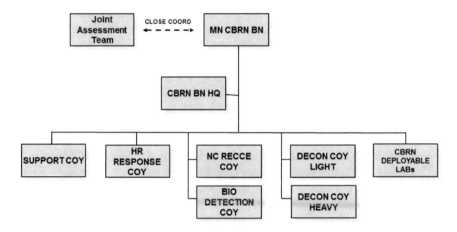

Fig. 2 Combined Joint CBRN Defence Task Force

Games in Athens 2004 or the NATO Summit in Riga 2006. To conclude, the CJ-CBRND-TF has to be considered as one of NATO's success stories in the area of CBRN defence. Currently ACO reviews the Task Force's Concept of Operations (CONOPS) to adapt its capabilities to future challenges.

The virtual Centre of Excellence for NBC Weapons Defence developed into the Joint CBRN Defence Centre of Excellence (JCBRND COE) in Vyškov/Czech Republic. The mission of the JCBRND COE is to assist and support NATO and its military transformation, partner nations, other international institutions and organizations in the area of CBRN defence, and to provide CBRN Defence related expertise and support in the areas of doctrines and concepts development, training and evaluation, and lessons learned processes. Through its Reachback capability, JCBRND COE supports NRF, as well as other allied units.

The JCBRND COE is not the one and only Centre of Excellence (COE) taking care of CBRN defence issues: in their special area of expertise at least the COE Defence Against Terrorism (DAT) in Ankara/Turkey, the COE for Military Medicine (MilMed) in Budapest/Hungary, the Counter Improvised Explosive Devices (C-IED) in Madrid/Spain, the Explosive Ordnance Disposal (EOD) COE in Trenčin/ Slovak Republic, and the Strategic Communications (StratCom) COE in Riga/ Latvia should not be neglected.

At the 2002 NATO Prague Summit among other issues, warnings were discussed about the shortfall of a Disease Surveillance System (DSS) to detect disease outbreaks and the determination of whether these outbreaks are to be attributed to the use of biological weapons or to natural causes. The COE for Military Medicine established the Deployment Health Surveillance Capability (DHSC) in Munich to contribute to a better preparedness of the deployed NATO Forces against the threats of infectious disease and bioterrorist attacks. Therefore, the DHSC is supporting operational readiness of NATO's deployed forces.

4 NATO's Comprehensive WMD/CBRN Defence Policy

On its way to "NATO's Comprehensive, Strategic-Level Policy for Preventing the Proliferation of WMD and Defending Against CBRN Threats" NATO conducted a couple of summits addressing WMD Non-Proliferation and CBRN Defence only as a side topic. However, at the 2004 Istanbul Summit [4] NATO's HoSG suggested to reviewing Operation Active Endeavour's mission. In accordance with international law, NATO addressed the risk of terrorist related trafficking in, or use of, CBRN weapons, their means of delivery and related materials. Terrorism and the proliferation of Weapons of Mass Destruction (WMD) and their means of delivery were assessed as key threats and challenges to Alliance and international security. In this regard, HoSG demanded a continued robust support for the CJ-CBRND-TF.

At the first glace, the Istanbul decision, to provide NATO's Partners with increased opportunities to enhance their contributions to NATO-led operations, and to assist in transforming their capabilities in-line with NATO's own evolving operational capabilities, including through enhancing the Operational Capabilities Concept (OCC). However, the OCC enabled a Ukrainian CBRN defence company to reach NATO standards, and to provide a decontamination platoon to the German led CJ-CBRND-TF (NRF 2010) as the first non-NATO contribution up to now.

In 2006, at the Riga Summit [5], NATO's HoSG endorsed the "Comprehensive Political Guidance". The Comprehensive Political Guidance provides a framework and political direction for NATO's transformation, setting out the priorities for all Alliance capability development until 2020.

Finally, at the 2008 Budapest Summit [6], HoSG decided to develop a comprehensive policy for preventing the proliferation of WMD and defending against CBRN threats. NATO continued contributing to policies to prevent and counter proliferation, mainly with a view to prevent terrorist access to, and use of, WMD.

At the 2009 Strasbourg/Kehl Summit [7], the summit on NATO's 60th anniversary, NATO's HoSG finally endorsed "NATO's Comprehensive, Strategic-level Policy for Preventing the Proliferation of WMD and Defending Against CBRN Threats". After 60 years of its existence, NATO got a policy on WMD non-proliferation and CBRN defence (Fig. 3).

The policy is influenced by a vision that the Alliance – its populations, territory and forces – will be secure from threats posed by weapons of mass destruction and related materials. Present and future security challenges require NATO to be prepared to protect and defend against both State and non-State actor threats. The Alliance's approach has to take into account every stage of an adversary's potential acquisition, intention and preparation to use, and employment of WMD. NATO responded to this challenge by addressing WMD proliferation, CBRN defence and consequence management within relevant NATO bodies. In addition, the policy introduces so-called "strategic enablers" that will allow the Alliance to prevent the proliferation of WMD, protect against a WMD attack, and recover should an attack take place. These enablers consist of intelligence and information sharing, international outreach and partner activities, as well as public diplomacy and strategic

NATO's Comprehensive, Strategic-Level Policy for Preventing the Proliferation of Weapons of Mass Destruction (WMD) and
Defending against Chemical, Biological, Radiological and Nuclear (CBRN) Threats

Fig. 3 NATO's Comprehensive, Strategic-Level Policy on WMD Non-Proliferation and CBRN
Defence

communication. To implement Alliance's policy, Allies were also encouraged to
reflect this policy in relevant national documents. NATO and NATO Allies will
accelerate their efforts to transform their capabilities to address WMD threats and
should develop an indicative roadmap for further consideration, which identifies
priorities for capability development. NATO's Military Authorities shall review and
revise, as appropriate, the relevant strategic-level guidance, analyses, defence and
force planning documents, as well as develop required concepts for preventing, protecting against and recovering from WMD attacks or CBRN events.

In 2010, NATO HQ's Deputy Permanent Representatives' Group on Committee
Review suggested where there are single-service groups working on capabilities in the
same field, they should be merged. The Military Committee Joint Standardisation
Board (MCJSB) CBRN Operations Working Group was merged with the Joint
Capability Group on CBRN of the Conference of National Armaments Directors
(CNAD), and the NATO Training Group (NTG) CBRN Defence Training Group. The
newly formed Joint CBRN Defence Capability Development Group (JCBRND-CDG)
is responsible for supporting the Alliance's prevention of the proliferation of WMD
and defence against CBRN Threats by supporting the development of CBRN defence
capabilities across all lines of development focusing on doctrine, material, and training. Seven Panels are subordinated under the JCBRND-CDG. One of the panels is the
Doctrine and Terminology Panel (DTP). DTP is responsible for monitoring the structure and hierarchy of NATO CBRN Defence standardization documents along with

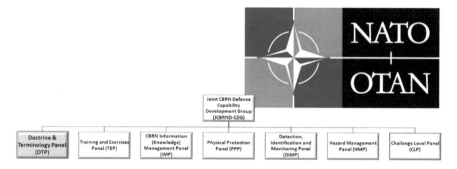

Joint CBRN Defence Capability Development Group (JCBRND-CDG)

Fig. 4 Joint CBRN Defence Capability Development Group and its subordinated panels

proposing and developing Terminology, CBRN Defence Doctrines and Procedures in support of capability requirements on behalf of the JCBRND CDG. Therefore, DTP could be seen as primus inter pares among all of the panels (Fig. 4).

5 Lisbon Summit Tasking

At the 2011 Lisbon Summit [8] the HoSG tasked the North Atlantic Council (NAC) to continue reviewing NATO's overall posture in deterring and defending against the full range of threats to the Alliance, taking into account changes in the evolving international security environment. This comprehensive review should be undertaken by all Allies on deterrence and defence posture principles agreed in the Strategic Concept, taking into account WMD and ballistic missile proliferation. In order to maintain, and develop further, appropriate consultations among Allies on these issues, NAC had to establish a Committee to provide advice on WMD control and disarmament in the context of the deterrence posture review.

Furthermore, the implementation of NATO's Strategic-Level Policy for Preventing the Proliferation of WMD and Defending against CBRN Threats continued. The North Atlantic Council (NAC), NATO's principal political decision-making body, was tasked to assess and report, before the meeting of Defence Ministers in June 2011, on how NATO can better counter the proliferation of WMD and their means of delivery. NATO's Defence Ministers endorsed that report.[2]

Based on the Lisbon Tasking, it should be ensured that NATO has the appropriate capabilities, including for planning efforts, training and exercises, to respond to CBRN attacks.

---

[2] The Report was classified as NATO RESTRICTED.

NATO's 2012 Policy Guidelines on Counter-Terrorism [9] stated that modern technology increases the potential impact of terrorist attacks employing conventional and unconventional means, particularly as terrorists are suspected seeking to acquire CBRN capabilities. However, NATO's work on responding to CBRN threats and the non-proliferation of Weapons of Mass Destruction has been well established.

6   Framework Nations Concept (FNC)

NATO's Heads of States and Government endorsed the Framework Nations Concept (FNC) during the NATO Wales Summit [10] in September 2014. The FNC focuses on groups of Allies working at multinational level for the joint and combined development of capabilities required by the Alliance. Facilitator will be a framework nation. Its implementation will contribute to providing the Alliance with coherent sets of capabilities. To implement this concept, a group of ten Allies, facilitated by Germany, have, through a joint letter, committed to working systematically together, to create, in various configurations, a number of multinational projects called "Clusters" to address Alliance priority areas across a broad spectrum of capabilities. One of these clusters focuses on creating coherent sets of capabilities in the area of CBRN protection.

A cluster focuses on groups of nations coming together coordinated by a nation that is willing to provide a framework to develop their national capabilities in the medium to long term, to come together for collective training and exercise purposes, and to produce operational groupings. With the Political guidance 2015, NATO further articulates the strategic direction set in the 2010 Strategic Concept and the subsequent defence capabilities. The implementation of the FNC will contribute to providing the Alliance with coherent sets of forces and capabilities, particularly in Europe. The Political Guidance linked FNC to the Readiness Action Plan (RAP) with a focus on the enhanced NATO Response Force (NRF) and its Follow-on Forces Group, as well as larger formations of the Follow-on Forces and related capabilities.

Germany offered to provide the framework for the Cluster CBRN Protection. It facilitates the development and the provision of specialized, sustainable and sufficient CBRN defence capabilities for NATO defence planning and operations. This Cluster intends to bridge the capability gap between defence planning and operations planning, and facilitates the provision of CBRN defence capabilities for both predictable and ad-hoc requirements (Fig. 5).

Participating Nations (PN) will support improving NATO's CBRN defence capabilities by coordinating and adapting national capabilities to NATO's current and future requirements.

In order to achieve its aim of supporting NATO's CBRN Defence, the Cluster has to coordinate capability requirements of the PN, to optimize education and training of CBRN specialists, to optimize and coordinate exercising in the area of CBRN Defence, and finally to provide trained, educated and harmonized CBRN Specialist Forces to NATO Article 5 Operations as well as to Non-Article 5 Operations. Within this context,

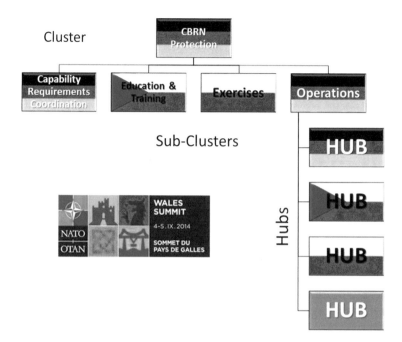

Fig. 5 FNC Cluster CBRN Protection

the Cluster CBRN Protection will respect the PN sovereignty, act within the framework of NATO's Defence Planning Process (NDPP), and will not duplicate existing NATO bodies, but support them as appropriate.

The Cluster CBRN Protection is subdivided into four main areas of work:

- Coordination of the capability requirements of the PN, based on the NATO Defence Planning Process, its capability targets, and the capability development assessment provided by the JCBRND-CDG.
- Education and training of CBRN specialists, thus improving interoperability, increasing the efficiency of existing education and training facilities, and enhancing and coordinating the efforts of training in national and NATO responsibility.
- National and multinational exercises in the area of CBRN Defence, in order to improve CBRN Defence support to operations in a multinational environment.
- Operations Integration in order to plan, and prepare the provision of trained, educated and harmonized CBRN Specialist Forces (and their enablers) to NATO Article 5 and Non-Article 5 Operations.

The four main areas of work will be conducted within the Cluster itself and in three Sub-Clusters. Capability Requirement Coordination will be realised at the immediate Cluster level, the others in subordinated Sub-Clusters, and four so called Hubs (within the Sub-Cluster Operations/Integration). The Sub-Cluster Education and Training (E&T) will be led by the Czech Republic, Sub-Cluster Exercises by Poland, and Sub-Cluster Operations/Integration by Germany.

A Steering Committee (SC) established by the PN, supported by a permanent multinational Cluster Support Cell (CSC), guides the Cluster. The CSC will be established in Bruchsal/Germany in 2017.

The NATO Army Armaments Group (NAAG) Joint CBRN Defence Capability Development Group (JCBRND-CDG) conducted five multinational workshops in order to develop a concept for the FNC Cluster CBRN Protection. Up to 21 nations, but also ACT and the Joint CBRN Defence Centre of Excellence (JCBRND COE) contributed to the concept development. As of now AUSTRIA (as a non-NATO nation), BELGIUM, BULGARIA, CZECH REPUBLIK, GERMANY, HUNGARY, ITALY, LATVIA, THE NETHERLANDS, POLAND, ROMANIA and SLOVAKIA declared their willingness to participate at the FNC Cluster CBRN Protection. Other nations are interested to follow.

Within this context, it will be ensured that NATO is postured to counter CBRN threats, including the Combined Joint CBRN Defence Task Force.

At the 2016 Warsaw summit [11], NATO HoSG expressed their concerns regarding the proliferation of WMD, as well as their means of delivery, by states and non-state actors. Despite the changing political landscape, NATO's populations, territory, and forces remain threatened. They welcomed all multinational and national initiatives, such as the FNC, contributing to capability development and strengthening the Alliance.

7 Conclusion

NATO's member nations will ensure that NATO continues to be both strategically and operationally prepared with policies, plans, and capabilities to counter a wide range of state and non-state CBRN threats. These efforts will be based on NATO's Comprehensive, Strategic-Level Policy for Preventing the Proliferation of WMD and Defending against CBRN Threats endorsed in 2009. In Warsaw NATO HQ was tasked to present a report on its continued implementation at the next Summit. NATO HQ will then have to answer a couple of important questions. What is the status of the implementation of NATO's WMD/CBRN defence policy? Is the policy still fit to meet future security challenges? Do the 2009 CBRN defence capability requirements enable the Alliance to address effectively current and future CBRN challenges? The 2017 NATO Summit will take place at NATO Headquarters in Brussels/Belgium, and provide then required answers.

References

1. Declaration of the Heads of State and Government participating in the meeting of the North Atlantic Council ("The Brussels Summit Declaration") in Brussels on 10–11 January 1994.
2. Washington Summit Communiqué Issued by the Heads of State and Government participating in the meeting of the North Atlantic Council in Washington, D.C. on 24 April 1999.

3. Prague Summit Declaration Issued by the Heads of State and Government participating in the meeting of the North Atlantic Council in Prague on 21 November 2002.
4. Istanbul Summit Communiqué Issued by the Heads of State and Government participating in the meeting of the North Atlantic Council in Istanbul on 29 June 2004.
5. Riga Summit Declaration Issued by the Heads of State and Government participating in the meeting of the North Atlantic Council in Riga on 29 November 2006.
6. Bucharest Summit Declaration Issued by the Heads of State and Government participating in the meeting of the North Atlantic Council in Bucharest on 3 April 2008.
7. Strasbourg / Kehl Summit Declaration Issued by the Heads of State and Government participating in the meeting of the North Atlantic Council in Strasbourg / Kehl on 4 April 2009.
8. Lisbon Summit Declaration Issued by the Heads of State and Government participating in the meeting of the North Atlantic Council in Lisbon on 20 November 2010.
9. Chicago Summit Declaration Issued by the Heads of State and Government participating in the meeting of the North Atlantic Council in Chicago on 20 May 2012.
10. Wales Summit Declaration Issued by the Heads of State and Government participating in the meeting of the North Atlantic Council in Wales on 5 September 2014.
11. Warsaw Summit Communiqué Issued by the Heads of State and Government participating in the meeting of the North Atlantic Council in Warsaw 8–9 July 2016.

Preventing Hostile and Malevolent Use of Nanotechnology Military Nanotechnology After 15 Years of the US National Nanotechnology Initiative

Jürgen Altmann

Abstract This chapter is to update the assessment of potential military applications of nanotechnology published in 2006. Expecting nanotechnology to bring the next industrial revolution, around 2000 many countries started research and development (R&D) programs. In particular in the US National Nanotechnology Initiative, military aspects have figured prominently, with a budget share around 30% until 2008, then slowly decreasing to about 10%. While applications are increasingly emphasized, most of the work is still at the research stage. Many other countries do military R&D of nanotechnology, too, but apparently at a much lower level of funding. Rough estimates of 2006 about when specific military applications would arrive have turned out about correct in one third of the cases, but for the others the expected arrival times have to be postponed by five to 10 years.

When re-considering the list of potential military applications, the 30 areas given in 2006 still seem appropriate. Also their evaluation under criteria of preventive arms control does not need to be changed. The eight applications that would be particularly dangerous to arms control and international humanitarian law, to international peace and military stability, or to humans and societies in peace time, remain the same: small sensors, small missiles, small satellites, small robots, metal-free firearms, implants and other body manipulation, autonomous weapon systems, and new biochemical weapons. They should be subject to specific international bans or limits as proposed in 2006.

Keywords Military • Nanoscience • Nanotechnology • Preventive arms control • Technology assessment • USA

J. Altmann (✉)
Experimentelle Physik III, Technische Universität Dortmund, Dortmund, Germany
e-mail: altmann@e3.physik.tu-dortmund.de

© Springer International Publishing AG 2017
M. Martellini, A. Malizia (eds.), *Cyber and Chemical, Biological, Radiological, Nuclear, Explosives Challenges*, Terrorism, Security, and Computation,
DOI 10.1007/978-3-319-62108-1_4

1 Introduction

Nanotechnology (NT) – sometimes grouped with biotechnology, information technology, cognitive science and other fields under the notion of converging technologies [1, 2] – is about the analysis and engineering of systems at the nanometers (1 nm = 10^{-9} m) size range: between about 1 nm (several atoms) and 100 nm (very large molecule). Following the projections and promises that NT will bring the next industrial revolution, with far-reaching consequences in how we communicate, produce and live, the highly industrial states have launched big NT research and development (R&D) programs around the year 2000. Since then, the USA with its National Nanotechnology Initiative (NNI, founded 2001), Japan, and the European Union have spent on the order of $1 billion per year each, the rest of the world combined at a similar level. By now, there are over 60 national NT programs [3].

From the beginning, military aspects have played an important role in the US NNI; until 2008, the Department of Defense (DoD) got about one third of the Federal funding (see Fig. 1). This is quite different from the other NT R&D programs. With respect to military R&D of NT (as in all other fields of military R&D), the USA sets a precedent for all other countries. For this reason and due to its unequalled transparency in military matters, potential future applications of NT in the military can be studied best by investigating the US work.

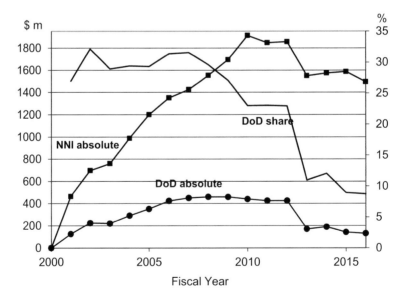

Fig. 1 Expenditure for the US National Nanotechnology Initiative (NNI, squares) and, within that, for the Department of Defense (DoD, *dots*) in nominal dollars (*left axis*) and percentage of DoD funding in the total (*thick curve*, *right axis*) (Numbers from the annual NNI Supplements to the President's Budget [29])

While there is considerable information available on concrete aspects of military NT R&D in the USA (overviews [4, 5], specific [6]) much of which takes place at universities with results published in academic journals, systematic studies of military applications are relatively scarce, in particular with a view to international security and preventive arms control. My study, based on work mainly done during years 2004–2005, appeared in 2006 [7]. It discussed 30 generic areas of potential military NT applications and gave recommendations for preventive limitations (see Sects. 3 and 4 below). A German book of the same year explained fundamentals of NT and presented the military potential of NT in more detail, in the categories: Intelligence and Command Systems, Platforms and Carriers, Weapons and Munition, Soldier, and Support [8]. This book neither treated international-security consequences nor preventive limitations. A much shorter overview was given in a brochure from the Netherlands [9]. NT Applications for chemical and biological defense were presented in 2009, again without consideration of arms control [10]; a short discussion was given by Kosal [11]. Nasu and Faunce [12] presented arguments about international regulation of "nano-weapons". Aspects of international humanitarian law were preliminarily assessed with respect to potential toxic effects of engineered nanomaterials and nanoparticles [13], with respect to enhanced thermobaric explosives, precision-guided, insect-size air vehicles and invisibility cloaking [14], and with respect to NT-enabled attacks on natural and potential future artificial photosynthesis [15]. A rather pessimistic assessment of the chances for preventive arms control of NT was given by Whitman [16], arguing that civilian and military applications would be too close to each other and that the dangers of malign uses of and an arms race in NT are not appreciated sufficiently.

After more than 15 years of R&D for military applications of NT, significant advances have been made, but the great general breakthrough with broad application throughout military systems has not yet appeared. Nevertheless, dangerous developments are arriving or appear on the horizon. Small armed drones (such as the TiGER [17] or the Switchblade [18]) have been deployed, but in these probably NT is still only involved in the computer chips on board. 3-D printing has been used to produce first firearms from plastics and metal without NT [19]; carbon-nanotube composite materials for gun barrels or small-missile casings have not yet arrived. Autonomous weapon systems do not yet exist, but could arrive – with or without NT – relatively fast; the international community is fully justified in being alarmed and discussing about a prohibition or limitation in the context of the Convention on Certain Conventional Weapons [20]. Scatterable small sensors, micro-robots and swarms of small vehicles are still under research, but prototypes are being demonstrated [21]. Genetic manipulation has become much easier with the CRISPR/Cas technique, again somewhat independent of NT [22]. Far-reaching concepts such as enhanced or superhuman soldiers have gained in traction [23, 24].

This chapter presents developments in military nanotechnology that have occurred since the early 2000s. My evaluations and assessments of the period 2004–2005 are revisited in the light of years 2016–2017. The focus is on near- to mid-term possibilities. Potential military applications of so-called molecular nanotechnology [25, 26] are not discussed here; the earlier general assessments [27] remain valid.

The next Sect. 2 presents developments in military R&D of NT since 2001, with the main focus again on the USA. Section 3 discusses whether the former assessment under criteria of preventive arms control is still appropriate. Section 4 re-considers the recommendations for preventive arms control. General conclusions are presented in Sect. 5.

2   Military Research and Development After 15 Years of the US National Nanotechnology Initiative

From the beginning of the US National Nanotechnology Initiative (NNI) in 2001, military research and development (R&D) figured prominently. Since the US has spent by far the most on military NT R&D – in 2006 my cautious estimates were 80–90% of the global funding– the US has set a precedent and certainly has acted as a role model for other countries, even if these could not or did not match the US spending level. It seems that the US is still setting the pace of military R&D of NT and is most advanced with actual application of NT-enabled systems in the military. Thus, it is sensible to give the US developments a strong emphasis. This is made easier by the relatively high degree of transparency of military R&D in this country.

2.1   Military NT in the USA

As Fig. 1 shows, the Federal funding for the National Nanotechnology Initiative was increased over 10 years to $ 1.8 billion per year, with reduction to $ 1.5–1.6 billion in 2013. In the first 7 years the military share was increased in proportion, at about 30% of the total. From 2008 the DoD part levelled off at slightly above $ 400 million, but was reduced from 2013 on to less than $ 200 million, bringing the DoD share to around 10%. The reasons given for these adjustments were on the one hand expiration of some projects, on the other hand competition with other military R&D programs [28, p. 11]. Obviously in the face of reductions of the overall R&D budget of the DoD[1] NT R&D was not deemed more important than other areas.

The balancing between the budget categories Basic Research ("Category 6.1"), Applied Research ("Category 6.2") and Advanced Technology Development ("Category 6.3") has been roughly the same as traditionally in DoD R&D: about 50%, 40%, and 10%, respectively [30].

In recent years, following the efforts of the NNI at large to improve lab-to-market transitions [31], also the DoD work has put a stronger emphasis on applications and transitioning. Already in 2010 the Air Force Research Laboratory published a list of

---

[1]The annual DoD R&D budget had been around $50 billion in the 1990s, was increased to above $75 billion 2005–2010, then was reduced again, passing through $60 billion in 2013. This was roughly parallel to the time course of the total defence budget with values of $400 billion, $760 billion and $600 billion, respectively (all nominal dollars) [33].

Table 1 Research results of the Air Force Research Laboratory that have led to applications (fielded, in use), transitions (being validated, scaled up and certified in system-level testing) and innovations (may enable future paradigm shifts) [32]

Applications

Nano-layer midwave infrared detectors and cameras

Nano-particles enable superconducting wire for compact power systems

Nickel nano-strands for aircraft lightning strike protection and electro-magnetic shielding

Nano-coating enables new class of high-power microwave devices

Commercialization of nano-composites for space systems insulation

Dualband quantum well infrared camera for target identification

Transitions

Semiconductor quantum well lasers for aircraft self-protection

Nano-taggants for biological threat detection

Self-protection and self-healing of polymer nano-composites

Nano-membrane flexible electronics for surveillance radar

Nano-jet micro-thrusters for space propulsion

DNA[a] Nano-taggants for covert watermarking, tracking and communication

Nano-aluminum alloy to replace titanium in liquid rocket engines

Nano-composite coatings for F-35 lift fan (STOVL)[b] and RL-10 rocket engine turbopump

Innovations

First electrically pumped semiconductor room-temperature terahertz radiation

Nano-additive demonstrates potential for JP-900 fuel

Processing of explosive formulations with nano-aluminum powder

Nano-column fabrication for gallium nitride radio frequency electronics

Photon-plasmon-electron conversion enables a new class of imaging cameras

Nano-particle lubricant produced by biosynthesis

Polymer nano-layer coatings for optical applications

Nano-crystalline zinc oxide for high performance thin film transistors

Quantum cascade laser nano-antenna for future identification of biological threats

Uncooled infrared detector made possible with controlled carbon nano-tube array

[a]Deoxyribonucleic acid

[b]Short take off vertical landing

research projects that have led to application (already in use), transitions (validated and certified in system-level testing), and innovations (that may enable future paradigm shifts) (Table 1). The Air Force also does research on environmental, safety and occupational health risks of nanomaterials.

There are far too many projects in military NT R&D in the USA to be reviewed or even only listed here. Thus, only a few exemplary activities are presented.[2]

Relatively early in the NNI, about 20 Centers of Excellence were founded, three of them within/by the DoD: the Institute for Soldier Nanotechnologies (ISN) at MIT, the Institute for Nanoscience (NSI) at the Naval Research Laboratory, and the Center for Nanoscience Innovation for Defense (CNID) at the University of California.

The ISN, founded 2002 by the Army with contributions from industry, at a budget of around $10 million per year, has been evaluated and prolonged every 5 years. The work is characterized as basic research (6.1) and as applied research (6.2).

---

[2] For more information see the annual reports to Congress on DoD NT R&D until 2009 [e.g. 39]. Later, overviews on DoD activities were given in the annual NNI Budget Supplements [e.g. 40].

Advanced technology development (6.3) is left to other institutions and companies by transitioning results to them. Over time, the goals and the structure have been adapted (Table 2).

The vision of a "21st-century battlesuit that combines high-tech protection and survivability capabilities with low weight and increased comfort", forming splints or releasing medications and providing energy [34, 35] has been less emphasized in recent years [36, 37]. In particular the concept of a powerful exoskeleton that had figured prominently in the reports about the founding of the ISN [38], seems to have been laid aside. Also the ISN stresses transitioning from laboratory innovations to more advanced stages of development; Table 3 lists a few examples.

The Institute for Nanoscience (NSI) at the Naval Research Laboratory was founded in 2001 and got a new laboratory building with many facilities in 2003 [49]. It does "multidisciplinary research at the intersection of the fields of materials,

Table 2 Research fields and structure of the ISN over time. Slight name changes are not listed

2004/2006 Seven research teams [41, 42]	1	Energy-absorbing materials
	2	Mechanically active materials and devices
	3	Sensing and counteraction
	4	Biomaterials and nanodevices for soldier medical technology
	5	Processing and characterization – the nanofoundries
	6	Modeling and simulation of materials and processes
	7	Systems design, hardening, integration, and transitioning
2009/2013 Five strategic research areas [43, 44]	1	Lightweight, multifunctional, nanostructured materials and hybrid assemblies
	2	Soldier medicine – prevention, diagnostics, and far- forward care
	3	Multiple blast and ballistic threats – materials damage, human injury mechanisms, and lightweight protective systems
	4	Hazardous substances sensing, recognition, and protection
	5	Nanosystems integration for protected communications, diagnostic sensing, and operational flexibility in complex environments
2016 Three strategic research areas [45]	1	Lightweight, multifunctional nanostructured materials
	2	Soldier medicine – prevention, diagnostics, and far-forward care
	3	Blast and ballistic threats: materials damage, injury mechanisms, and lightweight protection

Table 3 Examples of ISN Research Accomplishments and Transitions [46–48]

Examples of ISN research accomplishments and transitions
Optoelectronic fibers
Detection of hazardous substances
Blast and ballistic protection
Health monitoring, protection, and treatment
Wireless power
Nano-particles and quantum dots
Gentle chemical vapor deposition

Table 4 Programmes and topics of the NSI [50]

Nanoassembly
Nanofilaments: interfacial, interactions, manipulation and assembly
Directed self-assembly of biologically-based nanostructures
Chemical templates for nanocluster assembly
Assembly of laterally coupled molecular nanostructures
Interfacing electronics and biomolecular processes
Nano-optics
Organic and biologically conjugated luminescent quantum dots
Nanoscale-enhanced processes in a quantum dot structures
Nano-engineered photovoltaic devices
Integration of nanostructured light emitting devices
Photonic crystal optoelectronic components in advanced materials
Nanochemistry
Nanoelectronics
Coherence, correlation and control in nanostructures
Nanocluster electronics by macromolecular templating
Nanoelectronics physics research
Neural-electronic interfaces
Nanomechanics
Nanoscale measurement techniques
Nanochemical resonators and advanced nanodynamics

electronics, and biology", mostly at the fundamental level. Table 4 lists the programs and topics, there has been nearly no change since 2007–2009.

The Center for Nanoscience Innovation for Defense was founded end-2002 at the University of California at Santa Barbara, Los Angeles and Riverside. Concrete information about the activities of this center was always difficult to get. It seems that it is no longer active; at Los Angeles it is listed under "Previous Centers" [51]. Whether the ten industrial partners mentioned at the founding [52] are still active in this context is unclear.

The Army Research Laboratory and the Air Force Research Laboratory, while not among the NNI Centers of Excellence, were and still are very active in nano-technology. Universities are involved into military research by several funding bodies, among them the Defense Advanced Research Projects Agency (DARPA) and the research offices of the three services Army, Navy and Air Force.

2.2 Military NT in Other Countries

The US activities and their high budgets have certainly been noticed in other countries. In particular in countries that see themselves as potential adversaries the US efforts must have contributed to their own military-NT programs, and indications thereof had already been observed in 2003–2004.

The first part of Table 5 gives the countries with known or presumed military programs listed in [53].

Table 5 Countries other than the USA with nanotechnology activities listed in the 2006 book [55], in the 2009 DoD report to Congress [56], and found in other sources, in the sequence of the respective source. Where military activities including preparations were mentioned explicitly this is noted by (M); where a military program was presumed it is noted by (~M)

Germany, UK (M), France (M), Netherlands (M), Sweden (M), Russia (~M), China (~M), Australia (M), Israel (M), India (M)
China, Japan, Singapore (M), South Korea (M, 2 defense research centers), Taiwan, India (M, programs at several defense research institutes), Thailand, Vietnam, Iran (M, centers of defense-non-defense transfer, including work on nanoenergetics), Israel (M, with project on micro unmanned vehicles against individual terrorists), Australia (M, electronics, photonics, sensors, actuators, composites), Russia (M, energetics, aerospace, signature reduction, temperature and oxidation resistance, coatings), Germany (M, for protection, for equipping combat troops), France (M, composites, coatings, energies), Canada, Mexico, Brazil and others in the Americas, UAE, Pakistan
Russia (M)[a], India/Brazil/South Africa (M)[b], India (M)[c], China (M)[d], South Korea (M)[e]; European Union (M)[f]

[a]Ministry of Defence and military-industrial complex mentioned as partners in State NT program [57, p. 34].
[b]Energy & Defence mentioned as one area of cooperation [58, p. 28]
[c]More than 30 DRDO (Defence Research & Development Organization, India) laboratories are pursuing R&D in NT for defense [59].
[d]Technology for high energy fuel pyrotechnics, materials, coatings, sensors, integrated micro-electromechanical systems for micro/nano-type aircraft, micro/nano-satellite systems and special-purpose integrated technology platform [60].
[e]Defense Acquisition Program Agency among recipients of government funds [61].
[f]In the context of dual-use research [62].

It is probably safe to assume that many of the more than 60 national NT R&D programs have a military component. In the preparation of the present chapter, no detailed investigations about activities and budgets in other countries could be done. Instead, the second part of Table 5 shows the countries listed in the 2009 report to Congress of the US DoD; the reference explicitly mentions military activities only for a few countries.[3] The third part of Table 5 gives a few countries for which a military NT program was found by a cursory search.

Lacking concrete budget figures for countries other than the USA, my cautious estimate had been that in 2003 the combined expenditure outside of the USA was between one tenth and one quarter of the US one, then $243 million (current dollars) [54]. Assuming that in other countries military NT programs were formed and expanded, it is well possible that the share has grown to one third or one half of the present US spending ($130 million in 2016). If true, this would mean that the closest competitor spends no more than one or two tenths of the US expenditure, and the others considerably less.

---

[3]From 2010 on there was no longer an annual report on the DoD NT R&D program; instead the annual NNI budget supplements gave some detail on DoD activities. These documents do not discuss military NT in other countries.

2.3   Estimated Arrival Times

In my 2006 book, I had conceived of, or found in the literature, 30 fields of potential military applications of NT (Table 6) and made rough estimates of when they might be deployed [63]. The arrival times with respect to the time of writing (years 2004–2005) were classed in the following categories: A (0–5 years), B (5–10 years), C (10–20 years), D (more than 20 years), speculative, and unclear. The ensuing years thus were: A: 2006–2010, B: 2011–2015, C: 2016–2025, D: after 2025. Taking into account continued advance, some applications were assigned several such time periods.

An in-depth new assessment of the arrival times was not possible here, but a coarse estimate was done. In particular a fast check was made whether the applications estimated to arrive in the last 10 years, that is categories A and B, have actually been deployed. Table 6 shows the results of the comparison: the second column repeats the estimates published 2006, and the third one shows whether the estimate had been about correct.

Looking at what has actually been achieved in the last 10 years, the expectations of the respective first arrivals in categories A and B were roughly correct in many areas, in particular the generic ones. This applies to electronics, photonics, magnetics; computers, communication; software/artificial intelligence; materials; energy sources, propellants and explosives; light armor/garments.

There seems to be no application that has arrived earlier than estimated. However, for about two thirds of the areas the estimated first-deployment time has passed by now while the intended application is still under research or in development. Here the new estimate will postpone the first-deployment time by further 5–10 years. This holds for: lighter, faster and more agile vehicles; variable camouflage; distributed sensors for all purposes; heavy armor; metal-less arms, small guidance, armor piercing, small missiles; soldier systems; implants and body manipulation; autonomous systems; mini−/micro-robots; bio-technical hybrids; small satellites/space launchers; chemical weapons; biological weapons; chemical/biological protection/neutralization. With respect to nuclear weapons, auxiliary systems for safing, arming and fusing may have been built into them, but whether this has been done could not be looked at. The uncertainty about the expected advance in computer modelling persists, and very small (e.g. fusion-only) nuclear weapons remain speculative.

3   Military-Technology Assessment of NT Updated

The 30 fields of potential military applications of NT had been evaluated under criteria of preventive arms control. This concept is explained here in short terms, followed by the evaluation proper and the list of applications where the strongest dangers can ensue. The evaluations are re-examined to find possible changes.

Table 6 Potential military NT applications, starting with more generic ones. The second column gives the time to potential introduction as roughly estimated in 2004–2005[64]: A (next 5 years), B (5–10 years from now), C (10–20 years), D (more than 20 years); 'u' means unclear, '??' denotes speculative applications where a time frame could not be estimated. The third column indicates whether the earlier assessment had been about correct, or whether the (first) deployment as estimated in 2017 will likely come several years (y) later

Application area	Estimated time (2004–2005)					About correct? (2017)
Electronics, photonics, magnetics	A	B	C			Y
Computers, communication	A	B	C	D		Y
Software/artificial intelligence	A	B	C	D	??	Y
Materials		B	C			Y
Energy sources, energy storage	A	B	C			Y
Propulsion		B	C			Y
Vehicles		B	C			+10 y
Propellants and explosives	A	B				Y
Camouflage		B				+5–10 y
Distributed Sensors						
Generic	A					+5–10 y
Battlefield		B				+5–10 y
Verification	A					+5–10 y
Armor, protection						
Heavy armor		B				+5–10 y
Light armor/garments		B				Y
Conventional weapons						
Metal-less arms		B				+5–10 y
Small guidance		B				+5–10 y
Armor piercing	A	B				+5–10 y
Small missiles		B				+5–10 y
Soldier systems		B	C			+5–10 y
Implanted systems, body manipulation		B	C		u	+5–10 y
Autonomous systems		B	C			+5–10 y
Mini−/micro-robots		B	C			+5–20 y
Bio-technical hybrids		B	C			+5–10 y
Small satellites/space launchers		B	C			+5–10 y
Nuclear weapons						
Auxiliary systems	A	B				?
Computer modelling					u	?
Very small weapons					??	Y
Chemical weapons		B D	C			+5–10 y
Biological weapons		B D	C			+5–10 y
Chemical/biological protection/neutralisation	A	B				+5–10 y

3.1   A Short Look at Preventive Arms Control [65, 66]

The goal of preventive arms control is similar to the goal of "normal" arms control, namely to prevent dangers to international peace and military stability. Rather than capping numbers after new weapons technologies and systems have been widely deployed, such technologies and systems are to be prevented from being introduced from the beginning. Thus, preventive arms control prohibits deployment and use; often the earlier stages of development and testing are explicitly included, too. Examples of arms-control or disarmament treaties with preventive function are the Nuclear Non-Proliferation Treaty of 1968, the Chemical Weapons Convention of 1993 and the Comprehensive Nuclear Test Ban Treaty of 1996.

If a potential new weapon or other military technology becomes visible at the horizon, research for preventive arms control is advisable. Such research proceeds in four steps:

- Prospective scientific analysis of the technology in question. What would be the properties of the new weapon? How would the propagation to the targets work? What would be the effects on the targets?
- Prospective analysis of the military-operational aspects. What would be probable uses? Against which targets? How about unusual employment forms? What is the potential for collateral effects?
- Assessment of both under the criteria of preventive arms control.
- If preventive limitations seem needed: Devising possible limits and verification methods. At which stage in the "life cycle" (research, development, testing, deployment, use) should the prohibition work? Positive or legitimate uses of the technology need to be pondered, and a balance needs to be struck between these and the problematic military uses that one would like to curtail. Methods and procedures for verification of compliance need to be developed, balancing the need for transparency concerning the objects to be prohibited while protecting military and industrial secrets and civilian privacy. The same holds for confidence-building measures.

In the ideal case then the relevant nations take up the results of such research and start negotiating about limitations, finally concluding a new treaty or protocol. Proposals for preventive arms control with respect to problematic military applications of NT (see below) have so far not been taken up by countries. However, there is one important exception: autonomous weapon systems: This topic has been discussed since 2013 in the context of the Convention on Certain Conventional Weapons (CCW) [67].[4] Three informal expert meetings were held in Geneva in 2014, 2015 and 2016, and a formal group of governmental experts will meet in 2017 [68].[5] Principally,

---

[4] The full name is "Convention on Prohibitions or Restrictions on the Use of Certain Conventional Weapons Which May be Deemed to be Excessively Injurious or to Have Indiscriminate Effects". This framework agreement at present has five Protocols of which Protocol IV on Blinding Laser Weapons (1995) is the only preventive one; it could be a role model for a Protocol VI on AWS.

[5] The notion used in the CCW context is „Lethal Autonomous Weapons (LAWS)", but non-lethal AWS pose similar fundamental questions. Thus, optimally all AWS should be included in a future CCW protocol prohibiting or limiting weapon systems that select and engage targets without human control.

this is independent of the question whether NT is used in them, but NT would greatly enhance the capabilities of AWS, including the possibility to strongly miniaturize them. And since present-day computer chips use structure sizes markedly below 100 nanometers, NT would be inside by this token anyway.

The criteria of preventive arms control can be grouped under three headings:

I. Adherence to and further development of effective arms control, disarmament, and international law:

   1. Prevent dangers to existing or intended arms control and disarmament treaties.
   2. Observe existing norms of international humanitarian law.
   3. No utility for weapons of mass destruction.

II. Maintain and improve stability:

   1. Prevent destabilization of the military situation.
   2. Prevent arms race.
   3. Prevent horizontal or vertical proliferation/diffusion of military-relevant technologies, substances or knowledge.

III. Protect humans, environment, and society:

   1. Prevent dangers to humans.
   2. Prevent dangers to environment and sustainable development.
   3. Prevent dangers to the development of societal and political systems.
   4. Prevent dangers to the societal infrastructure.

Criteria group III applies to new military technologies in peace time (for example, if a new explosive is toxic for the environment). Whenever an envisioned new military technology can create problems under one or more of the criteria, considerations about preventive limitation are advisable.

3.2 Areas of Potential Military Application of NT

The 30 potential military applications of NT conceived of or found in the literature had been evaluated in 2004–2005 using the above criteria; Table 7 shows the summary results. Ten to twelve years later I do not see a reason why this evaluation needs to be changed. The arms race and proliferation criteria are flagged with nearly all applications. The more generic areas such as electronics, computers or materials pose no direct problems for arms control and international law and little or none to humans, environment and society; some could have positive effects in the environment. The more military-specific applications, on the other hand, can have negative effects with respect to several criteria, in particular if they contain a weapon function. Nuclear weapons much smaller than existing ones get negative signs in nearly all categories, but they are still hypothetical. On the other hand, new chemical or biological weapons get similar negative grades, but are quite possible, coming

Table 7 Evaluation of potential military NT applications under the criteria of preventive arms control; see [69] for more detail on the applications. Evaluation levels: –: danger 0: neutral +: improvement u: unclear; these refer only to the direct consequences.

Prev.-AC criterion Application area	I.1 AC	I.2 IHL	I.3 WMD	II.1 Dest.	II.2 AR	II.3 Prol.	III.1 Hum.	III.2 E/S	III.3 S/P	III.4 SI
Electronics, photonics, magnetics	0	0	0	0	–	–	0	0	0	0
Computers, communication	0	0	0	–0	–	–	0	0	–0	0
Software/artificial intelligence	0	0	0	–0	–	–	0	0	–0	0
Materials	0	0	0	0	–	–	u	u	0	0
Energy sources, energy storage	0	0	0	0	–	–	0	0+	0	0
Propulsion	0	0	0	–0	–	–	0	0+	0	0
Vehicles	0	0	0	–0	–	–	0	0+	0	0
Propellants and explosives	0	0	0	0	–	–	u	u	0	0
Camouflage	0	0	0	0	–	–	0	0	0	0
Distributed sensors										
Generic	0	0	0	0	–0	–0	0	0	–	0+
Battlefield	0	0	0	–	–	–	0	0	–	0
Treaty verification	+	0	0+	+	+	+	0+	0+	0	0+
Armor, protection	0	0	0	0	–	–	0	0	0	0+
New conventional weapons										
Metal-less arms	–0	0	0	0	–	–	–	0	–	–
Small guidance	0	0+	0	–	–	–	0	0	0	–
Armor piercing	0	0	0	0	–	–	0	0	0	0
Small missiles	–0	–0+	0	–	–	–	–	0	0	–
Soldier systems	0	0	0	0	–0	–0	0	0	0	0
Implanted systems, body manipulation	–0	–0	0	–0	–	–	–0	0	–	0
Autonomous systems										
Unarmed	0	0	0	0	–	–	–0	0	0	0
Armed	–0	–	0	–	–	–	–0	0	0	–0
Mini-/micro-robots incl. bio-technical hybrids										
No weapon function	–0	0	0	–	–	–	0	0	–	–
Target beacon/ armed	–0	–	0	–	–	–	–0	0	–	–
Small satellites/space launchers	–0	0	0	–	–	–	0	0	0	–
Nuclear weapons										
Auxiliary systems	0	0	–0+	0	–0	–0	0	0	0	0
Computer modelling	–	0	–	0	–	–	0	0	0	0

(continued)

Table 7 (continued)

Prev.-AC criterion Application area	I.1 AC	I.2 IHL	I.3 WMD	II.1 Dest.	II.2 AR	II.3 Prol.	III.1 Hum.	III.2 E/S	III.3 S/P	III.4 SI
Very small weapons	–	–0	–	–	–	–	0	–	–0	–0
New chemical weapons	–	–	–	–	–	–	–0	–0	–0	–
New biological weapons	–	–	–	–	–	–	–0	–0	–0	–
Chemical/biological protection/ neutralization	0	0	+	–0+	–0	–0	+	0+	+	+

Prev. preventive, *AC* arms control, *IHL* international humanitarian law, *WMD* weapons of mass destruction, *Dest.* destabilization, *AR* arms race, *Prol.* proliferation, *Hum.* humans, *E/S* environment/sustainability, *S/P* societal/political systems, *SI* societal infrastructure

closer with fast advances in the life sciences and nanobiotechnology. Distributed sensors deployed on the battlefield can act destabilizingly, but if used for treaty verification they can support arms control. Positive evaluation under many criteria resulted for protective or neutralizing materials against chemical or biological weapons. A special case is formed by non-medically motivated implanted systems and other manipulation of soldiers' bodies: Here some destabilization could occur, but more important may be that they could create facts so that a thorough discussion in society about what should be allowed in this respect would be undermined.

3.3   Most Dangerous Military NT Applications

Following from the unchanged assessment, the list of applications that raise strong concerns directly has not changed as well as shown in Table 8.

The arms race and the proliferation criteria apply throughout the list due to military advantages. Particular urgency will ensue where threats to one's forces or restrictions of their operations will be feared. The other criteria are affected in different ways, as explained in the next paragraphs.

Distributed small sensors, that is self-contained systems that can be scattered in high numbers and form a flexible communication network are in the list because if used on the battlefield, while difficult to be seen, they would help to locate targets, making easier fast precision attacks. In civilian life they would endanger privacy and industrial secrecy, as well as freedom, if used for eavesdropping and tracking – by state agencies, companies or criminals.

New conventional weapons with higher speed and more precise targeting would have some negative effect on stability. Strong destabilization would result if conventional attack against nuclear-strategic installations came into reach [71]. Dangers to humans and society would ensue if faster, more precise conventional weapons proliferated to criminals. Metal-free firearms that are not visible by x rays

Table 8 The potential military NT applications that pose the strongest dangers under the criteria of preventive arms control, together with the respective criteria group [70]

Area of concern Potential application	I. Arms control/ international humanitarian law/weapons of mass destruction	II. Military stability/ arms race/ proliferation	III. Humans/ environment/ society/ infrastructure
Distributed small sensors		X	X
New conventional weapons: Metal-free arms		X	X
Small missiles		X	X
Implants and other body manipulation		X	X
Autonomous fighting systems	X	X	
Small robots	X	X	X
Small satellites and launchers	X	X	
New chemical/biological weapons	X	X	X

or metal detectors would provide new options for terrorists. Gun control would become much more difficult if such weapons could be produced in masses by widespread 3-D printers.

Small missiles would make verification more difficult. Dangers to humans and society would result if they got into the hands of criminals, including terrorists.

Implants and other body manipulation that would reduce reaction times and link brains with computers could have some destabilizing effect. More relevant would be the effects on humans and society if military R&D and application contributed to general use in society, pre-empting a thorough debate of the pros and cons. Many negative secondary effects on body and mind are conceivable, similar to and beyond those of addictive drugs.

Autonomous fighting systems (today one would rather speak of autonomous weapon systems) could undermine arms control of conventional arms.[6] Strong dangers for compliance with international humanitarian law can be foreseen because discriminating between combatants and non-combatants and proportionality considerations in a complex combat situation require human levels of intelligence that computers and algorithms will not be able to achieve for a long time.[7] Because of very fast action and reaction, autonomous weapon systems would create strong pressures for fast attack if both opponents have got them. In a severe crisis, misunderstandings or system errors could lead to the start of hostilities; interactions between two partly automatic systems of command and control that could never be

---

[6] The main agreement in this category is the Treaty on Conventional Armed Forces in Europe of 1990; however, it is under severe stress at present and should be reinvigorated.

[7] In the 2006 book I was too cautious in stating that for "at least one decade –, artificial systems will not reach such a level" [73]. Despite remarkable advances in artificial intelligence, in 2017 it is obvious that this condition has not been reached; it is probably safe to state that this goal will not be achieved for at least two more decades.

tested together could fast escalate out of control [72]. Humans could be endangered in peacetime or under an occupation, if autonomous weapon systems were used by security forces, due to accidents or incorrect situation assessments by the control programs.

Small robots, if armed, could similarly undermine arms control, including the Anti-personnel Landmine Convention of 1997 in that they would not qualify as a mine but could function as such. Also here compliance with international humanitarian law would be at risk. Specific forms of destabilization would ensue if micro-robots could be sent covertly before armed conflict, ready to strike (or guide strikes) at important nodes from within at any time.[8] Together with a strong motive to send mini- and micro-robots for reconnaissance into an opponent's territory this would create high uncertainty and nervousness. Dangers to humans and society would result if small robots – for eavesdropping or attack – fall into criminals', especially terrorists', hands.

Small satellites, if used for anti-satellite attack, would counteract the general ban on space weapons that the international community has striven for since decades. Swarms of small satellites could improve military communication and detection of targets on Earth, resulting in higher threats of surprise attack. Much more destabilizing would be their capability to attack other satellites, either by kinetic impact or by rendezvous and docking, maybe with autonomous manipulation. Small launchers could be used to lift small satellites for such purposes, in particular on short notice. Such launchers would not depend on large spaceflight centers, but could take off from nearly anywhere, including ships and aircraft. Small satellites could also endanger the civilian infrastructure in outer space.

New chemical/biological weapons would contravene the Chemical and the Biological Weapons Convention. They would constitute weapons of mass destruction if capable of attacking many target humans, animals or plants. Even if NT and biomedical advances could provide mechanisms for very selective uses, maybe even against only one individual, the knowledge gained in the R&D could be used for agents and carriers targeting larger collectives. New chemical or biological weapons that would make them more selective and protect one's own forces and population, would undermine stability. New chemical/biological weapons would produce dangers to humans and society if criminals, in particular terrorists, would get hold of them; large groups of people or single individuals could be the targets. Accidental release from military facilities forms another risk. Biochemical weapons that would confuse large numbers of people for a sufficiently long time could lead to a breakdown of societal production and distribution systems with the indirect consequence of mass death.

Nuclear weapons are not listed because NT-enabled auxiliary systems (e.g. safing and triggering devices) would not change the fundamental characteristics, because it is unclear whether much more powerful computers would allow qualitatively

---

[8] Such a scenario had been described in a report by the US RAND Corporation [74], later wrongly assigned to Chinese military planning [75] while the Chinese source [76] just had reported to a Chinese military audience about the RAND report.

different modelling than present ones (e.g. rendering obsolete actual tests of radically new weapon designs), and because very small weapons (e.g. using pure fusion without a first fission state) are still hypothetical.

4   Recommendations for Preventive Arms Control

For several reasons, preventive limitations should not affect NT applications across the board. Many civilian NT applications will provide strong advantages, for example for more sustainable energy production and consumption or for health, and should not be hampered. Some military applications would be so close to civilian ones that limiting or outlawing them would simply be unrealistic, think for example of computers. NT-based better sensors or decontamination materials for biological- or chemical-warfare agents, once developed in the military, could help in preventing or mitigating terrorist attacks in civilian society. These and other sensors could be helpful in the co-operative verification of arms-control agreements.

As a consequence my recommendations for preventive arms control were directed at the most dangerous military NT applications (as listed in Table 8). Since the dangers arise rather from the respective applications, not from NT as such, the recommendations do not focus on NT, but rather on the military systems or missions. Thus, general prohibitions are advisable, independent of NT being used or not. This approach is also better for verification of compliance: if systems are categorized with the help of size and other characteristics observable from the outside, intrusive inspections of the components inside can be avoided.

Some dangerous NT applications would fall under existing arms-control treaties, for example the Biological Weapons Convention (BWC) or the Chemical Weapons Convention (CWC) with their respective general-purpose criteria, so that new agreements are not needed. However, the verification gap in the BWC should be closed.

With future widespread use of NT in civilian life, limits on military applications should be secured against circumvention by diverting civilian systems. Thus, for reliable prohibitions of certain military NT uses, some regulation of civilian systems is needed, striving to keep the restrictions to a minimum.

To prevent fast breakout, prohibitions should not only hold for the stages of deployment and use, but also for the earlier stages of development and testing.

A final consideration is that limitations should focus mainly on new, future applications where political, economic or military resistance will generally be lower than with systems that have already been deployed widely.

The most important recommendations in 2006 were [77]:

- A ban on self-contained[9] sensor systems below a certain size limit (3–5 cm), for the military and civilian sectors.
- A ban on small arms, light weapons and munitions that contain no metal, for the military and civilian sectors.

---

[9] Capable of functioning on their own, with power supply and communication link.

- A ban on missiles below a certain size limit (0.2–0.5 m), for the military and civilian sectors.
- A moratorium on body implants and other body manipulation that are not directly medically motivated, for the civilian and military sectors.
- A prohibition of armed, uninhabited vehicles[10] of normal size, with numerical limits on unarmed vehicles (exempting the already existing ballistic and cruise missiles).
- A prohibition of small, uninhabited vehicles below a certain size limit (0.2–0.5 m) in the military and civilian sectors.
- A ban on space weapons of all kinds – this would include nanotechnology-enabled small satellites used as weapons. Small satellites and launchers for other uses should be subject to regulation, notification and inspection.
- The Chemical Weapons Convention and the Biological Weapons Convention should be upheld, the latter should be augmented by a Compliance and Verification Protocol.

All these prohibitions and limits should include development and testing. For some of the rules, exceptions should be agreed upon for important positive civilian or military uses. For example, with respect to small missiles, firecrackers and life-saving apparatus for stranded ships should be exempted from the ban. Concerning small uninhabited vehicles, exceptions could be agreed upon for exploration of shattered or dangerous buildings or for surgical operations from inside the body. Abuse would need to be prevented by technical measures and licensing procedures. Verification of compliance could rely on various forms of on-site inspection, with allowed use of magnifying equipment to find indications of circumvention by systems of millimeter- or sub-millimeter size.[11]

Corresponding export controls could support the recommended prohibitions, but could not replace them, because most export controls are asymmetric in that some countries allow themselves the use of technologies while attempting to block access by others (the exceptions are the global bans of chemical and biological weapons).

It is noteworthy that, while non-state actors, in particular terrorist groups, cannot be parties to international agreements, their capability to develop sophisticated systems is lower by far than that of states. Thus, prohibitions among states would go a long way in limiting what types of NT-based weapons terrorists could use.

Already in 2006, I discussed the possibility that a complete prohibition of uninhabited armed vehicles might not be achievable. For this case, I recommended:

- Counting them under the CFE Treaty (and potential similar treaties for other world regions), adapting the treaty definitions where needed.
- No aiming and weapon release without human decision.
- Prohibition of qualitatively new types of nuclear-weapon carriers.

Similar recommendations were agreed in 2010 at the Berlin Workshop Arms Control for Robots [79]. With the present massive interest of states in armed drones

---

[10] On land, on/under water, in air.

[11] For more details on exceptions, inclusion of civilian uses and verification see [78].

(still remotely controlled at least in the attack function),[12] a general ban on armed uninhabited vehicles has become even more remote, even though from an international-security standpoint it would be the best solution, the same holds with respect to verification [80]. While a general ban seems unrealistic for the time being, a prohibition of autonomous weapon systems – that do not yet exist – is a possible outcome of the discussions among the CCW Member States. Here the term "meaningful human control" of attacks has become an important category [81, 82].

5   Conclusion: Responsible Use of NT in a Global Perspective

Special "nanoweapons" are not to be expected in the near to medium future. But NT will allow several new kinds of military systems and missions that could be dangerous for peace and international security as well as for humans and society. The recommendations show how such dangers could be contained. For the next one to two decades traditional forms of verification as they were agreed upon in nuclear, conventional as well as chemical arms control would suffice, namely on-site inspections in declared military installations and in some industry facilities. Over the longer run, however, this solution may run into problems, as production technologies will become smaller, cheaper and more widespread. 3-D printers, DNA synthesizers, chemical microreactors could be used in small, inconspicuous facilities including private basements. With easily available raw materials, miniature weapon systems or new biochemical agents could be made using software downloaded from the Internet. Verification would then need to be extremely intrusive, actually anytime, anywhere, in any country. That such intrusiveness would be difficult to accept for high-technology countries was already visible in the 2001 failure of the efforts to add a compliance and verification protocol to the BWC that foresaw much more limited inspection rights [83]. Anytime, anywhere would include all military installations, endangering the secrecy that armed forces want to maintain to not facilitate (surprise) attack by an adversary. Should the required intrusiveness turn out as too big a hurdle, then arms races in revolutionary technologies with unprecedented destabilization are the foreseeable outcome, unless humankind will learn to organize international security in a fundamentally different way, no longer mainly relying on national armed forces.[13]

The recommendations made above for certain potential military NT applications do not require fundamental change in the international system. Based on them, principally, the relevant states could start negotiations on appropriate limitations

---

[12] At present, 11 countries have got armed drones and further 10 are developing them [84].

[13] This could mean overcoming the security dilemma by creating a monopoly of legitimate violence resting with a democratized United Nations Organization, similar to the situation in many democratic states where citizens need not arm themselves because the state protects their security. Obviously getting there would be a long and complicated process, but first steps in this direction have been made already, with UN, International Tribunals, European Union etc.

with adequate verification. That they have not yet done so may have to do with the promise of military advantages from new technology. In particular the USA has the explicit goal of military-technological superiority [85–87]. That new weapons technology can be turned against it when potential adversaries will have caught up is seen as possible or even probable; the countermeasure is a permanent effort to stay far ahead. The price for the USA as well as for the world at large is ever increased threats and destabilization. Changing this fundamental approach will need a more enlightened view of national security that is not narrowly focused on one's own military strength but sees national security embedded in international security, where war prevention takes precedence over war preparation. As many arms-control treaties show, there is such a tradition in the USA, too.

Among the countries with active NT programs, there is an International Dialogue on Responsible Research and Development of Nanotechnology [88, 89]. Here questions of NT risks for the environment, health and society are being discussed and protective measures coordinated. Unfortunately, military NT applications are not part of this dialogue, even though they can produce the strongest dangers. An international dialogue on responsible handling of potential military NT uses, in particular on limitations of the most dangerous of such uses, is urgently needed. This needs insight into the dangers and subsequent political decisions. Awareness of the problem in science and the general public will be important to set such a process into motion.

References

1. Roco, M.C., Bainbridge, W.S. (eds.): Converging Technologies for Improving Human Performance – Nanotechnology, Biotechnology, Information Technology and Cognitive Science. Boston etc., Kluwer (2003), http://www.wtec.org/ConvergingTechnologies/1/NBIC_report.pdf
2. High Level Expert Group "Foresighting the New Technology Wave", A. Nordmann, Rapporteur: Converging Technologies – Shaping the Future of European Societies. Brussels, European Commission, Directorate-General for Research (2004), http://nanotech.law.asu.edu/Documents/2009/09/final_report_en_243_5158.pdf
3. Clunan, A., Rodine-Hardy, K. (with Hsueh, R., Kosal, M., McManus, I.): Nanotechnology in a Globalized World – Strategic Assessments of an Emerging Technology, PASCC Report Number 2014-006, Monterey CA, Naval Postgraduate School (2014), http://www.dtic.mil/dtic/tr/fulltext/u2/a613000.pdf
4. Director, Defense Research & Engineering: Defense Nanotechnology Research and Development Program. Department of Defense, Washington DC (December 2009), http://www.nano.gov/sites/default/files/pub_resource/dod-report_to_congress_final_1mar10.pdf (and earlier reports from 2005, 2006, 2007)
5. Department of Defense Sections in: NNI Supplement to the President's 2017 Budget, Washington DC: National Science and Technology Council (March 2016), http://www.nano.gov/sites/default/files/pub_resource/nni_fy17_budget_supplement.pdf (and earlier years)
6. Nano for Defense Conference series, https://www.usasymposium.com/nano/
7. Altmann, J.: Military Nanotechnology: Potential Applications and Preventive Arms Control. Routledge, Abingdon/New York (2006); Russian Version: Tekhnosphera, Moscow
8. Kretschmer, T., Wiemken, U. (eds.): Grundlagen und militärische Anwendungen der Nanotechnologie. Report, Frankfurt (Main)/Bonn (2006)

9. Schilthuizen, S., Simonis, F.: Nanotechnology İnnovation Opportunities for Tomorrow's Defence. TNO, Eindhoven (2009), http://publications.tno.nl/publication/13006495/s6VLVp/ Nanotechnology-book.pdf
10. Kosal, M.: Nanotechnology for Chemical and Biological Defense. Springer, New York (2009)
11. Kosal, M.E.: Anticipating the Biological Proliferation Threat of Nanotechnology: Challenges for International Arms Control Regimes. In Nasu, H., McLaughlin, R. (eds.): New Technologies and the Law of Armed Conflict. Asser/Springer, The Hague/Berlin etc. (2014)
12. Nasu, H., Faunce, T.: Nanotechnology and the International Law of Weaponry: Towards International Regulation of Nano-Weapons. Journal of Law, Information and Science 20 (1) 21–54 (2010)
13. Nasu, H.: Nanotechnology and Challenges to İnternational Humanitarian Law: A Preliminary Legal Assessment. International Review of the Red Cross 94 (886) 653–672 (2012)
14. Nasu, H.: Nanotechnology and the Law of Armed Conflict. In Nasu, H., McLaughlin, R. (eds.): New Technologies and the Law of Armed Conflict. Asser/Springer, The Hague/Berlin etc. (2014), pp. 143–157
15. Faunce, T.: Nanotechnology and Military Attacks on Photosynthesis. In Nasu, H., McLaughlin, R. (eds.): New Technologies and the Law of Armed Conflict. Asser/Springer, The Hague/ Berlin etc. (2014), pp. 175–190
16. Whitman, J.: The Arms Control Challenges of Nanotechnology. Contemporary Security Policy, 32, 99–115 (2011)
17. Crane, D.: MBDA TiGER (Tactical Grenade Extended Range) Small UAS/UAV ... Defense Review (October 2011), http://www.defensereview.com/mbda-tiger-tactical-grenade-extended-range-mini-flying-bombkamikaze-drone-for-tactical-reconnaissance-and-precision-kill-missions-is-low-observable-seriously-lethal-kittys-got-a-temper/
18. Switchblade, AV Inc. (2017), http://www.avinc.com/uas/view/switchblade
19. Jenzen-Jones, N.R.: Small Arms and Additive Manufacturing: An Assessment of 3D–Printed Firearms, Components, and Accessories. In King, B., Mcdonald, G. (Eds.): Behind the Curve: New Technologies, New Control Challenges. Occasional Paper 32, Small Arms Survey, Geneva (2015), http://www.smallarmssurvey.org/about-us/highlights/highlights-2015/highlight-op32. html
20. Background – Lethal Autonomous Weapons Systems. The United Nations Office at Geneva (2017), http://www.unog.ch/80256EE600585943/(httpPages)/8FA3C2562A60FF81C1257CE 600393DF6?OpenDocument; and 2014, 2015, 2016 Meetings of Experts on LAWS
21. Department of Defense Announces Successful Micro-Drone Demonstration (January 2017), https://www.defense.gov/News/News-Releases/News-Release-View/Article/1044811/ department-of-defense-announces-successful-micro-drone-demonstration
22. Sander, J.D., Joung, J.K.: CRISPR-Cas Systems for Editing, Regulating and Targeting Genomes. Nature Biotechnology 32, 347–355 (2014)
23. Lin, P., Mehlman, M.J., Abney, K.: Enhanced Warfighters: Risk, Ethics, and Policy. The Greenwall Foundation (January 2013), http://ethics.calpoly.edu/Greenwall_report.pdf
24. Kott, A. et al.: Visualizing the Tactical Ground Battlefield in the Year 2050: Workshop Report. ARL-SR-0327, US Army Research Laboratory, Adelphi, MD (June 2015), http://www.arl. army.mil/arlreports/2015/ARL-SR-0327.pdf
25. Drexler, K.E.: Engines of Creation – The Coming Era of Nanotechnology. Anchor/Doubleday, New York (1986/1990)
26. Drexler, K.E.: Nanosystems – Molecular Machinery, Manufacturing, and Computation. Wiley, New York (1992)
27. Altmann, J.: Military Nanotechnology: Potential Applications and Preventive Arms Control. Routledge, Abingdon/New York (2006); Russian Version: Tekhnosphera, Moscow, Sections 4.3, 6.2
28. NNI Supplement to the President's 2013 Budget (February 2012), http://www.nano.gov/ node/748
29. Via http://www.nano.gov, Search Publications and Resources, Type NNI Publications and Reports; e.g. NNI Supplement to the President's 2015 Budget (March 2014), http://www.nano. gov/node/1128

30. Director, Defense Research & Engineering: Defense Nanotechnology Research and Development Program. Department of Defense, Washington DC (December 2009), http://www.nano.gov/sites/default/files/pub_resource/dod-report_to_congress_final_1mar10.pd (and earlier reports from 2005, 2006, 2007), p. 20
31. NNI Supplement to the President's 2017 Budget, Washington DC: National Science and Technology Council (March 31, 2016), p. 20, http://www.nano.gov/node/1573
32. AFRL Nanoscience Technologies: Applications, Transitions and Innovations. Air Force Research Laboratory (2010), http://www.nano.gov/node/132
33. Altmann, J.: Military Research and Development (in German). In Altmann, J., Bernhardt, U., Nixdorff, K., Ruhmann, I., Wöhrle, D.: Naturwissenschaft, Rüstung, Frieden – Basiswissen für die Friedensforschung. Springer VS, Wiesbaden (2017)
34. Joannopoulos, J.D.: Institute for Soldier Nanotechnologies, MIT Reports to the President 2013–2014., http://web.mit.edu/annualreports/pres14/2014.18.06.pdf
35. Joannopoulos, J.D.: Institute for Soldier Nanotechnologies, MIT Reports to the President 2011–2012., http://web.mit.edu/annualreports/pres12/2012.08.08.pdf
36. Joannopoulos, J.D.: Institute for Soldier Nanotechnologies, MIT Reports to the President 2010–2011., http://web.mit.edu/annualreports/pres11/2011.08.08.pdf
37. Strategic Research Areas. ISN (2017), http://isnweb.mit.edu/strategic-research-areas.html
38. Army Selects MIT For $50 Million İnstitute To Use Nanomaterials to Clothe, Equip Soldiers. MIT News (March 13, 2002), http://news.mit.edu/2002/isn
39. Director, Defense Research & Engineering: Defense Nanotechnology Research and Development Program. Department of Defense, Washington DC (December 2009), http://www.nano.gov/sites/default/files/pub_resource/dod-report_to_congress_final_1mar10.pdf (and earlier reports from 2005, 2006, 2007)
40. Department of Defense Sections in: NNI Supplement to the President's 2017 Budget, Washington DC: National Science and Technology Council (March 2016), http://www.nano.gov/sites/default/files/pub_resource/nni_fy17_budget_supplement.pdf (and earlier years)
41. Thomas, E.L.: Institute for Soldier Nanotechnologies, MIT Reports to the President 2003–2004., http://web.mit.edu/annualreports/pres06/03.09.pdf
42. Joannopoulos, J.D.: Institute for Soldier Nanotechnologies, MIT Reports to the President 2005–2006., http://web.mit.edu/annualreports/pres06/03.09.pdf
43. Joannopoulos, J.D.: Institute for Soldier Nanotechnologies, MIT Reports to the President 2009–2010., http://web.mit.edu/annualreports/pres10/2010.08.08.pdf
44. Joannopoulos, J.D.: Institute for Soldier Nanotechnologies, MIT Reports to the President 2013–2014., http://web.mit.edu/annualreports/pres14/2014.18.06.pdf
45. Strategic Research Areas. ISN (2017), http://isnweb.mit.edu/strategic-research-areas.html
46. Joannopoulos, J.D.: Institute for Soldier Nanotechnologies, MIT Reports to the President 2011–2012., http://web.mit.edu/annualreports/pres12/2012.08.08.pdf
47. Joannopoulos, J.D.: Institute for Soldier Nanotechnologies, MIT Reports to the President 2012–2013., http://web.mit.edu/annualreports/pres13/2013.19.07.pdf
48. Joannopoulos, J.D.: Institute for Soldier Nanotechnologies, MIT Reports to the President 2013–2014., http://web.mit.edu/annualreports/pres14/2014.18.06.pdf
49. About the NSI, https://www.nrl.navy.mil/nanoscience/about
50. Programs, https://www.nrl.navy.mil/nanoscience/programs
51. UCLA Engineering, Research Centers, http://engineering.ucla.edu/research-centers/
52. UC Riverside: New Center For Nanoscale İnnovation Transfers Knowledge From Universities To İndustry (December 10, 2002), http://newsroom.ucr.edu/305
53. Altmann, J.: Military Nanotechnology: Potential Applications and Preventive Arms Control. Routledge, Abingdon/New York (2006); Russian Version: Tekhnosphera, Moscow, Section 3.2
54. Altmann, J.: Military Nanotechnology: Potential Applications and Preventive Arms Control. Routledge, Abingdon/New York (2006); Russian Version: Tekhnosphera, Moscow, Section 3.3
55. Altmann, J.: Military Nanotechnology: Potential Applications and Preventive Arms Control. Routledge, Abingdon/New York (2006); Russian Version: Tekhnosphera, Moscow, Chapter 4, Table 4.1

56. Director, Defense Research & Engineering: Defense Nanotechnology Research and Development Program. Department of Defense, Washington DC (December 2009), http://www.nano.gov/sites/default/files/pub_resource/dod-report_to_congress_final_1mar10.pdf (and earlier reports from 2005, 2006, 2007)
57. Kovalchuk, M.V.: Nanotechnology is the Basis for New Postindustrial Economy. In: Tomellini, R., Giordani, J. (eds.): Report on the Third International Dialogue on Responsible Research and Development of Nanotechnology, Brussels (March 2008), http://cordis.europa.eu/pub/nanotechnology/docs/report_3006.pdf
58. Raj, B.: Indo-Brazil-South Africa (IBSA) Nanotechnology Initiative – A Case Study of Effective Cooperation. In: Tomellini, R., Giordani, J. (eds.): Report on the Third International Dialogue on Responsible Research and Development of Nanotechnology, Brussels (March 2008), http://cordis.europa.eu/pub/nanotechnology/docs/report_3006.pdf, p. 128
59. Nanotechnology for Defence Applications. DRDO Newsletter (Defence Research & Development Organisation, India), vol. 32, no. 11 (November 2012), http://www.drdo.gov.in/drdo/pub/nl/2012/NL_November_2012_web.pdf
60. Ministry of Science and Technology, National Nanotechnology Development Strategy (2001–2010) (in Chinese) (2001), cited after Kosal, M.: China's Institutions and Leadership Policy in Nanotechnology Development. In: Clunan, A., Rodine-Hardy, K. (with Hsueh, R., Kosal, M., McManus, I.): Nanotechnology in a Globalized World – Strategic Assessments of an Emerging Technology, PASCC Report Number 2014-006, Monterey CA, Naval Postgraduate School (2014), http://www.dtic.mil/dtic/tr/fulltext/u2/a613000.pdf
61. Lim, H.: Overview on Nanotechnology in Korea: Policy and Current Status. In: 6th US-Korea Nano Forum, Las Vegas (April 2009), http://www.cmu.edu/nanotechnology-forum/Forum_6/Presentation/Hanjo_Lim.pdf
62. European Defence Agency: Dual-use research (2016), http://www.eda.europa.eu/what-we-do/activities/activities-search/dual-use-research
63. Altmann, J.: Military Nanotechnology: Potential Applications and Preventive Arms Control. Routledge, Abingdon/New York (2006); Russian Version: Tekhnosphera, Moscow, Chapter 4, summary in Table 4.1
64. Altmann, J.: Military Nanotechnology: Potential Applications and Preventive Arms Control. Routledge, Abingdon/New York (2006); Russian Version: Tekhnosphera, Moscow, Chapter 4, summary in Table 4.1
65. Altmann, J.: Military Nanotechnology: Potential Applications and Preventive Arms Control. Routledge, Abingdon/New York (2006); Russian Version: Tekhnosphera, Moscow, Chapter 5
66. Altmann, J.: Nanotechnology and Preventive Arms Control. Forschung DSF No. 3, Osnabrück, Deutsche Stiftung Friedensforschung (2005), http://www.bundesstiftung-friedensforschung.de/images/pdf/forschung/berichtaltmann.pdf
67. Convention on Prohibitions or Restrictions on the Use of Certain Conventional Weapons Which May be Deemed to be Excessively Injurious or to Have Indiscriminate Effects. Geneva, (October 1980), http://www.icrc.org/applic/ihl/ihl.nsf/INTRO/500?OpenDocument
68. Final Document of the Fifth Review Conference. CCW/CONF.V/10, advance version. (December 2016), http://www.unog.ch/80256EDD006B8954/(httpAssets)/AF11CD8FE21E A45CC12580920053ABE2/$file/CCW_CONF.V_10_23Dec2016_ADV.pdf
69. Altmann, J.: Military Nanotechnology: Potential Applications and Preventive Arms Control. Routledge, Abingdon/New York (2006); Russian Version: Tekhnosphera, Moscow, Table 6.1
70. Altmann, J.: Military Nanotechnology: Potential Applications and Preventive Arms Control. Routledge, Abingdon/New York (2006); Russian Version: Tekhnosphera, Moscow, Section 6.3
71. Miasnikov, Y.: Precision-Guided Conventional Weapons. In: Arbatov, A., Dvorkin, V., Bubnova, N. (eds.): Nuclear Reset: Arms Reduction and Non-Proliferation, Carnegie Moscow Center, Moscow (2012), http://armscontrol.ru/pubs/en/Miasnikov-PGWs.pdf and references
72. Altmann, J., Sauer, F.: Autonomous Weapon Systems and Strategic Stability, Survival 59 (5), (Oct./Nov. 2017) (accepted for publication)

73. Altmann, J.: Military Nanotechnology: Potential Applications and Preventive Arms Control. Routledge, Abingdon/New York (2006); Russian Version: Tekhnosphera, Moscow, Section 6.1.1.2

74. Brendley, K.W., Steeb, R.: Military Applications of Microelectromechanical Systems, MR-175-OSD/AF/A, RAND, Santa Monica CA (1993)

75. Pillsbury, M.: China Debates the Future Security Environment (Ch. 6), National Defense University Press, Washington DC (2000)

76. Sun Bailin: Nanotechnology Weapons on Future Battlefields. In: Pillsbury, M. (ed.): Chinese Views of Future Warfare, rev. ed., National Defense University Press, Washington DC, 1998. (Chinese original: National Defense (China), June 1996)

77. Altmann, J.: Military Nanotechnology: Potential Applications and Preventive Arms Control. Routledge, Abingdon/New York (2006); Russian Version: Tekhnosphera, Moscow, Ch. 7

78. Altmann, J.: Military Nanotechnology: Potential Applications and Preventive Arms Control. Routledge, Abingdon/New York (2006); Russian Version: Tekhnosphera, Moscow, Ch. 7

79. Berlin Statement, signed at the International, Interdisciplinary Expert Workshop Arms Control for Robots – Limiting Armed Tele-Operated and Autonomous Systems), Berlin, Sept. 2010, http://icrac.net/statements/

80. Altmann, J.: Arms Control for Armed Uninhabited Vehicles – An Ethical Issue. Ethics and Information Technology 15, 137–152 (2013)

81. Killing by Machine – Key Issues for Understanding Meaningful Human Control, Article 36, London (April 2015), www.article36.org/wp-content/uploads/2013/06/KILLING_BY_MACHINE_6.4.15.pdf

82. Report of the 2016 Informal Meeting of Experts on Lethal Autonomous Weapons Systems (LAWS). Submitted by the Chairperson of the Informal Meeting of Experts. CCW/CONF.V/2. (June 2016), https://documents-dds-ny.un.org/doc/UNDOC/GEN/G16/117/16/PDF/G1611716.pdf?OpenElement

83. Kervers, O.: Strengthening Compliance with the Biological Weapons Convention: The Protocol Negotiations. Journal of Conflict & Security Law 7 (2), 275–292 (2002)

84. World of Drones: Military, http://securitydata.newamerica.net/world-drones.html

85. 2007 Department of Defense Research & Engineering Strategic Plan. US Department of Defense, Washington DC (June.), http://www.dtic.mil/cgi-bin/GetTRDoc?Location=U2&doc=GetTRDoc.pdf&AD=ADA472100

86. Defense Manufacturing Management Guide for Program Managers. US Department of Defense, Washington DC (October 2012), Section 8.3, https://acc.dau.mil/docs/plt/pqm/mfg-guidebook-10-16-12.pdf

87. Secretary of Defense Speech, Reagan National Defense Forum Keynote, As Delivered by Secretary of Defense Chuck Hagel. Ronald Reagan Presidential Library, Simi Valley, CA (November 2014), http://www.defense.gov/News/Speeches/Speech-View/Article/606635

88. Tomellini, R., Giordani, J. (eds.): Report on the Third International Dialogue on Responsible Research and Development of Nanotechnology, Brussels (March 2008), http://cordis.europa.eu/pub/nanotechnology/docs/report_3006.pdf

89. US-EU dialogue, bridging nanoEHS research effort. (2017) [EHS: environmental, health, and safety questions], http://us-eu.org/

Chemical Challenges, Prevention and Responses, Including Considerations on Industrial Toxic Chemicals for Malevolent Use, CW Precursor Material for IEDs

Gaetano Carminati, Fabrizio Benigni, and Emanuele Farrugia

Abstract The large use of chemical substances in our normal life, doesn't give the real perception of the potential threats that these substances pose if used for a terrorist attack. For example chlorine or ammonia, that are some of the most common chemical substances in our life, may have a catastrophic impact if used for a malevolent action. To this end, OPCW, the International Organization responsible for the application of the Chemical Weapons Convention, exercises control over all chemical substances that can be used as Chemical Weapons. Another aspect to be considered is the response in case of intentional or unintentional chemical release in terms of preparation, training of first responders, decision-makers, and other stakeholders involved in case of a CBRN incident and sharing of best practices. To be fully prepared to respond to, and recover from, the consequences of a CBRN incident, the actors involved (i.e. first responders, CBRN specialists, decision-makers) must be fully aware of their responsibilities and duties in mitigating potential threats and related likely hazards. The current nature of a CBRN incident and its trans-border effects imply that civil-military interaction/cooperation is necessary. So the importance of developing and implementing a comprehensive approach and the link between civil and military preparedness, resilience, deterrence and defence is nowadays well acknowledged. This paper provides a general overview of chemical threats, of preventive actions to control the spread of chemical substances, of

G. Carminati (✉)
Ministry of Foreign Affairs and International Cooperation, Rome, Italy
e-mail: gaetano.carminati@esteri.it

F. Benigni
Direction of Politics and Security General Affairs, Rome, Italy
e-mail: fabrizio.benigni@esercito.difesa.it

E. Farrugia
Italian National Authority for the implementation of the Chemical Weapons Convention, Kinshasa, Democratic Republic of the Congo
e-mail: emanuele.farrugia@esteri.it

© Springer International Publishing AG 2017
M. Martellini, A. Malizia (eds.), *Cyber and Chemical, Biological, Radiological, Nuclear, Explosives Challenges*, Terrorism, Security, and Computation,
DOI 10.1007/978-3-319-62108-1_5

73

requirements to cope with cases of chemical release, highlighting the need for a national response organization in order to assure safety and security of populations.

1   Introduction to the Chemical Threat

Nowadays, we are living in an era where we are faced with many different types of threats and many degrees of insecurity. Unfortunately, the existence of chemical agents and the willingness of terrorists to use them for malevolent purposes presents us with an important dilemma, namely how to prepare the CBRN first responders for any contingency situation. Chemical threats come from the accidental or intentional release of chemical substances, that are normally used for various purposes by toxic vapours, gases, aerosols, liquids, and solids that are harmful to people, animals, or plants.

Chemicals substances have been used throughout the ages, and it is clear that such weapons have always been seen as a despicable means of warfare. The first use on a massive scale was during World War I which resulted in more than 100.000 fatalities and a million casualties. In this case, well-known commercial chemicals were put into standard munitions. Chlorine, phosgene (choking agent) and mustard gas (which inflicts painful burns on the skin) were among the chemicals used.

During the first half of the twentieth century many countries developed new Chemical Weapons (CW), in particular, the powerful nerve agent that renewed interest in chemical warfare, in fact Chemical Weapons were used by a number of countries in the inter-war period and, at the end of World War II, many Chemical Weapons were abandoned, buried or simply dumped at sea. During the Cold War, the U.S.A. and the Soviet Union maintained active chemical and biological warfare programs and held stockpiles of tens of thousands of tons of chemical weapons (Fig. 1).

"If I seek to acquire these weapons I am carrying out a duty."
Osama bin Laden, February 1999.

Fig. 1 Osama Bin Laden

In the modern era there have been various episodes where chemical weapons were used, in particular, Iraq used Chemical Weapons against Iran, during their 8-year war in the 1980s, and mustard gas and nerve agents against the Kurdish people in 1988. In Japan, the cult Aum Shirinkyo, having built up a sizable CW arsenal without detection and without access to large scientific resources, carried out a terroristic attack on the Tokyo subway on 20 March 1995 using the nerve gas sarin, killing 10 people, injuring thousands and terrifying millions.

After the 11 September 2001 attack, the security environment focused attention on the evolution of terror threat, because if terrorists are capable of acquiring the necessary knowledge to pilot a being 737, any terrorist group can easily plan to organize a terrorist attack envisaging the use of weapons of mass destruction. In particular toxic chemicals are very easy to find and to use, like chlorine or ammonia, normally used for civilian purposes, whose high toxicity makes them eligible for being used like chemical weapons.

According to the general definition, a Chemical Weapon is a toxic chemical contained in a delivery system, like a bomb, shell, missile or aerosol system and the term is also applied to any toxic chemical or its precursor that, when released into the environment, can cause death, injury, temporary incapacitation or sensory irritation through its chemical action.

The toxic chemicals that are used as Chemical Weapons, or are developed to be used as Chemical Weapons, are divided into several categories: choking, blister, blood, or nerve agents. The most well-known choking agents are chlorine and phosgene, the blister agents (or vesicants) are mustard gas and lewisite, blood agents are hydrogen cyanide, nerve agents are sarin, soman, VX.

It must be pointed out that some toxic chemicals, especially their precursors, are utilized globally in industrial processes for the manufacturing of materials that we normally use in our everyday life, especially basic raw materials, pharmaceutical products like anti-neoplastic agents, which prevent the multiplication of cells, or fumigants, herbicides or insecticides that are widely used in agriculture.

Below a definition is given of chemicals used directly as weapons (toxic weapons) and of chemicals that are useful for manufacturing chemical weapons (precursors):

- **Toxic Chemical** "any chemical which through its chemical action on life processes can cause death, temporary incapacitation or permanent harm to humans or animals. This includes all such chemicals, regardless of their origin or of their method of production, and regardless of whether they are produced in facilities, in munitions or elsewhere".
- **Precursor** "any chemical reactant which takes part at any stage in the production, by whatever method, of a toxic chemical".

This concept is the Dual Use. "*Dual-use items' shall mean items, including software and technology, which can be used for both civil and military purposes, and shall include all goods which can be used for both non-explosive uses and assisting in any way in the manufacture of nuclear weapons or other nuclear explosive devices*" [3].

The most traded dual-use items that can be used for both civilian and military applications and/or can contribute to the proliferation of Weapons of Mass Destruction (WMD) are subject to controls to prevent the risk that these items may pose in international security. The controls derive from the obligations of UN Security Council Resolution 1540, the Chemical Weapons Convention and from other treaties that are in line with the commitments agreed upon in multilateral export control regimes [1].

The analysis of the dynamic and rapidly evolving security environment, its emergent trends and future implications need to be supported by a common understanding within tomorrow's geo-political scenario. Today's security environment includes multiple emergent destabilizing factors and, moreover, the ongoing transition to a multi-polar and multi-dimensional world has created instability that is likely to continue. Therefore any innovative transformation solution needs to address these potential challenges.

The aim of this paper is to provide a wider and multidisciplinary approach toward CBRN attitudes and defence related activities to better respond to the new threats emerging from the mentioned instability era, including the use or even just the potential use of CBRN agents or Weapons of Mass Destruction (WMD).

To be fully prepared to respond to, and to recover from, the consequences of a CBRN incident, the involved actors (i.e. first responders, CBRN specialists, decision-makers) must be fully aware of their responsibilities and duties in order to mitigate any potential threat and related likely hazards.

This paper recommends a way to meet the challenges of the future, recognizes the need to develop a National comprehensive plan (directive) to respond to CBRNe incidents, and proposes an innovative, effective and "comprehensive" CBRN defence approach to cope with the CBRN threat that includes: common education, training and exercise of all the CBRN defence stakeholders, the establishment of a national "CBRN Guidance Committee" that will direct and harmonize the Italian CBRN Community, a common C4I response Centre at National level, and agile, prompt and resilient CBRN assets supported by proactive strategic communications and a cognizant national leadership networked with a wide range of Partners.

According to the aforementioned framework, national political leaders, operational and all planners/players should have a common perspective, focusing on the supreme objective: an effective, coordinated response. To this end, the ability to handle the unexpected is important.

2 Preventing Action to Control the Spread of Chemical Substances

In order to avoid the possibility that terrorists or non-state actors may acquire CBRN substances it's necessary to adopt a control regime based on a dedicated chemical substances control treaty, an implementation system and an inspection regime, in order to control the potential spread of chemical substances, and precursors, that may be potentially used for malevolent purposes.

For these reasons the *"Convention on the Prohibition of the Development, Production, Stockpiling and Use of Chemical Weapons and on Their Destruction"* also called Chemical Weapons Convention (CWC) [2], that is an arms control treaty which outlaws the production, stockpiling and use of chemical weapons and their precursors, comprises a preamble, 24 articles and three annexes on chemicals, verification and confidentiality.

The Chemical Weapons Convention (CWC) was the world's first multilateral disarmament agreement necessary to provide the elimination of an entire category of Chemical Weapons within a fixed time frame, headed by the Organization for the Prohibition of Chemical Weapons (OPCW).

At present, 192 States have joined the CWC that represent about 98% of the worldwide chemical industry and all Member States share the collective goal of preventing that chemicals will be ever used again for warfare, in order to strengthen international security.

The Convention contains all measures necessary to enhance confidence in treaty compliance, enable the detection and determination of any violations, provide for mechanisms to clarify compliance concerns, establish relevant facts, and compel State Parties to re-establish compliance, if this were to be necessary.

The CWC lays down some steps with deadlines within which to complete the process for the destruction of chemicals weapons; there is also a procedure for requesting timeline extensions in case a State Party has problems in implementing the CWC. At the present time, no country has reached the target of total elimination within the date set by the treaty, but several have already completed the process within the extended deadline. The order of destruction of Chemical Weapons is based on the obligations specified in the CWC, including obligations regarding systematic on site verification.

The CWC divides controlled chemical substances that can be used in modern industry and medicine on a daily and routine basis, or in the production of Chemical Weapons into three classes (Schedule 1, 2 and 3).

This classification is based on the quantities of the substance that can be produced commercially for the purposes permitted by the convention:

Schedule 1 Chemicals have few or no use for purposes not prohibited under CWC. These may be used for research, medical, pharmaceutical or chemical weapon defence testing purposes but production above 100 grams per year must be declared to OPCW.

Schedule 2 Chemicals are not produced in large commercial quantities for purposes not prohibited under CWC. Manufacture must be declared and there are restrictions on exports to countries which are not CWC signatories. An example is Thiodiglycol which can be used in the manufacture of Mustard Agents but is also used as solvent in inks.

Schedule 3 Chemicals may be produced in large commercial quantities for purposes not prohibited under CWC. Plants which manufacture more than 30 tons per year must be declared and can be inspected, and there are restrictions on exports to countries which are not CWC signatories. Examples of these substances are Phosgene, which has been used as Chemical Weapon but which is also a precursor in the manufacture of many legitimate organic compounds, and Triethanolamine, used in the manufacture of Nitrogen Mustard but also commonly used in toiletries and detergents.

To verify the application of the CWC, the OPCW have a complex verification regime based on site inspections of declared facilities in order to ensure that States Parties are fully complying with the objectives of the Convention and the relative declarations submitted every year to the OPCW to have true overview on the activities of the facility/industry declared. For these reasons, States Parties are obliged to accept all inspections at facilities that are involved in the production, trading and stockpiling of such chemicals.

The inspection regime is divided into three types of Inspection:

Routine Inspections

These inspections are carried out to confirm all the declarations and establish whether the activities and features of the chemical and pharmaceutical facilities are in line with CWC obligations. The selection of the facility to be inspected is made after the State Party and is determined and based on a weighted, random-choice, computerized model. The fact that a facility has been chosen for an OPCW inspection does not mean that the facility has done anything amiss. In terms of the CWC, the OPCW Inspection Team – OPCW IT- does not inspect the facility but it is the State Party that is being checked for compliance with the CWC.

Challenge Inspections

This kind of inspection regime was created to clarify and resolve disputes concerning the possible noncompliance with the CWC. The CWC Article IX, gives State Parties the possibility to request an on-site challenge inspection anywhere in the territory (or under its jurisdiction) of any other State Party. And the same article encourages, but does not oblige, State Parties to first of all offer information to resolve the noncompliance concerns also through a consultation before requesting the inspection.

It's also characterized by the 'any time, any place' concept and is launched at very short notice and can be directed at declared or undeclared facilities and locations. The complexity of this type of inspection requests a preliminary submission to the Executive Council; the Director-General and the Council have the possibility to block the inspection within 12 h of receiving the request and a three-quarter majority of Council members is required to confirm the inspection.

Investigations of Alleged Use of Chemical Weapons

During the late 1980s and early 1990s the United Nations created *ad hoc* teams to investigate the use of chemical weapons. After the entry into force of CWC, OPCW is actually the only international organization that has the legal authority and the requirements to maintain on standby a fully trained team equipped to investigate any possible of uses of chemical weapons.

Under the CWC, an investigation of alleged use of chemical weapons team can be activated by any of two ways; The first is on request for a challenge inspection in a situation in which another State Party is alleged to have used chemical weapons and the second is based on the request for assistance and protection in accordance with Article X directly to the Director-General in a situation where there is an alleged use of chemical weapons against the requesting State Party. The scope of this kind of investigation is to confirm and establish all facts related to the alleged use and provide the Executive Council with sound information for to decide to take further action to assist the requesting State Party.

If the case of alleged use of Chemical Weapons concerns a non-State Party or a territory not controlled by State Parties, the OPCW is also called upon to intervene and in such circumstances, it cooperates closely with the Secretary-General of the United Nations.

3   Requirement to Face a Chemical Risk and All Actions Required to React in Case of Malevolent Use of Chemical Weapons

An effective response to cope with the effects of a CBRN event requires a strong and balanced National organization, continuously trained and exercised.

Organizing the CBRN response on the National territory mainly involves the role of first responders (fire fighters) supported by military forces on request and other supporting Agencies [40, 41, 44, 45, 54].

While interaction between the two components is already in place and opportunities of cooperation have been increasing in recent times, there is still room for improvement in most sectors, especially in the field of command and control, interoperability and there is still a sort of "fragmentation" within the two main components: the civilian and the military one.

The objective of this paper is to pave the way towards cooperation between civilian and military CBRN assets so that they may act together in a coherent, effective and efficient manner.

The rise of new security threats are calling for a transformation of the Armed Forces, including CBRN organization, by evolving towards a single component approach and a new adherent joint vision, capable of facing the current security environment. From this standpoint, therefore, it is essential to create, firstly, a real integration within the military organization between the different components of the Italian Armed Forces and, then, establish (and enhance) an effective operational interaction between the military and the civilian side.

3.1   The CBRN Organization of the Italian Armed Forces

The CBRN organization of the Italian Defence is primarily focused on providing military forces with protection against the effects of CBRN incidents and to take recovery actions, aimed at accomplishing the mission and maintaining freedom of action in a CBRN environment.

Actively engaged in support of peace and security in many regions across the world, the Italian Army, through tactical specialist capabilities provided by the 7th NBC defence regiment, has the leading role in protecting military forces in operation abroad and reacting to CBRN incidents, not only in operations abroad but in case of necessity in support of civil defence across the national territory [45, 54]. This imposes that a variety of capabilities and competencies be attributed to the military in order for it to promptly react and respond to a CBRN event in the three traditional domains: land, air and maritime.

Current Italian military CBRN defence organization is centred on a unique joint pole responsible for the training (Joint NBC School), technical reference centres for C, B and R identification, a specialized unit at regiment level (7th NBC defence rgt.) and specific enhanced capabilities provided by the Navy and the Air Force (see Fig. 2). It is based on hierarchical (pyramidal) scale of proficiency and, according to NATO requirements [11, 16–19], on the following three capabilities levels required across all services (see Fig. 3):

(a) Basic level. It is the level requiring all military personnel to have basic CBRN defence proficiency in order to be able to survive, during and after a CBRN incident. It's basically guaranteed through individual protection equipment (IPE);

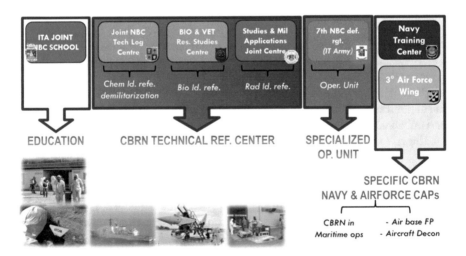

Fig. 2 Overview of the Italian CBRN defence current organization. Specialist and enhanced capabilities

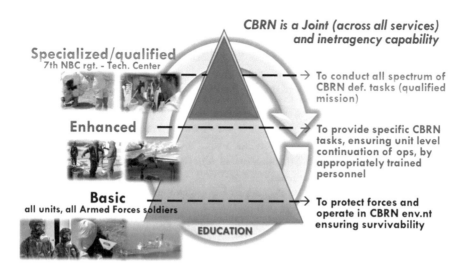

Fig. 3 The hierarchical scale of CBRN defence proficiency

(b) Enhanced level. It is the level required enhanced activities (operational decontamination, early warning of an attack and first detection) conducted by selected, generally trained and equipped CBRN defence personnel, to ensure continuation of operations by units following the threat of use or use of CBRN weapons and devices. Time is essential to appropriately respond to a CBRN attack, therefore necessary CBRN defence tasks need to be carried out by appropriately trained personnel from all arms and branches (usually at platoon/company/battalion levels) of all military components (Army, Navy, Air Force and Carabinieri);

(c) Specialist level. It is the level requiring qualified proficiency by specialist CBRN defence personnel in order to provide the whole spectrum of tactical military CBRN missions in the three prevention-protect-recovery pillars: render safe operations, SIBCRA activities, detection, identification and monitoring (DIM) activities, warning & reporting, thorough decontamination, and physical protection.

Italian Defence possesses consolidated, robust and efficient CBRN defence and niche specialist capabilities, but suffers a sort of "structural fragmentation" and a lack of coordination between different CBRN capabilities and education and technical reference Centres (Joint NBC School, CeTLI NBC, Centro SaNiVet, CISAM), which undermines the principles of efficiency and economy, as well as unity of direction, and harmonious development and effective use of such capabilities by the military and national Authorities. This fragmentation affects the proper implementation of interoperability criteria between all military units and centres involved in the CBRN defence and the interactions with other institutional agencies of the National CBRN organization, as well as in information management and capability development.

For the abovementioned reasons, coherently with the MoD "White Paper" and with financial constraints, in order to ensure the optimal use of the resources available to Defence, efforts need to be coordinated in order to guarantee a guiding direction and implement a coherent capability development process, introducing the necessary factors of change and driving the capacitive consequent evolution, in the direction of:

– organization for CBRN defence based on shared and integrated resources;
– strong joint integration and synergistic use of niche CBRN resources;
– cooperation and coordination at all levels with the various CBRN stakeholders at international and national level;
– coherent and balanced development of competences, resources, systems and materials dedicated to CBRN defence.

In compliance with the above mentioned objectives, Italian Defence has identified the following transformation projects in the short-medium term, some of which are underway, that consist of:

– completion of the military Information Management system (W&R organization) and setting up of its National Reach-back (RB) organization;
– development of new required capabilities (Forensic SIBCRA, CBRN-EOD, exploitation teams, etc.);
– providing personnel and units training in a more realistic scenario, adapting training capabilities including the possibility to create live agents training facilities (according to the actual requirements);
– international accreditation of the Defence identification reference Centre;
– participation, on an opportunity basis, in the main international and NATO CBRN initiatives (e.g. Smart Defence) and in the rotation of the CJ-CBRND-TF;
– bilateral cooperation activities with selected NATO and partner countries.

3.2   A Comprehensive and Multidimensional Innovative Approach Some Proposals

The "NATO Comprehensive Strategic-Level Policy for Preventing the Proliferation of WMD and Defending Against CBRN Threats" [9], as endorsed by Heads of State and Government in 2009, and the policy's prevent, protect, and recover framework represent and provide NATO with the necessary guidance and direction, including the need to improve integration between military, civilian and other stakeholders (IOs). The descending "CBRN Comprehensive approach" is aimed at informing both military authorities, political and civilian authorities of their role in the demanding effort of creating a CBRN Defence capability through a comprehensive civil-military cooperation [11].

The comprehensive approach is focused on cooperation between civilian and military institutions to ensure civil-military preparedness against attacks with CBRN materials. The risk of the disruption of governmental operations and services due to a targeted hybrid attack or a conventional or a CBRN attack have to be considered.

The importance of enhancing a comprehensive approach to crisis management in order to have the ability to plan for, employ, and coordinate civilian as well as military crisis management capabilities is the main objective of most initiatives undertaken by the CBRN community. All these efforts emphasize the importance of close civil-military cooperation in all areas of CBRN defence capability development:

– training and education;
– research & development;
– equipment, capability and structures.

In this context, it is important to ensure that military forces and first responders are able to operate cooperatively in a CBRN environment and that mechanisms exist to provide the required support in a timely manner.

In the changing security environment, the division between military and civilian responsibilities is less defined. The nature of CBRN incidents, in particular their trans-national effects, implies that civil-military cooperation is necessary. Particularly, planning for CBRN consequence management is a multi-dimensional effort, requiring coordination within the Alliance at all levels, as well as with civilian emergency planning authorities and other international organizations, as appropriate.

Military means are not sufficient on their own to meet the many complex challenges to international security. NATO and the Alliance Armed Forces recognized that it is vitally important to develop and implement a comprehensive approach and emphasized that it is essential to act in a coordinated way, applying a wide spectrum of civil and military instruments in a concerted effort that takes into account their respective strengths and mandates. This comprehensive approach addressed the need of enhancing joint training for civilian and military personnel with particular focus on planning of operations and exercises, lessons learned and sharing of best practices.

Therefore, coordinating and harmonizing the development of military and non-military CBRN Defence capabilities to the greatest extent possible and developing common recovery mechanism and procedures for a CBRN event is essential.

Both civilian and military CBRN communities acknowledge that training and exercises remain a key strategy for enhancing the interoperability with partners and fostering better CBRN preparedness of partners and host nations. Therefore it is mandatory to reinforce civil-military interaction and cooperation in CBRN Defence and Consequence Management. Its level, scope and depth have to be considered for every organization, at every stage of an operation and at all levels of command.

However, despite all the efforts, there is still room to better position our national domestic response to CBRN events, taking into account of recent lessons learned and deriving recommendations, including insights coming from some recent national interagency exercises.

The overall ambition is to reach an appropriate mix of military combat and non-military capabilities, reliant upon common tested operational concepts and doctrine, common task-specific equipment and personnel (decision makers, advisors, first responders, etc.) educated and trained at the required levels of proficiency. The full range of capabilities necessary to meet emerging requirements from a comprehensive approach to crisis response operations may be filled by military as well as non-military assets. Several nations have begun to use supportively civilian and military capabilities in a common effort to satisfy domestic requirements and such use may continue, according to the current CBRN environment in which we are involved.

To this end, adopting a DOTMLPFI capability line of development approach, the following proposed lines of action should be accomplished at National level.

Doctrine and Procedures A common national CBRN defence doctrine and procedures have to be developed across all the pillars of the comprehensive approach envisaged in this paper. The National CBRN doctrine should consider the key criteria:

- the national CBRN system should be addressed within the framework of a comprehensive political, military and civilian approach;
- this organization must be understood as a continuum of activities and measures aimed at stopping a CBRN incident as early as possible through advance planning, preparing and exercising appropriate preventive and protective measure.

Organization Assessment of national CBRN defence capabilities: Minister of Interior, in coordination with other relevant Minister (Defence, Health, etc.) and all national stakeholders, should undertake an analysis of the current capabilities to determine the existence of the diverse multipurpose capabilities necessary to provide the operational flexibility for a wide range of future CBRN response efforts in the national territory. Same rationale for the establishment of a comprehensive database or inventory that includes an overview of National civil and military capabilities, including emergency civil/military medical capabilities, is strongly envisaged.

An inventory of existing information sharing arrangements and identification of potential partners, that could lead or assist efforts to prevent or respond to CBRN events, are required. The Central Authority should promote the establishment of specific security agreements to assure exchange of information and suggest opportunities for exercising and training, including all different stakeholders and other IOs.

The integration of various existing warning and reporting systems on CBRN is necessary to improve civil and military preparedness. An effective civil-military interface can facilitate communication to national Civil Defence (and Civil Protection) authorities in order to warn the population, alert and/or mobilize response organizations and notify critical infrastructure operators. This could significantly reduce the loss of life and property and improve critical infrastructure resilience. The possibility to warn an area, through an appropriate and effective sharing of information and civil-military interface could have a great impact on the way of responding to a CBRN event, that is the case for example of a Ballistic Missile attack that in any case cannot be entirely ruled out. The differences between such an attack and a no-notice CBRN incident is clearly evident in case of good integration between all stakeholders.

Training Include new and evolving CBRN elements in all crisis response exercises. Ensure that CBRN contingencies and other CBRN elements are routinely included in all crisis response exercises. Including diverse CBRN scenarios can help ensure awareness of the threats by national leaders and decision makers and the potential responses needed at all National Government levels.

It is necessary to continuously assess and learn from recent CBRN-related events and take steps to strengthen the national CBRN defence capabilities.

The need to reinforce civil-military interaction and cooperation in CBRN defence and consequence management is widely known. Initial links between civil and military stakeholders have been established during past interactions/exercises but a new adaptive and flexible process focusing on synchronizing and unifying military and civilian WMD/CBRND training efforts is required at all levels, starting from the political-military up to the operational and tactical level. Military forces and first responders have to operate cooperatively in a CBRN environment.

Material All national CBRN defence units (military and civilian) must be provided with modern, robust CBRN equipment and systems that adheres, where possible, to NATO standards in order to guarantee a common understanding, interoperability and standardisation (at national and) international level.

Personnel and Leadership Adopting a comprehensive approach to CBRN defence requires decision makers, leaders, military commanders who are aware of the CBRN threat and understand the usefulness of adopting solutions that minimise risk, and who are educated and trained to deal with CBRN threats more effectively.

Facilities The availability of a national environmentally-secured crisis management centre is another key factor. This could be achieved by using and adapting the already existing DC-75 infrastructure, taking into consideration appropriate plans for its manning and activation.

Live Agent Training (LAT) facilities are required to train and certify CBRN specialist personnel. Efforts should, therefore, be focused on establishing a national LAT facility, i.e. supporting the proposal to adapt already existing facilities located in the "NuBiCh" training area of the Joint NBC School or creating a safe and confined space where to conduct such training.

Interoperability The establishment of a "CBRND Interagency Civil-mil Guidance Committee" that will drive the mentioned CBRN national transformation is envisaged. This body should be composed of representatives of all relevant CBRN national agencies and stakeholders. The objective of this national Committee is to:

- promote CBRN awareness, establishing a forum of discussion and harmonization for the national CBRN community;
- contribute and support the development of the national interagency CBRN defence organization and a National strategy within the 'prevent, protect and recovery' framework;
- support and develop standardization, doctrine and emergency response procedures and mechanisms of different National stakeholders, promoting interagency training and exercises at all levels and international exchanges of experience and personnel;
- develop and maintain coherent National CBRN response plans;
- monitor the progress of technology in the CBRN field and ongoing projects at industrial and academic levels;
- collaborate and maintain, as appropriate, working relationships with relevant International Organizations in the CBRN field.

This objective represents the key issue that can enable the CBRN (r)evolutionary process required.

4 Conclusions

The rapidly evolving security environment, particularly the rise of sophisticated hybrid threats, its emergent trends and future implications need to be supported by a common understanding of strategic fit within tomorrow's geo-political scenario. From this standpoint, "homeland defence" and civil-military cooperation have to be taken into account in order to strengthen preparedness and resilience.

Today we face multiple emergent destabilizing factors, mostly created by the transition to a multi-polar and multi-dimensional world, and new dimensions of the CBRN threat: the possible use of WMD and CBRN materials by States or, more likely, by terrorist groups is now associated with the CBRN incidents posed by Toxic Industrial Hazards (TIHs) or by natural disasters.

This new reality makes civilian populations and infrastructures the "real target" of such attacks and, in this view, a consistent and coherent CBRN defence posture should be able to defend the territory and the population along with own forces. These considerations represent the basis from which a stronger and enhanced civil-military cooperation and interaction has to be closely planned, organized, coordinated and carried out in the short-medium term. The current environment requires for joint military-civil response teams, dual use equipment, complementary logistics, CBRN response and resilience common planning, coordination in capability planning, coordinated capability planning and common interagency exercises.

The importance of developing and implementing a comprehensive approach and the link between civil and military preparedness, resilience, deterrence and defence is nowadays well acknowledged. The current nature of a CBRN incident and its trans-national effects imply that civil-military interaction/cooperation is necessary.

To be fully prepared to respond to, and to recover from, the consequences of a CBRN incident, the involved actors (i.e. first responders, CBRN specialist, advisors and decision makers) must be totally aware of their responsibilities and duties in order to mitigate any potential threat and likely hazards, enhancing interagency training for civilian and military personnel, with particular focus on planning of operations and exercises, and sharing of best practices.

Concluding, this paper has taken into consideration the current chemical threat, the relevant role of OPCW in preventing actions, and has proved the importance and the need for a wider and multidisciplinary national approach, in order to promptly respond to the new threats. In particular, it emphasizes the essential role of a comprehensive civil-military, political and diplomatic approach, applying the full spectrum of National resources in a concerted effort.

The analysis conducted in this paper paves the way towards the ability to meet the challenges of the future and the need to develop a comprehensive National response plan (directive) to respond to CBRN incidents on the national territory, and proposes new innovative, effective and "comprehensive" instruments required in the current CBRN environment in light of new evolving threats.

According to the aforementioned framework, national political leaders, advisors, planners and all operational players should have a common perspective, focusing on the supreme objective: a strong, appropriate and coordinated CBRN defence response that will contribute to deter potential adversaries, non-state and state actors, from considering the use of CBRN agents and materials against our Nation and the Western community.

References

1. United Nations Security Council Resolution 1540, Adopted by the Security Council at its 4956th meeting, on 28 April 2004
2. NATO Strategic Concept for the Defence and Security of the Members of the North Atlantic Treaty Organization (Ed. 2010)
3. Prague Summit Declaration, 21 November 2002
4. The Prague Summit and NATO's Transformation, 2003
5. PRICP (2010) 0155, Lisbon Summit Declaration, 20 Nov 2010
6. PRICP (2012) 0062, Chicago Summit Declaration, 20 May 2012
7. Wales Summit Declaration, 05 September 2014
8. C-M(2011) 0022, Political Guidance, 14 Mar 2011
9. C-M(2009)0058 (INV), NATO's Comprehensive, Strategic-Level Policy for Preventing Weapons of Mass Destruction (WMD) and Defending Against Chemical, Biological, Radiological and Nuclear (CBRN) Threats, 31 march 2009
10. MC-0511 Change 1 (Military Decision), MC Guidance for Military Operations in a Chemical, Biological Radiological and Nuclear (CBRN) Environment, Including the Potential Military Contribution to NATO's Response to the Proliferation of Weapons of Mass Destruction (WMD) (Jul. 2010) (NC)
11. MC 0603/1 NATO Comprehensive Chemical, Biological, Radiological, Nuclear (CBRN) Defence Concept, 11 June 2014 (NC)

12. MC-0590 NATO Chemical, Biological, Radiological and Nuclear (CBRN) Reachback and Fusion Concept, 12 August 2010
13. MC-0400/3, MC Guidance on military implementation of NATO's Strategic Concept, August 2013
14. AJP-1(D) Allied Joint Doctrine, 21 December 2010
15. AJP-3(B) Allied Joint Doctrine for the Conduct of Operations, 16 March 2011
16. AJP-3.8 (A) Allied Joint Doctrine for Chemical, Biological, Radiological and Nuclear (CBRN) Defence AJP3.8 Edition Aversion 1, 30 March 2012
17. ATP-3.8.1 vol. 1, CBRN Defence on Operation, 14 January 2010
18. ATP-3.8.1 (A) vol. 2, Specialist CBRN Defence Capabilities, 13 May 2014
19. ATP-3.8.1 vol. 3, CBRN Defence Standards for Education, Training and Evaluation, 5 April 2011
20. AJP-3.10 Allied Joint Doctrine for Information Operations, 23 November 2009
21. AJP-3.14 Force Protection (FP), 26 November 2007
22. AJP-5 Allied Joint Doctrine for Operational Planning, 26 June 2013
23. ATP 71 Allied Maritime Interdiction Operations
24. EU strategy against proliferation of Weapons of Mass Destruction, doc 15708/03, 10/12/2003
25. EU Concept for CBRN EOD in EU-led Military Operations, doc 8948/08, 29/04/2008
26. Council conclusions on strengthening chemical, biological, radiological and nuclear (CBRN) security in the European Union - a EU CBRN Action Plan - Adoption, doc 15505/1/09 REV 1, 12/11/2009
27. Communication from the Commission to the European Parliament, the Council, the European Economic and Social Committee and the Committee of the Regions on a new EU approach to the detection and mitigation of CBRN-E risks, COM(2014) 247 final, 5/5/2014
28. Chemical, Biological, Radiological and Nuclear (CBRN) Countermeasures Concept for EU-Led Military Operations, 11 July 2014
29. OPCW "Eliminating Chemical Weapons and Chemical Weapons Production Facilities" - https://www.opcw.org/fileadmin/OPCW/ Fact_Sheets/Fact_Sheet_6_-_destruction.pdf
30. The Chemical Weapons Convention: A Synopsis of the Text https://www.opcw.org/fileadmin/ OPCW/ Fact_Sheets/English/Fact_Sheet_2_-_CWC.pdf
31. "Lessons learned from recent terrorist attacks: Building national capabilities and institutions NRC conference (Ljubljana, Slovenia 27 June-1 July '15) – Chairman's Report. - http://www.nato.int/DOCU/conf/2005/050727/ index.html
32. CBRN: Lessons Learned from Fukushima, Ebola and Syria" - EU Non-Proliferation and Disarmament Conference, 12 November 2015 - https://www.iiss.org/en/events/eu%20 conference/sections/eu-conference-2015-6aba/special-session-1-a350/special-session-9-a5ac
33. AC/23(EAPC)N(2014)0013, Revised Guidelines for First Responders to a CBRN incident, 1 October 2014
34. AC/23(EAPC)N(2014)0017, Updated International CBRN Training Curriculum, 1 October 2014
35. Occasional Paper 8 - Defining "Weapons of Mass Destruction", W. Seth Carus Revised.
36. Blum, A., Asal, V., Wilkenfeld, J.: "Nonstate Actors, Terrorism, and Weapons of Mass Destruction". *International Studies Review*. (2005)
37. Mazzone, A.: "The Use of CBRN Weapons by Non-State Terrorists, Global Security Studies", Fall 2013, Vol. 4, Issue 4 (2013)
38. "Use of Chemical, Biological, Radiological and Nuclear Weapons by Non-State Actors. Emerging trends and risk factors". Lloyd's Emerging Risk Report – (2016)
39. L. 23 agosto 1998 n. 400 "Disciplina dell'attività di Governo e ordinamento della Presidenza del Consiglio dei Ministri"
40. L. 30 luglio 1999 n. 300 "Riforma dell'organizzazione del Governo a norma dell'art. 126 della legge 15 marzo 1997 n. 59"
41. L. 24 febbraio 1992, n. 225 "Istituzione del Servizio nazionale della protezione civile" e successive modificazioni e integrazioni

42. Concetto Strategico del Capo di Stato Maggiore della Difesa (Ed. 2004-2006)
43. Libro Bianco per la Sicurezza internazionale e la Difesa (Ed. 2015)
44. D.P.C.M. 5 maggio 2010, Organizzazione nazionale per la gestione delle crisi
45. Piano Nazionale di Difesa da attacchi terroristici di tipo Chimico, Biologico e Radiologico (Ed. 2010)
46. ANMD Approccio Nazionale Multi-Dimensionale alla gestione delle crisi (Ed. 2012)
47. CC-001 Implicazioni militari dell'ambiente operativo futuro (Ed. 2012)
48. ND-003 La dimensione militare della Comunicazione Strategica (Ed. 2012)
49. PID/O-3 La Dottrina interforze italiana per le Operazioni (Ed. 2014)
50. PID/O-5 (Vol. II) La pianificazione delle operazioni (Ed. 2012)
51. PID/O-7 (Vol. I) L'addestramento militare (Ed. 2009)
52. PID/O-3.14 La Protezione delle Forze (Ed. 2012)
53. SMD-DAS-001 Organizzazione del vertice militare interforze per la condotta delle operazioni (Ed. 2007)
54. COI-42-R, Predisposizioni per la difesa da attacchi terroristici di tipo Chimico, Biologico e Radiologico (Ed. 2012)
55. DM del 23 novembre 2010, Procedura nazionale relativa al trasporto di pazienti in alto bio-contenimento
56. NATO. Conferenza annuale su disarmo, non-proliferazione e controllo armamenti (Lubiana, 9-10 maggio 2016), MAE0096282, 15/05/2016

A Fresh Approach: Review of the Production Development of the CBRN/HAZMAT Equipment

Boban Cekovic and Dieter Rothbacher

Abstract Paper is describing basic postulates of product development of CBRN/ HAZMAT equipment, current status in development protocols of the different equipment with recognized reviews, with emphasis on how much feedback of end-user appears to have been used in finalization of the product for serial/commercial production. As well, special emphasis is placed on views based on comparison of the trials in the laboratory conditions to a full scope investigation in as much realistic conditions as possible, and by using real live agents-contaminants, performed by experienced end-users. Experiences and results of this paper are presented comparatively for several types of current products used by experienced CBRN operators, both in real operations and in live agent field trials and trainings/exercises. Proposals and recommendations are given for more complex systematic integration and cooperation of product developers with end-users of CBRN equipment in implementing feedback from early trials in as near as possible live agent conditions (and not from just more controlled laboratory environment).

Keywords CBRN • Equipment development • Live agent • Field validation

B. Cekovic (✉)
HotZone Technologies, The Hague, The Netherlands
e-mail: boban.cekovic@hotzonesolutions.com

D. Rothbacher
CBRN Protection GmbH, Vienna, Austria
e-mail: dieter.rothbacher@cbrn-protection.com

© Springer International Publishing AG 2017
M. Martellini, A. Malizia (eds.), *Cyber and Chemical, Biological, Radiological, Nuclear, Explosives Challenges*, Terrorism, Security, and Computation,
DOI 10.1007/978-3-319-62108-1_6

1 Introduction

As the CBRN terrorism is perceived to be one of the growing threats facing the world today [1], the field of CBRN[1] is considered an *"area of increasing concern for the international community. Responding to a major CBRN incident will entail a multi-faceted response, requiring coordination and cooperation at all levels to ensure the safety of the public"* [2, p. 1].

Some of the important observations made in certain case studies *"reveal that the mindset of the leadership, opportunities and technical capacity are some of the most significant factors that influence a terrorist group's propensity to seek, acquire and use CRBN weapons. But with the rise in the level of expertise of the terrorists' explosives experts, the possibility of terrorist organizations succeeding in obtaining CBRN materials that may be used for the construction of CBRN weapons is a hard reality that cannot be ignored"* [1, p. 6].

Also, *"when considering strategies for preparedness against CBRN/HAZMAT incidents, the possibility of a low-probability catastrophic outcome must be weighed against public health hazards of higher probability but smaller magnitude"* [3, p. 7]. Proper balance should be kept between complacency about the possible effects of deliberately released biological, radiological or chemical agents/materiel, and overestimation of the same effects [3].

From point of accessibility, *"the worldwide availability of advanced military and commercial technologies and information (including dual-use and emerging non-traditional threats), combined with commonly available transportation and delivery means, may allow adversaries opportunities to acquire, develop, and employ CBRN weapons without regard for national or regional boundaries"* [4, p. 8].

In the light of the assessed rising threat, some strategic reports [5] predict that the *"CBRN market will continue to steadily increase to meet emerging threats from the Middle East and AsiaPacific regions. The reports also predict that CBRN protective equipment will comprise most the CBRN market share in the upcoming decade"* [5, p. 1].

As the CBRN/HAZMAT related hazards are increasing with growing and spreading industries,[2] but also as the threat of terrorism using WMD (weapons of mass destruction) materiel or attacking industries containing/using harmful materiel, the development of appropriate equipment for detection, analysis, protection and consequences removal is becoming more complex and challenging.

At the same time additional factors add up to the more complex requirements in the development framework, such as:

- Increasing variety of hazardous materials (with industry growth)
- Worldwide/local economy crisis related budget cutbacks in past decade(s)

---

[1]Although latest etymology mostly uses CBRNe abbreviation, paper will use term CBRN as results/experiences described are related to CBRN/HAZMAT response equipment.

[2]Chemical, pharmaceutical, petrochemical, and others that can cause release of harmful CBR/HAZMAT materiel (TIC, radioisotopes, bio materiel…) in case of incidents/accidents.

- Safety and environmental legislations in relation to field open trials/training becoming more stringent,
- Fact that, unlike in case of some other products having non-satisfied customer in case of product failure, in case of CBRN equipment failure, harm can occur to personnel and/or equipment, etc.

"In the light of stricter environmental regulations, as well as financial cut-downs, more and more live agent validation protocols of newly developed CBRN/HAZMAT equipment are focused to the laboratory conditions"[6, p. 1], even with usage of simulants only. *"Nevertheless, the controlled and safe validation in the field conditions with live agents and with end-user perspective still provides insights that cannot be covered in laboratory"* [6, p. 1].

Providing more detailed experimental support to still non-replaceable field live agent validation/testing is the main focus of this paper, as well as is proposing modified model of development of CBRN/HAZMAT equipment.

2 CBRN/HAZMAT Equipment Development Practices

2.1 Terminology

As widely accepted terminology, slightly modified here:

1. CBRN threats *"include the intent and capability to employ weapons or improvised devices to produce CBRN hazards"* [4, p. 14].
2. CBRN hazards include *"CBRN material created from accidental or deliberate releases, TIMs, chemical and biological agents, biological pathogens, radioactive material, and those hazards resulting from the employment of WMD, or encountered-by during any organized field activity"* [4, p. 14].
3. A CBRN/HAZMAT incident is an *"occurrence involving the emergence of CBRN hazards resulting from the use of CBRN weapons or devices, the emergence of secondary hazards due to counterforce targeting, or the release of toxic industrial materiel/chemicals (TIM/TIC) into the environment"* [4, p. 14]. Exposure to CBRN hazards could occur *"from an attack with militarized CBR weapons or from releases of CBRN materiel due to accident or from attacks on infrastructure, including urbanized industrial areas"* [4, p. 16].

Other terms with combined view on distinguishing (non)intentional CBRN/HAZMAT incidents:

1. CBRN materials are *"weaponized or non-weaponized Chemical, Biological, Radiological and/or Nuclear materials that can cause great harm and pose significant threats in the hands of terrorists. Weaponized materials can be delivered using conventional bombs (e.g., pipe bombs), improved explosive materials (e.g., fuel oil-fertilizer mixture) and enhanced blast weapons (e.g., dirty bombs).*

Non-weaponized materials are traditionally referred to as Dangerous Goods (DG) or Hazardous Materials (HAZMAT) and can include contaminated food, livestock and crops" [3, p. 1].

2. An accidental CBRN incident is "*an event caused by human error or natural or technological reasons, such as spills, accidental releases or leakages. These accidental incidents are usually referred to as DG or HAZMAT accidents. Outbreaks of infectious diseases, such as SARS, or pandemic influenza are examples of naturally occurring biological incidents*" [3, p. 1].

3. An intentional CBRN incident includes:

   • criminal acts such as "*the deliberate dumping or release of hazardous materials to avoid regulatory requirements*" [3, p. 1],
   • malicious, "*but non-politically motivated poisoning of one or more individuals*" [3, p. 1],
   • terrorist acts that involve "*serious violence to persons or property for a political, religious or ideological purpose and/or that are a matter of national interest*" [3, p. 1].

The response to an intentional CBRN incident may be similar to an accidental CBRN/HAZMAT incident; however, "*intentional CBRN incidents differ because there are unique implications relating to federal/provincial/territorial responsibilities, public safety, public confidence, national security and international relations*" [3, p. 1].

2.2 Basics and Classification of CBRN/HAZMAT Equipment

CBRN/HAZMAT response is determined, among others, by the nature of the CBRN incident, that may include all or some of the following characteristics [3]:

• possibility for mass casualties,
• possibility for loss of life,
• potential for long term effects,
• establishment of an extremely hazardous environment,
• relative ease and cheapness of production,
• initial ambiguity and/or delay in determining the type of material involved,
• possible use of a mixture of CBRN materials each having different response requirements,
• short timeframe for administration of life-saving measures/interventions/treatments,
• need for immediate medical treatment for mass casualties,
• need for instant availability of specialized pharmaceuticals,
• necessity of specific detection equipment,
• need for timely, efficient and effective mass decontamination systems.
• Etc.

The affiliated CBRN/HAZMAT equipment is related to every aspect of CBRN/HAZMAT incident response. In order to reflect all response aspects, core CBRN equipment can be grouped in main/standard categories[3] [4, 7–9]:

- Protective Equipment (personal protective equipment and mass shelter/protection),
- Detection & Monitoring Equipment,
- Decontamination Equipment,
- Sampling & Analysis Equipment,
- Health & Safety Equipment,
- Scene Management Equipment,
- Transport & Waste Management Equipment,
- Training/Instructional Equipment,
- Integrated Systems, ...

Also, some reports based on survey of (federal/state or local) officials with hazardous materials (HAZMAT) expertise, attempt to categorize the equipment through "*various levels of response capability*" to a CBRN incident [10]. These items were categorized to "*represent different levels of capability – basic and modest, moderate, and high in comparison to the basic level. A modest increase over basic HAZMAT would include additional detection and decontamination equipment. A moderate increase would include a greater array of detection and identification equipment than the modest level. The high level of increased equipment capability would include additional and more expensive detection*" [10, p. 4] and identification equipment, automated and/or integrated systems in reconnaissance and consequence removal as well as in scene management [10].

2.3   CBRN/HAZMAT Products Development Recent Protocols

The development of CBRN Equipment is being done in same manner as any other equipment. Many publications made on principles of product development [11–19], where product development is considered as "*set of activities beginning with market opportunity and end with production, sale, and delivery*" [15, p. 2].

Successful product development in a generic way is the one that delivers a product from production line that [14–16] (see Fig. 1).

The New Product Development (NPD) process is often referred to as *The Stage-Gate* innovation process, and "*when teams collaborate in developing new innovations, having the eight ingredients mixed into team's new product developmental repertoire will ensure that it's overall marketability will happen relatively quick, and accurately – making everyone productive across the board*" [11, p. 1].

---

[3]Although slight differences exist in classification on different national, regional or treaty's levels.

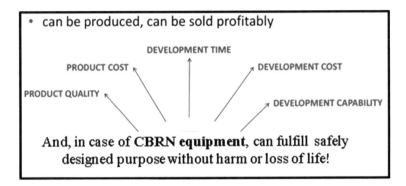

Fig. 1 Successful product development

As main steps in generic commercial NPD usually mentioned are [11–16]:

- Step 1: Generating – *"Utilizing basic internal and external SWOT (strength, weakness, opportunity and threat) analyses, as well as current marketing trends"* [14, p. 3]. But next to generated idea, basic research needs to be done before submitting for (pre)approval (to see how many people/companies *"are searching to solve the same problem"* [12, p. 1] …).
- Step 2: Screening the Idea – Do you go/no go? Setting particular parameters *"for ideas that should be continued or dropped. "Prescreening product ideas" – means taking top three competitors' new innovations into account, how much market share they're taking, what benefits end consumers could expect etc."* [11, p. 1].
- Step 3: Concept Design and Testing – Besides the *"patent research, design due diligence, and other legalities involved with new product development, knowing where the marketing messages will work best is often the biggest part of testing the concept. Does the consumer understand, need or want the product or service?"* [11, p. 1].
- Step 4: Business Analytics – During the NPD process, establishing a system of measurables to monitor progress. *"Including input metrics, such as average time in each stage, as well as output metrics that measure the value of launched products, percentage of new product sales and other figures that provide valuable feedback"* [11, p. 2].
- Step 5: Beta/Marketability Tests – *"Arranging private tests groups, launching beta versions, and then forming test panels after the product or products have been tested will provide valuable information allowing last minute improvements and tweaks"* [11, p. 2].
- Step 6: Technicalities and Product Development – The production division/section/department will make plans for product manufacture. *"The marketing department will make plans to distribute the product. The finance department will provide the finance for introducing the new product"* [11, p. 2]. As an example, in manufacturing, *"the process before sending technical specs to machinery involves printing MSDS sheets, a requirement for retaining an*

ISO9001 certifications (the organizational structure, procedures, processes and resources needed to implement quality management), etc." [11, p. 2].

- Step 7: Commercialize – At the commercialization phase, "*new product development has gone mainstream, consumers are purchasing good/service, and technical support is consistently monitoring progress. Keeping distribution pipelines loaded with products is an integral part of this process too, as one prefers not to give physical (or perpetual) shelf space to competition*" [11, p. 2], at the same time caring not to overstock [13]."*Refreshing advertisements, advising retailers, etc. during this stage will keep product's name firmly supplanted into the minds of those in the contemplation stages of purchase*" [11, p. 2].
- Step 8: Post Launch Review and Perfect Pricing-Is focused on "*reviewing the NPD process efficiency and looking for continuous improvements. Most new products are introduced with introductory pricing, in which final prices are nailed down after consumers have "gotten in." In this final stage, overall value relevant to COGS (cost of goods sold), is gauged, making sure internal costs aren't overshadowing new product profits*" [11, p. 2]. Needing to continuously distinguish consumer needs as products age, forecast profits and improve delivery process whether physical/digital, products are being continuously perpetuated [11].

The process is loose – The entire NPD process is an "*ever-evolving testing platform where errors will be made, designs will get trashed, and loss could be recorded. Having the entire team working in tight synchronicity will ensure the successful launch of goods/services*" [11, p. 2].

Other concepts consider product development as phases that typically include planning, concept design, product design and testing, and production start-up [17], which still correspond to steps 1–7 of the NPD phases discussed above.

> "*Productivity during product development can be achieved if, and only if, goals are clearly defined along the way and each process has contingencies clearly outlined on paper.*" [11, p. 2]

Of course, during new product development, many considerations and factors need to be considered, some of them being [12, 13]:

- Cost assessment of conceptual model and prototype development/manufacturing,
- Early feedback from potential customers reduces final development /modifications costs,
- Design/packaging investment pays off by customers' quick attention,
- Going for quality adds to the developer and company reputation,
- In-house production capability or outsourcing significantly influences the cost and communication/coordination lines,
- Consideration of ideas protection should be considered in cost/product planning,
- Etc.

Also, two main point on responsibilities within the company – who is responsible for design and who for product development process? Three core processes are reported in Fig. 2 [15].

Based on responsibilities, recommended product development team would be represented by Fig. 3.

MARKETING	*DESIGN*	*MANUFACTURING*
• Mediate interaction between firm and customers. • Identify product opportunities and market needs. • Oversee launch and product promotion.	• Define physical form of product to best meet customer needs. • Engineering design. • Industrial design.	• Designing, operating and coordinate the production system. • Also includes purchasing, distribution and installation

Fig. 2 NPD core processes

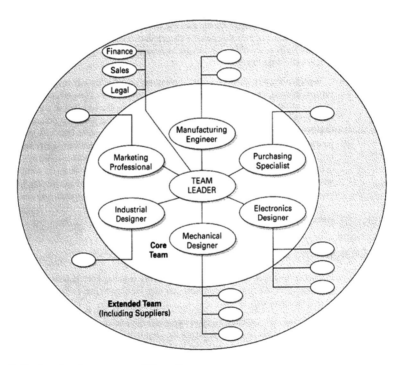

Fig. 3 Product development team [15, p. 4]

And the usual duration of product development can be grouped [15, 16]:

• Less than 1 year (very few);
• 3–5 years (commonly found);
• 10 years (possible).

Cost estimate would depend on [15, 16]:

- Number of people in the project;
- Duration of project;
- R&D activity;
- Tooling investment;
- Production cost, etc.

Number of factors are also influencing initial product concept design [18], having already in early design implementation of customers' requirements through various methodologies – surveys, trials, interviews, and even using social media [19] that allows for widespread group representation variety (depending also on relevance of the product to certain professional/consumer groups).But essential, as obvious, is also to have product development, in as early in development as possible, incorporated with customer needs, satisfaction and budget.

> *"Product, being a yield of producer including service or goods, is essential for satisfying customer needs. It is also shown for producer technical ability. Producer's revenue is fluctuated due to change customer trends by age, income, education, automation and technological advancement."* [20, p. 11]

"Delft innovation model of product development is conceived from SWOT analysis of the organization...After SWOT analysis, the organization has developed a new product with compliance to their standard. Evaluation of product is done by the information of consumers. So, delft innovation model is applied for product development, market introducing, and evaluation of product use" [20, p. 13]. Figure 4 is shown delft innovation model, part for product development.

A design structure matrix (DSM) is presented as a simple, condensed, and visual representation of a complex system that supports advanced solution to decomposition and integration problems for product development [20]. Figure 5 is showing a process of new product development (NPD), also in compliance to other references [14–16, 20].

2.4   Specifics of CBRN/HAZMAT Equipment Development

As focus in the paper is on development of CBRN/HAZMAT equipment in pre-commercialization stages (Steps 1–6 described above in this paper) and establishing end-user feedback and field validation data/results incorporation in the final product/prototype manufacture, examples will be discussed on certain CBRN defense industry R&D and product development mechanism as such representing one of common protocols being followed. For each, observations with mind on fresh perspective CBRN equipment NPD will be made by authors.

One of main noticeable differences/specificity is that NPD in area of CBRN/HAZMAT equipment can be in relation to the unique purposes of that equipment – directly/indirectly aimed to avoiding harm/protecting/saving/preserving lives of exposed to CBRN/Hazardous material. Thus, instead of most commercial product

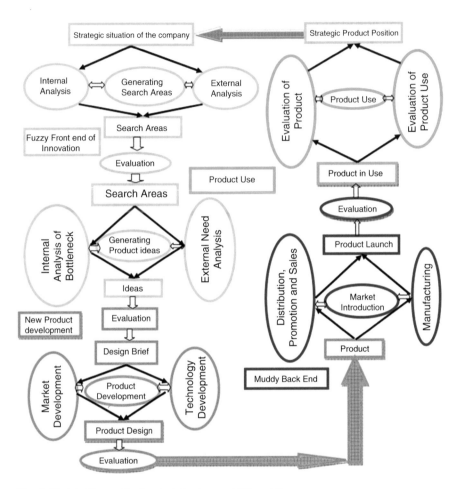

Fig. 4 Model of innovation/product development [20, p. 14]

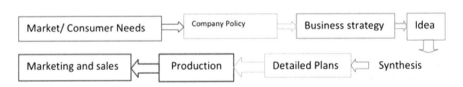

Fig. 5 NPD process [20, p. 16]

development strategies (that are trying to create new needs based on esthetics, user friendliness, comfort of everyday-life customers, overcoming competition), initial/new ideas for new CBRN/HAZMAT products are more coming from:

- Established/recognized new threats,
- Recognized gap in the response capability pending on the new/non-existing equipment,

- Recognized need for enhancing existing products with new/expanded capabilities
- Validated feedback on poor performance of existing equipment,
- Documented need for developing equipment to increase safety of potential affected "civilians" or professionals responding/working in hazardous environment, etc.

Off course if, at the same time, CBRN/HAZMAT NPD can bring esthetics, user friendliness, miniaturization and/or increased profits share from the competition, then the better for the manufacturer. Several examples to be mentioned:

(a) Example of development of CBRN equipment based on driving factors mentioned above (especially emerging threats and need for enhancing equipment capabilities) is well given in description of development of CBRN/HAZMAT intervention robot [21], trying to develop the product that has obvious advantages over existing products on the market, among others [21]:

  - having been sealed thus allowing efficient dealing with decontamination unlike the competition products, making it also more economical on the long run,
  - having enlarged tolerance to increased radiation of various types,
  - increasing modularity of attachments and their variety,
  - increasing mission range to a larger scope,
  - also, the whole R&D of the project including development of the basic and expanded capabilities was interloped with continuous validation/testing of the enhanced/introduced capabilities.

(b) Another example is of emerging threats that can be influencing overall development strategies of new CBRN/HAZMAT equipment for the whole countries:

  - thus having development and/or procurement focused on certain areas recognized as needing capability enhancement against new threats [22–25]:

  • Diagnostics. Investment in chemical, biological, and/or radiological (CBR) detectors and diagnostics to protect by "*quickly and effectively identifying, characterizing, and diagnosing to inform timely and appropriate treatment*" [24, p. 3].
  • Bio-surveillance. Investment in a "*bio-surveillance capability that will mitigate the threat from CBR events (intentional, accidental, or naturally occurring) by informing leadership with essential information to support decision making in a timely manner*" [24, p. 3].
  • Medical Countermeasure (MCM) Development. Investment in MCMs to pre-treat, protect, and treat against chemical and biological (CB) threats (both endemic in theater of operations and weaponized CB threats) [22, 24].
  • Non-Traditional Agent (NTA) Defense Capabilities. Investment in NTA defense capabilities in support of efforts to provide and enhance abilities to "*mitigate the threat from non-traditional and emerging chemical threats in priority areas of detection, MCMs, decontamination, and protection*" [24, p. 3].

Also, point is that development of certain types of equipment such as CBRN/HAZMAT one has a direct effect on safety and that malfunctioning equipment can cause negative health or even life threatening effects.

Next chapter of this paper will discuss experiences from several areas of CBRN/HAZMAT equipment development and present comparatively with experiences and results acquired by authors, with stressing suggestions and proposals for future NPD methodologies improvements, at the end suggesting the protocol of early integration of end-user feedback and live field validation in R&D and NPD.

3 Results and Discussions of Selected Laboratory and Field Validations/Testing

The main focus of the developmental or operational tests (DT or OT) is intended to be a realistic representation of how the system under test will be used by its intended operators in the projected operating environment. An OT includes actual responders executing response missions and using the system under test in the same manner that they would use it in reality. Realistic testing of CBRN defense/HAZMAT systems requires the use of an actual warfare agent (hazardous material). However, in most countries, because of treaties, public laws, and a desire not to harm (test participants, testers, the public, or the environment) – no real CBR/HAZMAT materiel/agents are released during operational tests or in any field test. Testing with an actual warfare agent is mostly restricted to the laboratory (controlled containment chambers) and field testing is performed with simulants studied/established as closest approximation [26–31].

This chapter focuses to give extracts of:

- Trials/tests performed under defined laboratory or field conditions,
- Parameters of the experimental conditions established/used,
- Relevant results indicating complemental data collected from laboratory and field testing,
- Benefits from combined laboratory and field testing,
- Emphasis given on additional data received from field testing done by end-user, etc.

There are three primary groups to consider during the development of CBRN/HAZMAT equipment performance standards, specifically:

1. Responders/Operators must have confidence that equipment meets the standard and are able to compare generated data [26],
2. "*Industry requires an equitable means to demonstrate that products meet responders' needs*" [26, p. 1], and
3. "*Vendors require a means to test new technologies and compare capabilities*" [26, p. 1].

"In similar situations, where underperforming products are potentially dangerous or place people at risk, a standards-based conformity assessment process can provide an effective method to quality-assure the products." [26, p. 1]

3.1 Detection Equipment Validations/Trials/Testing

In case of a toxic chemical release, *"either through an act of terrorism, industrial accident, or natural disaster, effective incident management requires accurate real-time chemical analysis of the materials in question. To help ensure that proper evacuation and decontamination procedures can be initiated, it is critical for first responders and soldiers to have chemical detection equipment which enables the identification of the chemical hazard, the threat level, and the boundaries of the contaminated area. Furthermore, the detection equipment must operate reliably and accurately, and the users must have confidence in the equipment"* [26, p. 1].

To ensure that detection equipment performs as required, the need to create a comprehensive structure for the coordination, establishment, and implementation of CBRNE/HAZMAT detection equipment standards has been recognized and worked on many national regional and treaty levels [26–30].

In case of detection equipment among most standard laboratory testing are [26–31]:

- Detection limits – relevant as main detection and/or monitoring indicator (having detectors mostly focus on detecting concentrations closely bellow and alarming above IDLH concentrations, and monitors mostly focused on low concentration alarm, for values between TWA and IDLH, for non-protected personnel, implying the need to escape from the area or don the protective equipment in order to continue working);
- Response time – indicating how long detector needs to be in certain concentration to indicate/show the response signal corresponding to current concentration;
- Recovery time – indicating how long detector needs to clean out if over exposed before being able to perform measurement to current/reduced concentration;
- Other relevant parameters to the operational use (cross-sensitivity, interference rejection, false alarm rate, reliability and shelf-life, temperature and humidity influence) can be controlled during chemical exposure experiments with the detectors under evaluation.

Additionally, there is a great need to certify the detector performance under more realistic conditions where uncontrolled environmental parameters and rugged field usage can impact the chemical measurements.

Regretfully, due to economical and safety/legislative restrictions, in many cases detection products find place on the market with validation performed to limited DT and OT conditions with a number of extrapolations made.

3.1.1 Example 1/3.1 [31]: As a Positive Example on Field Validation of Pre-production Prototype Detectors Using Experienced End-User Professionals/Operators

More than 200 CBRN defense professionals *"responded to simultaneous mock chemical attacks during joint training on land... Working with test officers and data collectors of the Operational Test Command... they tested the Next Generation*

Chemical Detector (NGCD)... The NGCD is a multi-service system used to assess CW Agents, TICs, and Non-Traditional Agents. The detectors are intended for use by the military for a wide range of missions ... and focused on collecting specific data necessary to inform the program manager and the services with information on the effectiveness, suitability, and survivability of the NGCD systems" [31, p. 2].The test also took in account *"human factors and the assessment of tactics, techniques and procedures used, with data collection efforts aimed to assist in the manufacturing, development, and production of critical chemical detection equipment"* [31, p. 3]. At the same time this next generation equipment was challenged with need to *"rely on the combined team approach to identify any joint operational gaps between the armed services"* [31, p. 3].

CBRN/HAZMAT Product Development Observation

The Example 1/3.1 demonstrates exceptionally important step of early integration in development process of experienced end-user professionals/responders, enabling continuous and real-time feedback. It would complement initial real CW agent laboratory testing in controlled static/dynamic chamber and being followed by next step – live agent testing in open air field conditions. Of course example given is on product development project being conducted jointly – from cooperation between governmental agency and commercial company, thus enabling resources that standalone commercial entity probably would not have.

3.1.2   Example 2/3.1 [32]: An Example of Vendor Being Organization with Number of Professionals with Extensive CBRN/HAZMAT Background and with Resources to Extend Test Done by Commercial Companies Producing and Selling Detectors

International Disarmament Organization (IDO)[4] has very strict protocols of procurement and approval of new equipment for their Inspectors performing various verification/investigation/inspection tasks worldwide. So, selecting functional and equipment fulfilling all necessary requirements is essential (otherwise replacement of non-satisfactory equipment with different model/type would also take some time).

Thus, in preparation for procurement of new detection equipment it was recognized that, in some cases, developers/manufacturers demonstrate detector performance with limited number of target chemicals in a strictly controlled laboratory environment [32].

Operational field tests with highly toxic chemicals, such as chemical warfare agents (CWA), are normally expensive and difficult to perform. Very few facilities in the world perform these kinds of tests as part of assessment for domestic military use and almost none of them commercially. Contrary to the laboratory testing, it is

---

[4] Due to confidentiality restrictions, actual name not disclosed in the paper. Sanitized non-restricted data used only.

almost impossible to achieve controlled conditions during the open field tests. Environmental conditions as well as concentrations of test chemicals (targets and interferences) can vary greatly in space and time making it very difficult to monitor them quantitatively. At the same time, no better conditions to evaluate the behavior of the detector as in as-real-as-possible live agent environment.

For the IDO needs, comparative testing using reference detection devices was selected as an alternative to the expensive and complicated field tests. The reference detection devices were chosen between detectors already in standard use by the IDO and based on the same detection technology as the candidate device to replace the existing one. Performance of a candidate detection devices was compared to performance of selected reference detection device(s) under identical conditions. The IDO Live Agent Training (LAT) offers unique and cost effective platform for conducting the comparative tests in the operational placement. The IDO inspectors, wearing adequate PPE, exercise in a toxic chemical environment (hot zone) protocols for detection, sampling, decontamination, etc. The same training platforms, set with authentic CWA, common potential interferons and/or simulants in both the laboratory and the field conditions, can be used for comparative and qualitative evaluation of chemical detectors, at the same time collecting also exploitation data from experienced end-users.[5] As this paper authors participated (as members of the IDO at the time of the tests performed) in referenced tests [32], they also took part in results review and final decisions on other candidate devices with focus on assessment of usability and adequacy of the devices for the IDO purposes.

Scope

CWA used were: GB, HD, VX, L.

- Group 1 (IMS based detectors): Reference detectors (RID1 & RID2), and three test detectors (TID1 to TID3)[5]
- Group 2 (FPD based detectors): Reference detector (RFD1) and one test detector (TFD1)[5]

The test concept for the IMS based candidate devices was as follows:

(a) Comparative evaluation of two current IDO chemical detection devices based on the same underlying technology as the candidate detection devices in order to choose the referent detection device.
(b) Comparative evaluation of candidate and referent chemical detection devices performed in both, the laboratory and the field conditions.
(c) Comparative evaluation of detection devices in respect to the sensitivity, sample processing time, throughput rate, and detection probability.

---

[5] As this paper not aiming to be used to discredit or support different commercial products, but to discuss development/testing protocols using experiences from references and from authors' own experimental studies/results, the exact names of detectors will be replaced with serial numbers and detection principle.

(d) Assessment of influence of limited number of interfering materials/chemicals on the false positive response.

(e) Assessment of usability of the detection devices for the IDO use, including human factors.

The test concept for other candidate devices was focused on assessment of usability and adequacy of the devices for IDO purposes.

Protocols (Extract) [32]

The candidate chemical detection devices were evaluated in the "as received" condition, with the optimized settings given by the manufacturer. No attempt was made to optimize their detection capability. The detection equipment was operated following the operating procedures outlined in the accompanying manuals.

All tests were performed first in controlled laboratory environment, then in the field, with parallel operation and timing of both reference and tested devices. All environmental measurements (laboratory controlled and in the open air) were recorded.

Results (Extract) [32]

Extract from number of results presented here is for response time for one reference and two test detectors for different amounts of agent density (HD) on reference surface (Fig. 6).

Relative standard deviation (RSD) data in respect to both (response time and signal strength) were more favorable for TID1 (Table 1). However, the RSD data should be used cautiously. The values are influenced not only with variation in the detectors performance(repeatability), but also with variation of the agent vapor concentration at the sampling point.

At 2.5 µl of HD applied, TID2 detector failed in detection. Very high false negative rate (defined as the ratio of the number of non-alarmed samples and the total number of samples) was observed for 5 µl of HD applied (0.8 or 5 out of 6 missed) and 10 µl of HD applied (0.3 or 2 out of 6 missed). For 12.5 µl of HD applied, TID2 completely failed to detect at all data point sets.

Conclusions (Extract) [32]

As a supplement to critically studied comprehensive test reports provided by the manufactures, comparative tests with a candidate chemical detection device run against a detection device in standard IDO use and based on the same underlying technology present probably were the simplest and cheapest way to evaluate suitability of new detection devices for IDO uses. The tests were conducted with authentic chemical agents in confined (laboratory) and open (field) environments under different climate conditions. For this, provisions of the already organized IDO

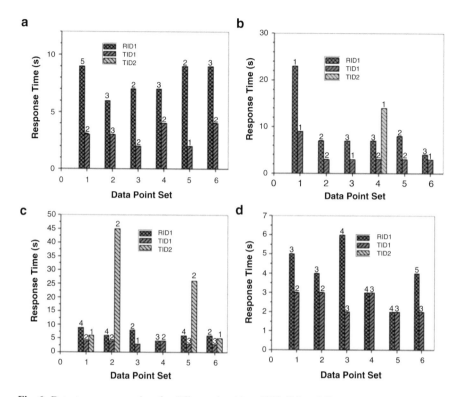

Fig. 6 Detectors response time for different densities of HD [32, p. 51]

Table 1 Relative standard deviation (RSD) data in respect to response time and signal strength [32, p. 52]

Volume of HD applied(μl)	RID1 RSD (%)		TID1 RSD (%)		TID2 RSD (%)	
	Response time	Signal strength	Response time	Signal strength	Response time	Signal strength
2.5	36	36	30	32	–	–
5	73	35	61	36	245	245
2 × 5	27	31	16	35	133	89
2 × 5 + 2.5	35	20	22	19	–	–

inspectors LAT were utilized cutting significantly testing costs. This included logistic and site support. The test environments and scenarios were chosen to simulate an operational situation. The test scenario was focused on a survey of contaminated spots, as this is deemed to be the most probable detector use in the IDO service. The detection devices were tested against neat agents Sarin (GB), Mustard (HD), Lewisite (L) and VX, respectively. These Schedule 1 chemicals are considered to be of the highest risk for the aim and purpose of IDO. They belong to the different classes of CWA, they contain different hetero atoms (phosphorus, sulfur, nitrogen, oxygen, arsenic) and they cover range of different physicochemical properties. For example, GB and VX produce in IMS detectors ions of a positive charge, while HD and L produce ions of a negative charge. That way, the IMS system and its drift tube(s) can be tested at both polarities. Similarly, compounds with different hetero atoms produce in flame emissions on different wavelengths challenging different channels on FPD based detectors. Coming from different sites, the same types of testing agents were of different quality (purity/degradation products composition/polymerization level/etc.) challenging the detection devices further. The tests were aimed at probing the sensitivity, sample processing time/through output rate, and detection probability of the devices. This was achieved by running the detection devices alternately at the same sampling point providing temporally changing agent vapor concentrations. The sampling point was spatially defined by covering the agent source (Petri dish/ watch glass with the agent) with a glass funnel, narrow end pointing upwards. Funnel also provided space for initial vapor build-up and sheltered the vapor flux to the sampling point from ambient air movements. Range of vapor concentrations was formed by a temporal decrease in the agent flux and by applying different amounts of the agent at the source dish, as it is important to challenge detection devices in a relatively low as well as in a relatively high concentration region. In the case of highly volatile agent GB, decrease in the vapor concentration at room temperature was too fast. The solution was found in applying the agent onto a piece of filter paper placed in the sample dish. Overall detector sampling time was restricted to 60 s, as the realistic average sampling time for surveying a spot for contamination. Failing to detect in this time window was regarded as a miss. Time to the alarm and signal strength were recorded, as they relate to a sample processing time and sensitivity of the instrument. Number of data points were collected in time to provide for changing sample concentration and better statistical coverage. The IDO PPE ensemble with APR and air permeable suit was worn during the tests, as the most probable initial personal protection during the IDO investigative/mission events. This allowed proper assessment of usability of the detection devices when placed in the operational environment, including human factors such as ergonomics of the hardware, display, controls, portability and robustness. Overall, of the three tested IMS based detectors, detector TID1 was found acceptable, TID2 not acceptable and TID3 favorable (above expectations).

The TID2 Detector Test Unit Conclusions (Extract) [32]

When tested in the laboratory conditions with HD and GB, TID2 displayed very high to unacceptable false negative rates. Eventually, the unit failed to pass the confidence tests for negative polarity ions. The problem was found to be beyond common user troubleshooting, so no further tests were performed with the unit. In the field, TID2 performance was comparable to TID1 with GB as the test agent. However, it had to be noted that the vapor concentration of test agent was relatively high. With L and HD, TID2 displayed higher false negative rate than the referent detection devices. Overall impression was that the instrument is not adequately designed for survey applications. This especially applies for the inlet system. The display was also found to be badly placed and it showed bad visibility/readability in the field conditions. On the other hand, the instrument is portable and with field favorable power supply solutions. Batteries used with the system are not restricted for transport, but the instrument itself contains radioactive source amenable to transport regulations.

The TID3 Detector Test Unit Conclusions (Extract) [32]

Under both, the laboratory and the field conditions, TID3 either outperformed or at least equaled performance of the referent detection devices in the standard IDO use for years. The instrument displayed fast response and clearing times, high sensitivity and steady performance. The device is truly portable, user friendly and suitably designed for survey applications. Together with the batteries, it weighs 5 times less than the referent detection device RID2 or 2.4 times less than RID1. The batteries are commercially available and the instrument's energy consumption is low. Unlike the other IMS instruments tested (TID1 or TID2), TID3 does not utilize radioactive source, so there are no transport restrictions applicable. For the IDO use, this feature is of outmost importance.

CBRN/HAZMAT Product Development Observation

Testing protocols applied (in supplement to the critical analysis of comprehensive laboratory testing results provided by the producer) allowed either confirmation or not of manufacturer's declared technical/operational capabilities of the detectors in testing.

Addition to that, usage in as close to real scenarios of operational deployments, with same PPE ensemble with actual end-users (inspectors) allowed for very realistic feedback on the usability of the devices being tested and was crucial in final selection of the new replacement device.

Although testing described in Example 2/3.1 was more in relation to supporting procurement final selection, obvious is that if manufacturers would have utilized similar type of feedback from open field live agent testing/validation (not just the laboratory tests) by experienced end-users during the development of the products, some of characteristics observed needing improvement could have been resolved in the first place.

3.2 Protection Equipment Validations/Trials/Testing

In case of personal protective equipment among most standard laboratory testing parameters/criteria are [33–35]:

– Breakthrough/permeation time of the material when challenged with CW/TIC agents – relevant as main protection indicator, implying the time of safe exposure in the defined contaminated environment,

- Decontamination of PPE materiel efficiency – indicating how easy PPE material can be decontaminated after potential exposure and, based on that, then (re)used or discarded,
- Physical-mechanical characteristics of the PPE materiel,
- Durability (on climate conditions, fungi, physical strains, etc.),
- Etc.

3.2.1 Example 1/3.2 [33]: Personal Protective Equipment (PPE) Decontamination Protocols and Their Efficiencies Challenged

CBRN/HAZMAT operators returning from the hot zone (HZ – potentially contaminated area) are passing with their equipment and samples (if collected) through contamination reduction station (CRS – Decontamination station, Contamination Control station, etc.). On CRS, usually decontamination of the potentially contaminated PPE, equipment and sample containers is performed as a risk mitigation measure before entering clean/cold zone. Although CBRN/HAZMAT professionals/operators are trained and apply contamination avoidance behavior in the HZ, the contamination reduction protocol on exit is mandatory in most services. At the same time, *"such reduction should be effective over a short period of time considering that the person leaving the exclusion zone may be on supplied air with a limited air supply remaining in their self-contained breathing apparatus (SCBA)"* [33, p. 2745]. Even when, alternatively, PPE may be doffed through a series of steps in the CRS without immediate decontamination of the PPE, protocol would have to be exercised promptly and with extreme care as to avoid contact with the chemical agent and to reduce exposure of the staff doffing the equipment.

As biggest focus in most studies of PPE is on protection time for selective representatives and from decontamination perspective on full decontamination efficiency. Very few studies have examined the effectiveness of decontamination technologies against variety of chemical or biological agents on PPE-related surfaces with realistic evaluation of hazard reduction on responder PPE. *"Further, decontamination studies have not examined the efficacies at dwell times that are representative of those used in the CRS prior to PPE doffing. Dwell times (contact between decontaminant and agent/surface) during the decontamination of PPE are inherently shorter (2 min or even less, depending on the urgency to doff the PPE) than typical building/equipment material dwell times (30 min or longer). In addition, the ratio of decontamination solution active ingredient amount over chemical contaminant amount may not be as high as during building/object/materiel surface decontamination. Laboratory studies are also often conducted using (stirred) solutions, facilitating the anticipated reaction chemistries. Information on short (less than 5 min) neutralization reaction times is limited"* [33, p. 2746].

Therefore, realistic studies (although one described in study [33] is with TICs and surrogate CW agents only), simulating real operational conditions, provide more relevant information in comparison to the pure laboratory environment usually used for validity testing.

Scope and Results (Extract)

The study [33] came to measured neutralization efficacies following a 2.0-min dwell time varying strongly by chemical with no/very minimal efficacy observed for decontaminants against materials contaminated with nitrobenzene, chlordane, and phenol. Higher efficacies up to 60% were observed for full strength bleach, RSDL and EasyDECON® DF200 products against malathion, carbaryl, and 2-chloroethyl ethyl sulfide. Other decontamination solutions like detergent and water, 10% diluted bleach, and pH-amended bleach were found to be non-efficacious (less than 20%) against any of the chemicals (Fig. 7).

   "The short dwell time and limited amount of decontaminant on the contaminated surface limits the expected efficacy. Therefore, the contribution of neutralization to the decontamination process while responders are preparing to doff personal protective equipment may be limited" [33, p. 2745], unless additional physical removal actions (brushing/scrubbing) would not be added (thus also reducing the processing time through the CRS).

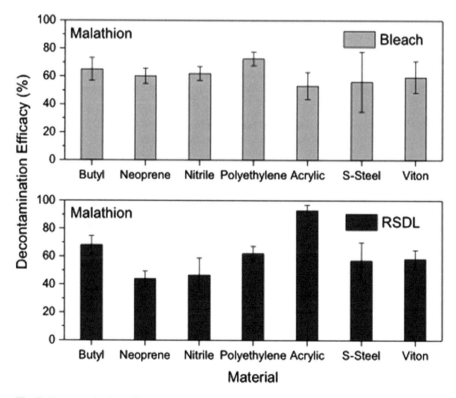

Fig. 7 Decontamination efficiency for Malathion (Bleach and RSDL) [33, p. 2750]

CBRN/HAZMAT Product Development Observation

This result, on first look, contradicts other study [36] done about exploring possibility to use household products for decontamination, where one of statements claims that *"for the decontamination of first responders' protective clothing and equipment, 1 min soaking is suggested for emergency decontamination using undiluted chlorine bleach, and 15 minutes soaking for normal decontamination using diluted bleach (1:22), before rinsing with large quantities of soapy/tepid water"* [36, p. 3]. Then it becomes obvious that, due to different experimental conditions (spraying vs. soaking dwell/contact time) comparison can be tentative only.

Nevertheless, example given is just one more indication that many claims unless field-tested instead just extrapolated to field environment from laboratory tests, have to be taken with reserve (not misinterpreted), even when made by a recognized institution/body.

3.2.2   Example 2/3.2 [35, 37]: Protective Equipment Standards and Their Application/Interpretation by Different Units/Levels

"The protection of first responders (FRs) against the threat of CBRN events has raised significant concern due to potentially fatal consequences. Fire, police, paramedics and other medical first responders/receivers most commonly represent the initial, front-line response to an uncharacterized CBRN event. These "front- line" FRs are most likely to be amongst the first casualties, thus calling for a more stringent CBRN response readiness program to prevent unnecessary loss of life." [37, p. 2]

In the referenced guidance, *"issues such as ensemble selection, integration, equipment, and response readiness, as well as the training and interim procedures necessary to address existing capability gaps, were explored as part of the standard's development and implementation"* [37, p. 2].

In majority situations, CBRN responders need much higher level of protection compared to other responders, thus having some of the accepted practices for selecting and fitting respirators insufficient [37]. *"Several methods were developed and researched so that a responder organization selecting a particular PPE ensemble will have more assurance that the RPD (respiratory protection device) will provide adequate protection in CBRN event"* [37, p. 2].

Scope and Results (Extract)

Overall conclusion/recommendation was that *"these methods, applied at various stages in the respiratory protection program, are necessary for routine implementation in selection, PPE sizing, and donning training procedures"* [37, p. 2]:

Firstly, *"the sensitivity of the methods normally used to measure total inward leakage needs to be considerably improved in order to measure the required levels of protection, 2 and these improved methods need to be implemented in routine respirator sizing/fitting activities"* [37, p. 2].

Then, "*it becomes paramount to evaluate the level of respiratory protection that may be achieved in an operational setting, by performing simulated workplace protection factor (SWPF) measurements. These measurements take into account operational activities that involve specific movements (bending, jumping, crawling, etc.) which may lead to dislodging of the RPD, thus compromising protection. Other SWPF issues include integration problems from interference with PPE components such as hoods or blast-protective helmets*" [37, p. 2].

Conclusions (Extract) [37]

At the same time, when developing an RPD, considerations must be made for the parameters listed below, that can lead to degradation of respirator performance in use:

- exposure to simultaneously occurring/present hazards;
- user factors such as poor integration with other PPE items that may affect the respirator to dislodge, such as weight change/shift, changes in facial hair, or dental features;
- compromised respirator integrity (damage/leakage under unexpected circumstances);
- aging (in storage) and/or wear and tear;
- strain from temperature, moisture and wind;
- contamination from environmental factors (dust, mold or similar);
- degradation of materials due to thermal and/or chemical stress.

> "*All of these factors must be taken into consideration when developing a concept of use for the respirator, as well as respirator inspection and maintenance programs in concert with life cycle management.*" [37, p. 9]

Virtually all jurisdictions in country in question have formulated some form of regulation related to occupational health and safety at the workplace. "*However, there appears to be a difference in how the standards and guidelines are applied from one jurisdiction to another*" [37, p. 14].

CBRN/HAZMAT Product Development Observation

Example is stressing the utmost importance on:

- end-user full ensemble compatibility and protection/performance testing (just having APR tested in laboratory conditions on sophisticated dummy has proven not to be enough),
- standardization of the PPE on all levels from local to federal, and
- as treaties and borders when comes to emergencies are envisioned to be expanded/ crossed in light of need for cooperation/coordination, interchangeability of equipment, translation of one to another quality criteria and other issues of such have to be worked out before responders actually deploy.

Manufacturer producing elements of PPE (just protective boots or just an APR), although having maybe high quality products, should focus also on having early in development end-user feedback of equipment compatibility with other parts of the PPE ensemble and finding ways to expend in which protection levels and scenarios their product can fit.

3.3   Decontamination Equipment Validations/Trials/Testing

In case of decontamination equipment among most standard laboratory testing criteria are [36, 38–42]:

– Decontamination rate – relevant as main effectiveness indicator, implying the time of agent half/full degradation,
– Decontamination stability – indicating how long after preparation for usage the decontamination formulation keeps active, based on that, then (re)used or discarded,
– Decontaminant compatibility with other chemicals and surfaces,
– Decontaminant environmental fate,
– Etc.

3.3.1   Example 1/3.3 [36]: Partial Study of Household Based Decontaminants and Their Efficiency

Sometimes, following a terrorist attack or industry accident, when chemical agents including chemical warfare agents (CWAs) and/or toxic industrial chemicals(TICs) are released, *"immediate response and long-term recovery will require the decontamination. In the event of a large scale chemical emergency, although standard/ specialized decontamination products are available, they may either not be readily accessible at affected sites or they may be quickly consumed. Thus, first responders and local authorities need as much information as possible on potential decontamination technologies subjected to specific chemical agents"* [36, p. 2].

Scope (Extract) [36]

Report mentions other study ("Evaluation of Household or Industrial Cleaning Products for Remediation of Chemical Agents" – EPA 600/R-11/055, 2011) where *"four commercial cleaning products, full strength K-O-K ® liquid bleach (5.25% aqueous solution of NaOCl), dish-washing detergent Cascade ® with Extra Bleach Action Gel, OxiClean ® Versatile Stain Remover Powder, and ZEP ® Industrial Purple liquid cleaner (proprietary caustic cleaner containing surfactants) were evaluated for the chemical agents including thickened sulphur mustard, thickened* *soman, V-series nerve agent (VX), and sulphur mustard contacted with four indoor building materials (galvanized metal, laminate, wood flooring and carpet). Among the commercial cleaning products tested, full strength K-O-K ® liquid bleach generally had the highest decontamination efficacy against all four chemical agents from all materials. Toxic by-products were generated when ZEP ® cleaner and K-O-K ® bleach were used for the decontamination of sulphur mustard and VX. No toxic by-products were found in the decontamination of VX using full strength K-O-K ® bleach (PH > 12) instead of 10% K-O-K ® bleach (PH < 10)"* [36, p. 5].

In another study "Decontamination of Residual VX on Indoor Surfaces Using Liquid Commercial Cleaners" (EPA/600/R-10/159, December 2010),*"two liquid decontaminants, full strength K-O-K ® liquid bleach (5.25% aqueous solution of NaOCl) and a 25% aqueous solution of ZEP ® Industrial Purple liquid cleaner (proprietary caustic cleaner containing surfactants), were tested on three building material coupons (galvanized metal, laminate, and carpet) treated with VX aging from 1 day to 21 days. In contrast to ZEP ® Industrial Purple resulted in moderate decontamination, full strength K-O-K ® liquid bleach showed high efficacy on all samples, regardless of the aging period and coupon substrate type. It was concluded that the full strength K-O-K ® liquid bleach can be effective for both porous and non-porous materials and that decontamination efficacy may be more dependent on the initial level of contamination rather than on surface characteristics"* [36, p. 5].

In case of both studies, contact times (from applying the decontamination chemical/solution to post-rinse) were 30 and/or 60 min.

Results/Recommendations (Extract) [36]

Although household cleaning products are already suggested as alternative to use for decontamination in many countries, there are still significant gaps to be addressed before using such commercial cleaning products for those purposes [36]:

– Risk assessment of potential toxic by/side-products,
– Tool aiding selection of right household chemical for different chemical contaminants,
– Evaluation of usage safety including potential personal injury, corrosiveness to equipment, pollution to the environment, PPE requirements,
– Temperature effect on decontamination efficiency (hot/cold weather),
– Existence/applicability of regulations (at a federal, provincial, or municipal level),
– Effectiveness on different contaminated materials including building materials, vehicle materials, soils, vegetation, etc.,
– Waste collection and disposal protocols,
– Shelf life study,
– Existence of standard operation procedures (SOP),
– Guideline/training for first responders, decontamination contractors, and/or civilian.

CBRN/HAZMAT Product Development Observation

Example is already recognizing that is partial study of laboratory limited effectiveness and that it did not take into consideration other factors mentioned above, that also sure need to be tested more comprehensively in the laboratory but also in open field live environment.

3.3.2 Example 2/3.3 [6, 38, 39, 40, 41, 42]: Live Agent Exploitation Validation/Testing of the Various Decontamination Equipment in the Field Conditions

Example describes comparative results (extracts) from several projects (this paper author's lead) on the live agent (CBR) exploitation testing of newly developed CBR equipment for decontamination. Projects were (in some cases) for validation as part of new product development and in some cases as tests in support of procurement process selection. During the development/validation of the equipment, usually the agent removal and/or destruction has been tested through:

– Initial testing of removal/destruction kinetics in homogenous conditions in the laboratory;
– Laboratory testing of other characteristics of decontaminant chemical/mixture (physical properties, storage stability, environmental fate, hazard levels, etc.);
– Laboratory testing of the decontamination equipment hardware (used for application/dispersion of decontamination mixture) and its optimal application (test table trials on ruggedness – vibration/drop, etc.);
– Exploitation testing of the final formulation – mostly done in controlled/laboratory environment, attempting to approximate some aspects of the field conditions (pressure control and values, temperature, water/fuel consumption, etc.).

Scope (Extract)

The full field live agent testing of different initial (personal, chemical) and thorough (secondary/final, radiological and chemical) decontamination systems have been conducted on multiple occasions. Agents and surfaces used are reported in Table 2.[6]
    At same time, exploitation observations/measurement were made with systematic approach, following defined parameters [42]:

– Users making the observations possessed documented long-term background/experience in all aspects of CBRN response/training/testing
    • End-user perspective was being addressed in the observation approach (training for usage and operation by the end user of low/high experience and skills),

---

[6] Table represents list of all live agents used on different tests/trials on different test locations, values of initial contamination densities and aimed limit values of decontamination efficiency.

Table 2 Experimental scope for decontamination efficiency tests [6, 38–42]

RAD contaminant	Half-life	Surface. act. S (Bq cm⁻²)	Limit values (Bq cm⁻²)	CHEM contaminant	Initial contamination (g cm⁻²)	Limit values (mg m⁻² h⁻¹)	Used surfaces (applicable for all contaminants)
U-238/235 (α) [38, 39, 42]	$4.49 \cdot 10^9$ /$7.1 \cdot 10^8$ year	8.163	<0.4	GB [40]	8.7	<10	Bare metal
							CARC/alkyd-carbam.
Tc-99 m (γ) [42]	6.0058 h	40.816	<4	GD [38, 39]	40.816	<10	Painted metal
Y-90 (β, (γ)) [42]	64.1 h	40.816	<4	VX [38–40]	40.816	<10	Concrete
							Butyl rubber
I-131 (β,γ) [38, 39]	8 days	21	<4	HD [38, 39, 42]	21	<10	Fabric (various.)
Cs-137/ Ba-137 m (β,γ) [38, 39]	30 0.17 year / 2.5 min	43	<4	M [42]	43	<10	Glass
							Full vehicle
							…

Two ways in case of chem. sampling were compared (rad. measurements completed on-spot):
- Wipe sample (defined wipe with DCM as solvent)
- Timed desorption collection (with impregnated filter paper)

Influence of pressure and decontaminant temperature were compared (where applicable):
- Ambient and elevated (40 ± 4 °C) temperature
- Low and elevated (22 ± 3 bar) pressure

BIO contaminant	*Bacillus subtilis* spores (*B. anthracis* sim.) [38, 39]	**Initial contamination** 0.1 mL (min. 10^5 density) with CFU > 25 from surface	**Limit value** Germicidal Effect GE ≥ 5

– Operational aspects covering the observations included [42]:

  • Packaging, carrying and maintenance,
  • Accessibility and prompt usage,
  • Basic usage parameters,
  • Endurance to intensive field usage,
  • Control and disposal,
  • Consumption and waste generation,

– Miscellaneous operational observation.

Examples extracted from obtained results are shown on following graphs, concerning decontamination efficiency (% of radionuclide surface activity removed for R-decontamination, and degradation half-life in reaction mixture or on surface for C-decontamination) as one of crucial validation parameters [6, 38–42] (Figs. 8 and 9):

CBRN/HAZMAT Product Development Observation

In overall, the results obtained were [6, 39–42]:

– Presenting 5–15% "*lower C-decontamination efficiency in the field in comparison to the laboratory conditions*" [6, p. 1];
– Presenting 6–27% "*higher R-decontamination efficiency in the field in comparison to the laboratory conditions (due to physical brushing and rinse-off effects)*"[6, p. 1];
– Indicating B-decontamination efficiency in the field in comparison to the laboratory conditions depending on the type of the solvent media used for decontamination mixture (app. 2–10-fold lower in water based solution to emulsion or organic based mixture) [38, 39];
– Demonstrating difference in "*ruggedness when challenged on laboratory/test table and on the field by end-user*" [6, p. 1];
– Demonstrating difference in "*ease of operation and training when tested in laboratory and field conditions by (non) trained users with(out) field background*" [6, p. 1].

Fig. 8 Comparative R-Decon efficiency (laboratory-field conditions) [6, p. 1]

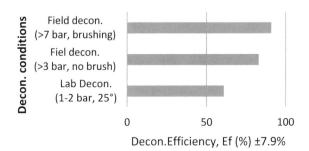

RAD Decon Efficiency comparison for U-radioisotope mixture

Fig. 9 Comparative C-Decon efficiency (laboratory-field conditions) [6, p. 1]

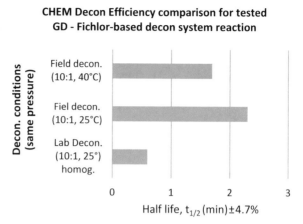

CHEM Decon Efficiency comparison for tested GD - Fichlor-based decon system reaction

Although all tested decontamination equipment has shown satisfactory results for all laboratory/test-table required challenges, the field tests in open ambient and controlled live agent contamination, with engagement of experienced end-users of CBRNe equipment have indicated the need for additional tests and/or improvements, both in [6, 39–42]:

- the validation protocols (making them more rugged and challenging and as close as possible to real exploitation/usage conditions; and at the same time remaining scientific reproducibility and safety),
- product development methodologies (having end-user input on continuous basis and from earlier development stages).

3.4 Scene Management Equipment Validations/Trials/Testing

Aiding professionals/operators as well as decision makers in response coordination, communication and in making prompt and right choices and decisions in as short time as possible are, with increasing variety and weight of threats, becoming growing area of CBRN/HAZMAT equipment development. Therefore, interaction with end-user, feedback and field trials are essential part even in initial development and capabilities parameters definition.

3.4.1 Example 1/3.4 [43]: Systems for Predicting Atmospheric Dispersion

To respond effectively to CBRN/HAZMAT attack/incident, "*rapid decisions need to be made concerning the transport, dispersion, deposition, and fate of the CBRN agent and its concomitant effects on the exposed population. This includes the need to address the specific threat of atmospheric releases of hazardous materials, for real-time emergency response, pre-event planning, and post-incident assessment*" [43, p. 1].

Systems for assessing atmospheric dispersion can provide "*planners, first-responders, and public health officials with information on which to base life-and-death decisions*" [43, p. 1]. These include [43]:

– definition of safe zones for the set-up of incident command posts, sheltering-in-place, or
– recommendations on evacuation, the need for protective equipment, and
– the use of hospital and healthcare resources, etc.

To achieve these objectives, this project implemented three principal activities [43]:

– providing further advanced modeling capabilities,
– development of the vital supporting infrastructure,
– merging all the components in a quasi-operational computing environment and demonstration of the operational prototype system over a number of cities in the country.

Scope and Results (Extract)

This project was accomplished within the allocated budget and with only manageable delays. One of key factors in succeed was in having "*decision makers and emergency management officers involved throughout the course of the project to make sure the new system would meet their operational needs. It was also a way to build a network of contacts in the urban dispersion world. This network will be instrumental to the success of a follow-up project to install and run a truly operational CUDM system*" [43, p. 16].

To achieve this, the following additional work was recognized to be needed: "*full integration of the urban capabilities in the operational CMC suite, integration of a capability to treat complex inflow conditions, integration of orographic capabilities, and development of tools to rapidly initiate an operational response*" [43, p. 16].

CBRN/HAZMAT Product Development Observation

Example presented is already recognizing the need for further development with inclusion of more complex factors simultaneously influencing the movement and concentration range. As well, clear is that legitimacy of the model can only be fully proven/validated by physical measurements (preferably in controlled live field testing environment than in real incident situation).

4 Conclusions and Recommendations for System Integration Between Product Development and Selected End-Users Enhancements

Companies/entities involved with NPD of CBRN/HAZMAT Equipment, in order to improve the efficiency of production and trade and to ensure the required quality of products (necessary to ensure safety of the users and environment), have to follow ever more stringent concept of quality management based on a set of quality standards. The concept of validation of product quality based on protocols where compliance is being cross-checked between declared features of the product with the mandatory (safety and other) requirements. Proof of quality through evaluation is performed during the total life of a product starting from the initial development, production, maintenance, during storage, for the purposes of extension of the shelf/usage time, destroying or recycling [34, 44]. During the development of the new product, from

- conceptual model,
- working model,
- prototype,
- prototype batch,
- initial production batch,
- serial production batches

evaluation of *compliance is a "complex process that is carried out preset methods, standards, procedures and criteria in the competent laboratories and consists of testing, control and certification. Supporting functions, and no less important, for evaluation of products conformity were metrology, standardization and nomenclature, which, with testing, controlling and certification, assemble the basis of quality infrastructure"* [34, p. 1].

Next to the establishment of quality infrastructure, basics fundaments in NPD of CBRN/HAZMAT equipment (and thus, proposed algorithm for CBRN equipment NPD), should be strongly focused on:

- early integration of potential/existing end-user feedback, especially from users that are deploying in real CBRN/HAZMAT situations (responders, live agent instructors, equipment testers, etc.)- if possible even while developing the initial concept and technical specifications
- Implementing validation protocols through different stages of NPD through both laboratory and field testing protocols, serving dual purpose,

  • providing feedback for performance/usability improvements during the development,
  • providing quality assurance/certification of the product,

- Validation/testing protocols having several levels:

  • Initial laboratory testing/validation of core characteristics in controlled/simulated environment,

- Basic quality parameters field testing protocols (both in simulated and live agent environment),
- Advanced testing protocols in live environment with as real as possible usage scenarios with combined quality parameters challenged by experienced end-users in a systematic and safe manner,
- Final laboratory testing of product quality after stresses of the field testing.

At the same time, the need for live-agent testing will remain a safety, security and commercial advantage of companies using it, *"as will the need for testing against real threats early in the development process"* [44, p. 4]. Table 3 presents proposal of enhanced Small-to-Medium Company Protocol for NPD of CBRN/HAZMAT equipment. Enhanced protocol is general and would have to be additionally modified/adopted per the specifics of different products, production lines, end-user requirements and number of other factors. Comments corresponding numerically to different stages of the Table 3:

1. Idea for new product can be coming out from different sources:

   – Established/recognized new threats and updated threat analysis results/recommendations;
   – Recognized gap in the response capability pending on the new/non-existing equipment;
   – Completely new/original idea of employee/partner;
   – Recognized need for enhancing existing products with new/expanded capabilities
   – Validated feedback on poor performance of existing equipment;
   – Documented need for developing equipment to increase safety of potential affected "civilians" or professionals responding/working in hazardous environment (based on governmental, regional or treaty funds/projects/tenders, etc.).

2. Already in this initial stage getting systematic feedback (defined questionnaires, interviews, analysis of already existing feedback, including social media and reviews off/on-line, etc.) from end users, as well as: – Analyzing potential competitor's solutions, scientific supporting references, market pre-screening, etc.

3. Before funds are obligated, pre-assessment of the market, preliminary cost projection and timeline (among other parameters of the product advantages/benefits…) have to be objectively and critically evaluated.

4. Including elements from above, but also human resources, logistic, financial and marketing drafts (as well as independent end-user feedback), The team is starting parallel actions on different activities of the NPD. Very important – technical specifications ("wish list") are also being pre-approved.

5. Demonstrational & materiel concept piece designed/manufactured (to prove the model is working). Besides main (physical design), design of marketing, pricing and other supporting strategies start.

Table 3 Proposed enhanced example of small-to-medium company protocol for NPD (new product development) of CBRN/HAZMAT equipment

No.	Phase/activity	Timeline
1.	Idea generating	
2.	Screening the idea with feedback from potential end-users	
3.	Submitting initial concept for pre-approval	
4.	Developing preliminary NPD plan	
5.	Design of conceptual model	
6.	Core function testing and NPD approval	
7.	Developing main NPD plan	
8.	NPD analytics/monitoring	
9.	Working model design	
10.	Logistical & financial planning and actions	
11.	Marketing planning and actions	
12.	Working model beta validation/testing	
13.	Re-evaluation & update of main NPD	
14.	Prototype (piece and/or batch) design & manufacture	
15.	Full operational and exploitational testing & validation	
16.	Validation report analysis and main NPD plan update	
17.	Establishing production line	
18.	Initial/serial production batch	
19.	Functional testing and quality validation & certification	
20.	Further commercialisation steps …	

1. Idea for new product can be coming out from different sources:
  – Established/recognized new threats and updated threat analysis results/recommendations
  – Recognized gap in the response capability pending on the new/non-existing equipment
  – Completely new/original idea of employee/partner
  – Recognized need for enhancing existing products with new/expanded capabilities
  – Validated feedback on poor performance of existing equipment

(continued)

Table 3 (continued)

– Documented need for developing equipment to increase safety of potential affected "civilians" or professionals responding/working in hazardous environment (based on governmental, regional or treaty funds/projects/tenders…)

2. Already in this initial stage getting systematic feedback (defined questionnaires, interviews, analysis of already existing feedback, including social media and reviews off/on-line, etc.) from end users, as well as: – Analyzing potential competitor's solutions, scientific supporting references, market pre-screening, etc

3. Before funds are obligated, pre-assessment of the market, preliminary cost projection and timeline (among other parameters of the product advantages/ benefits…) have to be objectively and critically evaluated

4. Including elements from above, but also human resources, logistic, financial and marketing drafts (as well as independent end-user feedback). The team is starting parallel actions on different activities of the NPD. Very important – technical specifications ("wish list") are also being pre-approved

5. Demonstrational & materiel concept piece designed/manufactured (to prove the model is working). Besides main (physical design), design of marketing, pricing and other supporting strategies start

6. Mostly done in the laboratory conditions – preferably (in case of CBRN/HAZMAT equipment) with live agents, to prove/confirm the core functionality of the device. Final approval of full NPD project is then expected

7. Based on above – main NPD plan is being submitted and approved. Off course, pending the further developments of the project, NPD plan will be constantly updated/modified if necessary

8. Establishment of the "project LoogBook" – during the NPD process, building a system of metrics to monitor progress (physical development/design, financial, etc.) – 1st material stage of the product development – noncommercial manufactured product (in rough stage, but with full functionality that final product should have

9. Actually, starting since adopting NPD Plan

10–11. Starting since adopting NPD Plan

12. Full laboratory and preliminary field testing, with live agent environment and potential end-user engagement for feedback on exploitational/operational usability

13. Based on reports from previous stage, improvements of the design, update of technical specifications and other aspects of NPD are made

14. 2nd material stage of the product development, manufactured product in close-to-final stage for functionality and design testing, (if satisfactory – can be made commercially available – can be sold as commercial prototype)

15–16. As already said, full laboratory and field live agent testing with representative sample of end-users for final feedback, quality validation and certification as well as for further (if any small improvements of the final design) financial, marketing and enhancement purposes

17. In-house or outsourced production line (used for prototype batch, if any) establishment with quality control measures

18–19. 3rd material stage of the product development, out-of-production line product batch for quality of serial production testing (if satisfactory – can be made commercially available/can be sold as commercial final product) and serial production can commence. Serial Production – Fully established production line batch output with formalized quality & quantity control

20. Further measures, marketing, pricing refreshing the offer/product with new packaging, upgrades, production costs reduction…

6. Mostly done in the laboratory conditions – preferably (in case of CBRN/ HAZMAT equipment) with live agents, to prove/confirm the core functionality of the device. Final approval of full NPD project is then expected.

7. Based on above – main NPD plan is being submitted and approved. Off course, pending the further developments of the project, NPD plan will be constantly updated/modified if necessary.

8. Establishment of the "project LoogBook" – during the NPD process, building a system of metrics to monitor progress (physical development/design, financial, etc.) – 1st material stage of the product development – noncommercial manufactured product (in rough stage, but with full functionality that final product should have.

9. Actually, starting since adopting NPD Plan.

10. Starting since adopting the NPD Plan.

11. Same comment as above (or as for 10).

12. Full laboratory and preliminary field testing, with live agent environment and potential end-user engagement for feedback on exploitation/operational usability.

13. Based on reports from previous stage, improvements of the design, update of technical specifications and other aspects of NPD are made.

14. 2nd material stage of the product development, manufactured product in close-to-final stage for functionality and design testing, (if satisfactory – can be made commercially available – can be sold as commercial prototype).

15. As already said, full laboratory and field live agent testing with representative sample of end-users for final feedback, quality validation and certification as well as for further (if any small improvements of the final design) financial, marketing and enhancement purposes.

16. Same comment as above (or as for 15).

17. In-house or outsourced production line (used for prototype batch, if any) establishment with quality control measures.

18. 3rd material stage of the product development, out-of-production line product batch for quality of serial production testing (if satisfactory – can be made commercially available/can be sold as commercial final product) and serial production can commence. Serial Production – Fully established production line batch output with formalized quality & quantity control.

19. Same comment as above (or as for 18).

20. Further measures, marketing, pricing refreshing the offer/product with new packaging, upgrades, production costs reduction, etc.

Proposed enhanced protocol, besides the usual NPD steps, places significance as already concluded, on important integration of end-user systematic feedback and live agent laboratory and field environment validation and testing (in as close to reality scenarios as safely possible) in early development stages. Off course, depending on the complexity of the product some steps could be, more or less, merged or overlapped.

Also, due to costs of:

- the developmental and operational testing,
- end-user systematic testing, and
- live agent laboratory and field environment validation and testing, budget of small to medium companies could not maintain on staff required variety of professionals and testing capabilities. Therefore, either:
- in partnership with potential big customers (governmental agencies, armed forces, etc.), or using services of specialized companies for live agent validation/ testing on a time limited basis for particular stage of development, might be worth the overall product investment return, embodied in a satisfied and safe customer.

References

1. Basumatry, J.: CBRN Terrorism: Threat Assessment. Centre for Land Warfare Studies, 76, http://www.claws.in (2016)
2. Shatter, A.: Opening Address – CBRNe World Convergence – All Hazards Response, http://www.defence.ie/ (2013)
3. The Centre for Excellence in Emergency Preparedness. CBRN Intro Sheet, http://www.ceep.ca/education/CBRNintrosheet.pdf
4. Operations in Chemical, Biological, Radiological, and Nuclear Environments. Joint Publication 3–1, http://www.dtic.mil/doctrine/new_pubs/jp3_11.pdf (2013)
5. SDI predicts CBRN market will surpass $13.69 billion by 2023, http://bioprepwatch.com/stories/510512028-sdi-predicts-cbrn-market-will-surpass-13-69-billion-by-2023
6. Sutulović, L.J., Karić, S.,Nikolić, V., Milovanović, D., MarčetaKaninski, M., Ceković, B., Rothbacher, D.: Research of Methodology Improvements of Live Agent (Chemical and Radiological) Exploitational Validation of the Decontamination Equipment in the Field Conditions, P19, 3rd Int. CBRNe Workshop, University of Rome Tor Vergata, (2016)
7. Standardized Equipment List (SEL), https://www.interagencyboard.org/sel
8. HotZone Solutions Catalogues, http://hotzonesolutions.com/solutions.html, http://hotzonesolutions.com/technologies.html
9. Capability Profiles., https://www.cbrneworld.com (2014)
10. Analysis of Potential Emergency Response Equipment and Sustainment Costs. Report to Congress. Requesters - GAO/NSIAD-99-151, http://www.gao.gov/assets/230/227601.pdf
11. Brands, R.F.: 8 Step Process Perfects New Product Development. http://www.innovationcoach.com/2013/05/8stepprocessperfectsproductdevelopment/
12. Le Beau, C.: 5 Steps to Take Your Product from Concept to Reality. https://www.entrepreneur.com/article/227390
13. Goodman, M.: The 7 Steps of Effective Product Development. https://www.scribd.com/article/327136071/The7StepsOfEffectiveProductDevelopment
14. Duval, J.: Eight Simple Steps for New Product Development. http://www.business2community.com/product-management/eight-simple-steps-for-new-product-development-0560298#IybbGms0zfzE7wEj.97
15. Ulrich, K.T., Eppinger, S.D.: Product design and development. McGraw Hill, (2008)
16. Loch, C.H., Kavadias, S.: Handbook of New Product Development. Elsevier Oxford, (2008)
17. Gopalakrishnan, M., Libby T., Samuels, J.A., Swenson D.: The Effect of Cost Goal Specificity and New Product Development Process on Cost Reduction Performance. Accounting, Organizations and Society 42, 1–11 (2015)

18. Eimecke, J., Baier, D.: Preference Measurement in Complex Product Development: A Comparison of Two-Stage SEM Approaches, Chapter Data Science, Learning by Latent Structures, and Knowledge Discovery, Part of the series Studies in Classification, Data Analysis, and Knowledge Organization. Springer-Verlag Berlin Heidelberg, 239–250 (2015)
19. Klein, A., Spiegel, G.: Social Media in the Product Development Process of the Automotive Industry: A New Approach, Human-Computer Interaction. Users and Contexts of Use, Volume 8006 of the series Lecture Notes in Computer Science, Springer Berlin Heidelberg, 396–401 (2013)
20. Rashid, M.: Compliance of Customer's Needs with Producer's Capacity: A Review and Research Direction, A study report, Department of Mechanical Engineering, Faculty of Engineering, Kitami, Hokkaido (2010) https://www.scribd.com/document/56314038/A-Study-on-Product-Development
21. Guzman, R., Navarro, R., Ferre, J., Moreno, M.: RESCUER: Development of a Modular Chemical, Biological, Radiological, and Nuclear Robot for Intervention, Sampling, and Situation Awareness,J. Field Robotics, 33, 931–945 (2016)
22. Biological Defense: DOD Has Strengthened Coordination on Medical Countermeasures but Can Improve Its Process for Threat Prioritization, http://oai.dtic.mil/oai/oai?verb=getRecord&metadataPrefix=html&identifier=ADA601867
23. Stewart, S.: The Jihadist CBRN Threat, https://www.stratfor.com/weekly/20100210_jihadist_cbrn_threat
24. Department of Defense (DoD) Joint Chemical and Biological Defense Program's (CBDP) 2014 Annual Report to Congress, http://www.acq.osd.mil/cp/docs/home/Final%202014%20DoD%20CBDP%20ARC_signed%2021%20Mar%202014.pdf
25. Ackerman, G.A., Pereira, R.: Jihadists and WMD: A Re-evaluation of the Future Threat, 2015, CBRNE World, http://www.cbrneworld.com (2014)
26. Chua P.M, Laljer, C.E.: Voluntary Consensus Standards for Chemical Detectors, https://www.mitre.org/sites/default/files/pdf/12_2554.pdf
27. Chemical Detection and Identification, https://www.tno.nl/media/7702/chemical_detectors_flyer.pdf
28. Instrument Standards for Detection of Hazardous Chemical Vapors, https://www.nist.gov/programs-projects/instrument-standards-detection-hazardous-chemical-vapors
29. DHS Science and Technology Directorate Office of Standards — Chemical Detection Standards Program, https://www.dhs.gov/sites/default/files/publications/Office%20of%20Standards%20-%20Chemical%20Detection%20Standards%20Programs-508_1.pdf
30. Holman, C., Loerch, A.G.: Chemical and Biological Test and Evaluation—Detector Agent Simulant Relationship, TEA J., 31, 525–530 (2010)
31. Next Generation Chemical Detector Tested During Joint Services Training, https://cbrnecentral.com/nextgenerationchemicaldetectortestedjointservicestraining/10293/
32. Terzic, O., et al.: Testing and Evaluation of New Chemical Detection Devices for The IDO Use During Toxic Chemical Trainings 2013–2014. IDO Internal Document, (2015)
33. Oudejans, L., O'Kelly, J., Evans, A.S., Wyrzykowska-Ceradini, B., Touati, A., Tabor, D., Snyder, E.G.: Decontamination of personal protective equipment and related materials contaminated with toxic industrial chemicals and chemical warfare agent surrogates. JECE, 4, 2745–2753 (2016)
34. Karkalić, R., Radulović, J., Blagojević, M., Popović, R.S.: Complex Approach in Testing of protective equipment for body protection from highly toxic chemicals. Kongres inženjera plastičara i gumara K-IPG, 5 (2008)
35. European Standards and Markings for Protective Clothing, http://www.hse.gov.uk/foi/internalops/oms/2009/03/om200903app7.pdf
36. Kuang, W.: Commercial Cleaning Products for Chemical Decontamination: A Scoping Study.http://oai.dtic.mil/oai/oai?verb=getRecord&metadataPrefix=html&identifier=AD1004327

37. Gudgin Dickson, E.F.: Implementation of Individual System Qualification (ISQ) in a CBRN Respiratory Protection Program. Part A: Guidance, http://cradpdf.drdc-rddc.gc.ca/PDFS/unc199/p800779_A1b.pdf
38. Ceković, B.: Plan and Program of Internal Testing of Decontamination Emulsion ED-1 Prototype. Mil. Tech. Institute - Internal Document, (2002)
39. Cekovic, B., Mladenovic, V., Lukovic, Z., Karkalic, R., Krstic, D.: Comparative Research on Chemical, Radiological and Biological Decontamination Efficiency of Present Decontaminants and of Multipurpose Emulsion-Based Decontaminant - Scientific Technical Information.Mil. Tech. Institute, 6(2006)
40. Cekovic, B., Stacey, P., Rothbacher, D.: Comparative Field Testing of Decontamination Agents – Summary Report. IDO - Internal Document, (2009)
41. Cekovic, B., et al.: Comparative Field Testing of Decontamination Agents Compatibility with Chemicals Commonly Used/Encountered on IDO Deployments. IDO - Internal Document, (2010)
42. Cekovic, B., Schmidt, O., Lebedinskaya, L., Sutulovic, L.J., Dvorak, J.: Report on the Preliminary Validation of the Efficiency of C/R-Decontamination in Field Conditions for Decontamination Trailer. HZS/TEST REPORT/DECONTAMINATION, 1 (2016)
43. Bourgouin, P.: Towards an Operational Urban Modeling System for CBRN Emergency Response and Preparedness, http://oai.dtic.mil/oai/oai?verb=getRecord&metadataPrefix=html&identifier=AD1004011
44. Early Live-agent Testing: The Competitive Edge for Manufacturers Specializing in Chemical/Biological Defense Equipment, http://www.battelle.org/lat-white-paper- download /

Radiological and Nuclear Events: Challenges, Countermeasures and Future Perspectives

Marco D'Arienzo, Massimo Pinto, Sandro Sandri, and Raffaele Zagarella

Abstract Over the last few years a broad array of organizations have practiced terrorism with the aim to achieve political, criminal, religious, and ideological goals. These acts have revitalized awareness of the threat of attacks involving chemical, biological, radiological or nuclear weapons. In particular radiological and nuclear methods are likely to be pursued by well organised terrorist groups, particularly those which have access to financial resources. The objective of this paper is to provide the reader with basic knowledge of possible radiological and nuclear events and the potential risks they pose. The document focuses on the characteristics of radiologic and nuclear agents as well as on the basics of response. Ultimately, this article explores how emerging technology has been infusing additional complexity into the global radiological and nuclear threat scenario.

Keywords Radiological and nuclear agents • CBRN • Terrorist attack • Weaponization • Radiological and nuclear response

M. D'Arienzo (✉) • M. Pinto
ENEA Casaccia Research Center, Department of Fusion and Technology for Nuclear Safety and Security, National Institute of Ionizing Radiation Metrology, Rome, Italy
e-mail: marco.darienzo@enea.it; massimo.pinto@enea.it

S. Sandri
ENEA Frascati Research Center, Radiation Protection Institute, Frascati, Rome, Italy
e-mail: sandro.sandri@enea.it

R. Zagarella
Centro Interforze Studi per le Applicazioni Militari (C.I.S.A.M.), Pisa, Italy
e-mail: raffaele.zagarella@esercito.difesa.it

© Springer International Publishing AG 2017
M. Martellini, A. Malizia (eds.), *Cyber and Chemical, Biological, Radiological, Nuclear, Explosives Challenges*, Terrorism, Security, and Computation,
DOI 10.1007/978-3-319-62108-1_7

1 Introduction

Chemical, biological, radiological and nuclear (CBRN) terrorism is a form of terrorism involving the use of weapons of mass destruction (WMD). A growing concern among homeland security professionals, is that terrorist groups will someday release CBRN materials in an attack against civilian populations.

Following 11th September 2001 and the anthrax attacks in America of the following month, the international community came to believe that the reality of a CBRN event has to be accepted. Furthermore, recent developments in the Middle East[1] have revitalized awareness of the threat of attacks involving CBRN weapons. Therefore it is generally accepted that there is a realistic possibility of some form of unconventional terrorist attack in the western world and that this could involve CBRN material. The authorities need to consider how the community is likely to react and collaborative efforts are needed to identify what steps would be required to mitigate the effects of such an event.

CBRN is a term that covers four distinct groups of hazards. *Chemical* hazard involves poisoning or injury caused by chemical substances. *Biological* agents are microorganisms (dangerous bacteria, viruses or fungi, or biological toxins) that may cause infection, toxicity or allergy in humans. *Radiological* agents cause illness by external or internal exposure to ionizing radiation and other harmful radioactive materials contaminating the environment. In particular, there has been extensive concern among security analysts that terrorists might steal medical or industrial radionuclides (such as radium, cesium or cobalt) to manufacture "dirty bombs", i.e. radiological weapons that combines radioactive material with conventional explosives with the aim to generate psychological harm through mass panic and terror. Ultimately, *nuclear* agents include exposure to ionizing radiation following the detonation of a nuclear device. The main differences between radiological and nuclear agents are related to their different origin. Broadly speaking, radiological agents are radioactive materials generated as by-products and waste from industrial processes or medical facilities. Nuclear agents are radioactive materials generated from nuclear fission or fusion reactions, such as those produced by detonation of a nuclear weapon or released from nuclear power plants.

According to the University of Maryland's Global Terrorism Database, from 1970 to 2014 CBRN weapons have been extensively used across the world, for a total of 143 attacks (of which 35 biological, 95 chemical, and 13 radiological). This information is depicted in Fig. 1. In particular, over the last 20 years the fear of terrorist groups using dirty bombs or other radiological agents has increased remarkably. Indeed, radiological incidents have already occurred in the past. The plausibility of a radiological attack became evident in November 2006 when former Soviet agent Alexander Litvinenko was murdered in the United Kingdom by poisoning

---

[1] Nerve agent sarin was used in an attack on the Ghouta agricultural belt around Damascus on the morning of 21 August 2013. The sarin attack claimed between 350 and 1400 lives and was a critical moment in the Syrian war, provoking a debate about who was responsible.

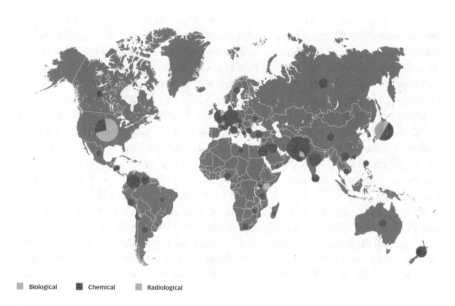

Biological   **Chemical**   **Radiological**

Fig. 1 Chemical, biological and radiological attacks across the world from 1970 to 2014 (From Unal et al. 2016 [2]. Reproduced with the permission of Chatham House, the Royal Institute of International Affairs)

with radioactive polonium. A small amount of ^{210}Po was placed in his drink at a restaurant. He was not the only person exposed to radiations. In the process, murderers irradiated hotel rooms, bars, restaurants, toilets, aircraft and cars. At least 12 people were contaminated as of January 2007. Tissue analyses showed the presence of ^{210}Po at lethal doses in several organs and post-mortem samples were used to estimate intake as 4.4 GBq [1].

Due to the relative technical simplicity of a dirty bomb, the probability of such an event depends almost entirely on the acquisition of radioactive materials. There have only been only a few cases of dirty bombs and none of them ever detonated. According to a UN report, Iraq tested a dirty bomb in 1987. The radiation levels were considered too low to achieve the objectives and the project was abandoned [3]. In 1995 a group of Chechen separatists buried a dirty bomb consisting of dynamite and cesium-137 in Moscow's Izmailovsky Park. The bomb never detonated and neither the Chechens who buried the bomb there nor the original source of the cesium were ever identified. A few years later, in 1998, the Chechen Security Service discovered a container filled with radioactive materials and attached to an explosive mine. The bomb was hidden near a railway line in Argun (Chechen Republic). The bomb was defused, but the radioactive materials involved were never identified. Finally, in 2003 British intelligence agents discovered documents in Afghanistan suggesting that al-Qaeda might have succeeded in building a dirty bomb.

On the other hand, the events of September 11, 2001 raised new concerns about the vulnerability of nuclear facilities to internal sabotage or terrorist attack. According to International Atomic Energy Agency (IAEA) spokesman David Kyd, "[Nuclear reactors] *are built to withstand impact, but not that of a wide-bodied passenger jet full of fuel. A deliberate hit of that sort is something that was never in any scenario at the design stage. These are vulnerable targets and the consequences of a direct hit could be catastrophic*" [4]. The actual likelihood of such an event is open to debate. However, it is worth mentioning that over the past three decades nuclear reactors have been repeatedly attacked during military occupations and campaigns. On 30 September 1980 a surprise airstrike carried out by Islamic Republic of Iran Air Force damaged an almost-complete nuclear reactor 17 km south-east of Baghdad, Iraq (Operation Scorch Sword). One year later, in June 1981, an Israeli air strike completely destroyed Iraq's Osirak nuclear research facility. Between 1984 and 1987, Iraq bombed Iran's Bushehr nuclear plant six times, while in 1991, the U.S. bombed three nuclear reactors and an enrichment pilot facility in Iraq. In 1991, Iraq launched Scud missiles at Israel's Dimona nuclear power plant. Ultimately, in September 2007, Israel bombed a Syrian reactor under construction [5]. As a consequence, Institutions, policy makers and relevant security agencies need to be on alert, and should consider addressing the possibility of a terrorist attack using radiological or even nuclear materials.

The objective of this document is to provide the reader with basic knowledge of radiological and nuclear agents and the potential risks they pose. It is worth noting that CBRN agents may be released in a variety of situations, whether incidentally (unintended CBRN mishaps) or maliciously (deliberate use of CBRN weapons in a conflict). This synthesis report is mainly focused on the intentional malevolent use of radiological and nuclear agents.

2   Basics of Radiation Exposure and Injury

2.1   Types of Ionizing Radiation and Units of Radiation

There are four distinct types of radiation, each with very different properties: alpha, beta, neutron particles and photons. Conventionally photons can be referred to as gamma radiation (if produced by a nuclear reaction) or x-rays (if photon emission originate in the electron shell). *Alpha* radiation occurs when an atom undergoes radioactive decay, giving off a particle (alpha particle) consisting of two protons and two neutrons (essentially the nucleus of a helium-4 atom). Therefore alpha particles are massive, about 4 times the mass of a neutron. Because of their size, alpha particles cannot travel far (only travel a few centimeters in air) and are fully stopped by the dead layers of the skin. As a general rule, alpha particles constitute a negligible external hazard. However, alpha particles pose significant hazard if they are ingested or inhaled. In fact, when inside the body, they are able to travel just enough distances into tissues to cause considerable damage. *Beta* particles are electrons ejected from the nucleus of a decaying atom and can travel a short distance in tissue. Although they can be stopped by a thick piece of plastic, or even a stack of paper, beta particles can penetrate the dead skin layer, potentially causing burns. They pose serious external and internal radiation threats and can be lethal depending on the amount received. *Neutrons* are uncharged particles and are emitted during nuclear reactions. They have significant mass and interact with the nuclei of atoms, severely disrupting atomic structures. Compared to gamma rays, they can cause 20 times as much damage to tissue. Neutrons can generally be stopped by a hydrogen-rich material, such as water, concrete or polyethylene. Photons have no mass, travel at the speed of light and can be very penetrating. Hence, photonic radiation require more shielding to reduce their intensity than do beta or alpha particles. Both gamma radiation and x-rays consist of photons. The main differences between gamma radiation and x-rays is related to their different origin. Gamma radiation is emitted by certain radionuclides during nuclear transitions from a higher to a lower energy state, while x-rays is electromagnetic radiation produced by deflection of electrons from their original paths, or inner orbital electrons changing their orbital levels around the nucleus. X-rays, like gamma rays can travel long distances through air and most other materials.

The energy per unit of mass imparted by ionizing radiation to matter is expressed by the physical quantity *absorbed dose*. In the SI system of units, the unit of measure of the absorbed dose is joules per kilogram, and its special name is gray (Gy). A dose of 1 Gy is equivalent to a unit of energy (joule) deposited in 1 kg of substance. The unit for dose, gray, is not restricted to any specific radiation but can be used for all forms of ionizing radiation. The former unit of absorbed dose is called *rad* (radiation absorbed dose) and is equivalent to an absorption energy of 10^{-2} J/kg, i.e. 1 Gy = 100 rad.

Beside the physical quantity absorbed dose, there are two types of quantities defined for use in radiological protection: *protection quantities* (defined by the ICRP and used for assessing the exposure limits) and *operational quantities* (defined by the ICRU and intended to provide a reasonable estimate for the protection quantities).

Protection quantities are *equivalent* and *effective dose*. It is worth noticing that different types of radiation have different effects on biological tissue. In order to account for these differences, each type of radiation is assigned a radiation weighting factor (w_R) as a measure of its biological effect. *Equivalent dose (H_T)* is calculated by multiplying the absorbed dose to the organ or tissue (D_T) with the radiation weighting factor, w_R and it is expressed in sieverts (Sv).

$$H_T = w_R \cdot D_T \qquad (1)$$

In this form equivalent dose measures the biological effects of ionising radiations in that it takes account of the type of radiation, the energy carried by the radiation and how much tissue absorbs the energy. In 2007 ICRP published an updated set of radiation weighting factors [6] (Table 1).

Table 1 Tissue weighting factors according to ICRP 103 [6] Tissue and continuous function in neutron energy, E_n (MeV), for the calculation of radiation weighting factors for neutrons

Radiation type	Radiation weighting factor w_T
Photons, all energies	1
Electrons, muons, all energies	1
Protons and charged pions	2
Alpha particles, fission fragments, heavy ions	20
Neutrons	A continuous function of neutron energy (see equation below)

$$w_R = \begin{cases} 2.5 + 18.2 \cdot e^{-\frac{[\ln(E)]^2}{6}} & E < 1 MeV \\ 5.0 + 17.0 \cdot e^{-\frac{[\ln(2E)]^2}{6}} & 1 MeV \leq E \leq 50 MeV \\ 2.5 + 3.25 \cdot e^{-\frac{[\ln(0.04E)]^2}{6}} & E > 50 MeV \end{cases} \tag{2}$$

Biological effects also depend on the type of tissue or organ that has been irradiated. Therefore a new radiological protection quantity, the *effective dose* needs to be introduced. The effective dose, E, is calculated as the sum of the weighted equivalent doses in all the tissues and organs of the body:

$$E = \sum_T w_T \cdot H_T = \sum_T w_T \sum_T w_R \cdot D_{T,R} \tag{3}$$

The factor by which the equivalent dose in tissue or organ is weighted is called "tissue weighting factor", w_T, and represents the relative contribution of individual tissues to total detriment of stochastic effects resulting from uniform irradiation of the whole body. Tissue weighting factors according to ICRP 103 are reported in Table 2. The unit for the effective dose is the same as for absorbed dose, J/kg, and it is expressed in sieverts (Sv). Broadly speaking, the effective dose attempts to provide a single numerical value that is proportional to the radiobiological detriment from a particular radiation exposure. In this context detriment represents a balance between carcinogenesis, life shortening and hereditary effects.

It is important to recognise that the ICRP protection quantities, *equivalent* and *effective* dose, do not relate to individuals but to reference persons. They are calculated using reference models and defined radiation and tissue weighting factors [7].

The human body-related protection quantities can't be measured directly as they are the result of a calculation. Therefore they are not appropriate for area and individual dose measurements. To overcome these practical limitation ICRU has introduced a set of *operational quantities*, which can be measured and which are intended to provide a reasonable estimate for the protection quantities. Furthermore *operational quantities* allows calibration of area and individual monitoring instruments. The operational quantities are based on point doses determined at defined locations

Table 2 Tissue weighting factors according to ICRP 103 [6]

Tissue	Tissue weighting factor w_T	Σw_T
Bone-marrow (red), colon, lung, stomach, breast, remaining tissues[a]	0.12	0.72
Gonads	0.08	0.08
Bladder, oesophagus, liver, thyroid	0.04	0.16
Bone surface, brain, salivary glands, skin	0.01	0.04
	Total	1.00

[a] Remaining tissues: Adrenals, extrathoracic region, gall bladder, heart, kidneys, lymphatic nodes, muscle, oral mucosa, pancreas, prostate ($\male$), small intestine, spleen, thymus, uterus/cervix ($\female$)

in defined phantoms. One such phantom is the ICRU-sphere, a sphere having a diameter of 30 cm and a density of 1 g/cm³ (mass composition of 76.2% oxygen, 11.1% carbon, 10.1% hydrogen and 2.6% nitrogen). For calibration purposes, the ICRU-sphere can be replaced by a 30 cm × 30 cm × 15 cm square block with the same composition.

Operational quantities are: ambient dose equivalent, $H^*(d)$, directional dose equivalent, $H'(d, \Omega)$ and personal dose equivalent, $H_p(d)$. Each quantity is specified at an appropriate depth d (and oriented in the fixed direction Ω for the directional dose equivalent).

The operational quantities were defined for area monitoring and for individual monitoring. Area monitoring generally characterizes radiation fields with respect to their relevance for radiation protection measures. For area monitoring ambient dose equivalent, $H^*(d)$, and Directional dose equivalent, $H'(d, \Omega)$ are used. Individual monitoring is used for determining the individual exposure of persons. For individual monitoring the individual dose equivalent penetrating, $H_p(d)$, is used.

It is worth noticing that regulatory dose limits (ICRP 103) are given in terms of:

- effective dose, E
- equivalent dose to the skin, H_{skin}
- equivalent dose to the lens of the eye, $H_{eye lens}$
- equivalent dose to the hands and feet

The relationship between regulatory dose limits and operational quantities is reported in Table 3. For a comprehensive treatment and for references to the extensive literature on the subject one may refer to [8–10].

2.2 Exposure to Ionizing Radiation

Radiation exposure may be classified as either external or internal. *External radiation exposure* occurs when all or part of the body are exposed to penetrating ionizing radiation from an external radiation source. Exposure from an external source stops when an individual leaves the area of the source, the source is shielded

Table 3 Operational dose quantities for external radiation dose exposure

Task	Area monitoring	Individual monitoring
Monitoring of ambient dose personal dose	Monitoring of ambient dose personal dose, H* [11]	Monitoring of ambient dose personal dose, H_p [11]
Monitoring of equivalent directional dose personal dose	Monitoring of equivalent directional dose personal dose, $H'(0.07, \Omega)$	Monitoring of equivalent directional dose personal dose, $H_p(0.07)$
Monitoring of equivalent directional dose personal dose	Monitoring of equivalent directional dose personal dose, $H'(3, \Omega)$	Monitoring of equivalent directional dose personal dose, H_p [4]

completely, removed or switched off. External contamination is an example of external exposure. It occurs when radioactive material (for example in the form of dust or liquid) comes into contact with a person's skin, hair, or clothing.

There are three general guidelines for minimising exposure to ionising radiation: reducing exposure time, maximising distance from the radiation source, and shielding yourself from the radiation source.

Time is an essential factor in limiting exposure to the public and to radiological emergency responders, especially following a radiological or nuclear attack. It is of utmost importance to carefully plan the work to be done prior to entering the radiation environment. Working as quickly as practicable and rotating workers who are in the radiation area will help minimise exposure of individuals.

A significant reduction in radiation dose can be obtained by increasing the distance between the exposed individual and the radiation source as the decrease in exposure rate follows the inverse square law, i.e. doubling the distance from the source (supposed to be point like) the exposure rate decreases to 1/4 of its original value.

Finally, the placement of a shield between the exposed individuals and the radiation source will also dramatically help minimise radiation dose. As a general rule, shielding of gamma and x-radiation is accomplished using high-density materials. Lead is particularly well-suited for reducing the exposure to photon radiation due to its high atomic number. On the other hand, materials with low atomic number are appropriate as beta particle shields. This is because beta radiation can produce secondary gamma and x-radiation when passing through elements with a high atomic number and density. This is called bremsstrahlung radiation and is produced during slowing down of beta particles while they travel in a very dense medium. As a consequence plastic and other low atomic number elements (e.g. aluminum) are generally preferred when dealing with high-energy beta radiation. Shielding of alpha radiation does not pose a difficult problem. Alpha radiation can be stopped after traveling a few centimeter of air and a single centimeter of plastic is sufficient for shielding against alpha particles. Even the dead cells in the outer layer of human skin provides adequate shielding because alpha particles can't penetrate it. Finally, neutrons are generally well attenuated by hydrogen and hydrogen-based materials. Compounds with a high concentration of hydrogen atoms (such as water, polyethylene and paraffin) form efficient neutron barriers. However, it is important to notice that an absolute barrier may not be possible in many situations.

Internal contamination can occur when radioactive material is airborne (inhalation), is present in contaminated food, drink or other consumable items (ingestion) or is spilled or aerosolizes onto the skin and absorbed or enters through cuts or scratches (skin contact). Internal deposition may also result from contaminated hands, with subsequent eating or rubbing of eyes. Internal exposure stops when the radionuclide is completely removed from the body, either spontaneously (such as through excreta) or as a result of a medical treatment. It is worth noting that a radiological attack could also be the cause of radioactive contamination of water, food, and other widely consumed commodities [11]. However, this possible outcome is considered unlikely to lead to significant internal contamination of a large number of people owing to the large amounts of radioactive material that would be required to cause high levels of contamination [11].

Although external exposure is primarily caused by x-rays, gamma rays and neutrons, all forms of radiation can cause internal radiation exposures. This is particularly true for alpha particles, which pose a significant internal hazard. Because of this double positive charge and their size, alpha particle create a high concentration of ions along their path, and can cause severe damage to internal organs and tissues when they are inhaled, ingested or are present on the skin.

In case of a nuclear explosion, over 400 radioactive isotopes are released, of which about 40 are likely to pose serious hazard to individuals [12]. In fact, it is known that many elements taken into the body will accumulate in certain target organs.

Iodine is present in the form of ten radionuclides produced during a nuclear explosion. Iodine ^{131}I is a major cause of concern for internal contamination because of its volatility and hence its ability to enter the body via inhalational pathways. Since the body cannot distinguish stable iodine from radioactive iodine, a significant portion of radioactive iodine take into the body will be deposited in the thyroid gland. Polonium tends to be a soft-tissue seeker and accumulates mainly in liver and kidney. On the other hand radium and strontium, which are alkaline earth elements, follow the metabolic pathways of calcium and are deposited almost exclusively in the bones. Another key radionuclide in case of internal contamination is cesium (with 21 radioisotopes), which is a metabolic homolog of potassium. The most important radionuclide is cesium ^{137}C, a product of nuclear fission. Ultimately, Uranium has three alpha emitters isotopes with long half-lives that pose significant hazards of internal contamination: ^{234}U, ^{235}U, ^{238}U. The osteotropic properties of uranium isotopes are of particular concern due to their final retention in the bone structure.

However, in the event of an internal contamination, the most important hazard is likely to be plutonium, which is osteotropic. In fact, plutonium is frequently referred to as *the most toxic substance known to man*. It accumulates in the bone crystals causing chromosomal aberrations, genotoxicity, malignant alterations, and cellular death. Plutonium can stay in the bones for decades, continuously delivering dose to the body [12, 13].

Exposure to ionizing radiation can cause both immediate and long-term health effects. Health effects resulting from radiation exposure are categorized as either deterministic or stochastic. Deterministic effects only occur once after a threshold dose is reached. Furthermore, the severity of deterministic effects increases as the dose of exposure increases. Stochastic effects occur without a threshold level of dose and the probability of having the effects is proportional to the dose. Furthermore, the severity of stochastic effects is independent of the absorbed dose. In the context of radiation protection, the main stochastic effect is cancer.

Immediate health effects to the general population are of great concern in case of a radiological or nuclear event. Therefore the following section will be dedicated to early effects of radiological and nuclear agents.

2.3 Early Effects of Radiological and Nuclear Agents

Even very low amounts of energy absorbed from ionizing radiation can cause severe consequences because of the interaction between radiation and the basic mechanisms of human cells. Anyway, the effects are not immediate like for chemical agents, because these processes require a certain time interval to damage the cells and so, depending on the occurring moment of the harmful outcomes, we usually speak of "*deterministic*" or "*stochastic*" effects.

Stochastic effects are long-term consequences and often appear several years after the exposure; at the moment it is hypothesized, at least for the regulatory purpose, that it's not possible to establish a threshold for such effects. The carcinogenic or teratogenic consequences are very severe and irreversible.

Deterministic effects are instead acute biological consequences that occur within several hours to several days following the exposure to ionizing radiation. Examples of deterministic effects are provided in Table 4.

After an acute event involving exposure to radioactive material (such as a terrorist attack, or a severe nuclear accident) mass measures of triage and medical care must be put in place following the emergency procedures planned by the civil and military authorities in charge. It can be useful for the first responders to know at least the basic details of the deterministic effects of the radiological and nuclear agents, to be able to identify early signs and turn the injured patients to the correct paths of assistance.

Deterministic effects in an organ/tissue arise when the absorbed radiation kills or prevents from reproducing and functioning a sufficient number of cells, and hence an exposure threshold can be determined. No harmful effect happens below this value while above it, the damage is directly proportional to the number of cells affected and hence to the radiation dose. Although the degree of pathological severity increases rapidly with dose, the deterministic effects can be reversed if proper medical care is administered promptly.

Detailed data on the deterministic effects are incomplete, especially as far as the lethal dose is concerned. Of course, it is not possible to perform experiments on real persons and on the other hand the results found for nonhuman primates cannot be directly transposed to humans, because of the difference in the DNA, the biologically relevant target molecule of ionizing radiation. Data can be estimated from computer simulations or from the monumental studies on the cohorts of survivors of the Hiroshima and Nagasaki bombings of 1945; many available data come from

Table 4 Operational dose quantities for external radiation dose exposure

Effect	Description
Erythema	Erythema occurs 1–24 h after the exposition, and the breakdown of the skin surface occurs approximately after 4 weeks. Epilation occurs 3 weeks following exposure and can be reversible or irreversible according to the absorbed dose
Sterility	Sterility is due to the damage of the germinal cells, leading to impaired or temporary non-fertility. The radiation dose required to cause this effect decreases with age because of the falling of the total germinal cells present. The higher the dose, the higher the probability the sterility might become permanent
Cataract	Cataract occurs due to the accumulation of damaged or dead cells within the lens, the removal of which cannot take place naturally

studies conducted during the cold war when a nuclear attack on large scale was conceived possible.

The effective dose range of interest for a responder to a radiological emergency should be, in the worst scenario, up to 1 Sv. Larger effective doses might be involved only in a severe nuclear accident or for a highly unlikely terroristic attack with nuclear weapons. A complete set of deterministic effects is summarized in the following for effective doses ranging from 0 to 100 Sv:

- below 0.15 Sv: no effect takes place
- between 0.15 and 1 Sv: different organs/tissues can be damaged
- between 1 and 3 Sv: radiation sickness occurs

Radiation sickness (correctly termed Acute Radiation Syndrome) involves nausea, vomiting, and diarrhoea developing within hours or minutes after intense radiation exposure.

- Between 3 and 5 Sv: 50% of exposed people will die in an average period of 60 days without any medical care (this quantity is called $_{60}LD^{50}$ in some textbooks)
- Between 5 and 10 Sv death occurs for secondary infections. The population of white blood cells is completely depleted and even normally harmless viruses cause lethal infections
- Between 10 and 100 Sv gastrointestinal death occurs. The walls of the GI tract are irreversibly damaged, and death occurs for the infections after the body is invaded by the bacteria coming from the intestine
- Above 100 Sv CNS death occurs. The dose is high enough to destroy directly the cells of the Central Nervous System

3   Radiological and Nuclear Agents

It is generally believed that radiological agents and nuclear materials are more accessible now than at any other time in history. This problem is exacerbated by a number of circumstances. Among these: (I) the widespread and increasing use of medical radionuclides (II) the diffusion of modern nuclear technologies (III) the reduced protection of the fissile material in the former Soviet Union. Surveillance of this material has become more difficult as many of the institutional mechanisms that once curtailed the spread of nuclear materials, technology, and knowledge no longer exist (IV) knowledge of radiological and nuclear weapons design is today sufficiently widespread and secrecy on this technical knowledge no longer offers adequate protection.

Radiologic and nuclear agents are radioactive materials or radiation sources that have the potential to pose severe adverse health effects. In particular they present both an external and an internal hazard: (I) emission of highly penetrating radiation leading to external irradiation or contamination of individuals (*external exposure and external contamination*). (II) Irradiation of tissues and organs in the body if the agents are inhaled, absorbed through the skin by direct contact or contact with contaminated matter, or ingested in contaminated food or water (*internal contamination*).

Radiological agents include by-products and waste radioactive materials from industrial processes or medical facilities. In recent years, the safety and security of these sources has gained considerable attention due to the possible risk of acquisition and manufacture of a radiological dispersal device by terrorist organizations. Furthermore, tracking and recording radiological sources is not always easy as they are used in a wide variety of activities.

Medical radionuclides are used in a broad range of applications like oncology, cardiovascular and neurological disorders. Approximately 140 radioisotopes are currently used worldwide in medical applications. In particular, the use of radioactive sources for the treatment of cancer and other pathological conditions has been undertaken clinically for over 50 years. This practice, generally referred to as "molecular radiotherapy" (MRT), is based on the delivery of radiation to malignant tissue through the interaction of an agent with molecular sites and receptors. MRT is today rapidly establishing as an additional treatment modality in oncology. Radionuclides generally used in MRT include Iodine-131, Iridium-192, Iodine-125, Strontium-89, Samarium-153, Rhenium-186, Yttrium-90, Lutetium-177 and Radium-223. In addition, high activity Cobalt-60 and Cesium-137 sources can be used for external irradiation of solid tumours. Diagnostic procedures in nuclear medicine also use radionuclides which emit gamma rays from within the body. Among these Technetium-99 m and Fluorine-18 are widely used for SPECT and PET acquisitions, respectively. Such radioactive sources merit real concern as they are vulnerable to theft, to black-market sale and can be possibly used to make a radioactive dispersal device.

In industry, gamma radiation sources are widely used for sterilising medical products and supplies such as syringes, gloves, clothing and instruments, many of which would be damaged by heat sterilisation. In many countries high activity gamma sources are widely used in the form of large-scale irradiation facilities. Cobalt-60, with a half-life of 5.27 years, and caesium-137, with a half-life of 30.1 years, are the best gamma radiation sources because of the high energy of their gamma rays and their relatively long half-lives. Caesium-137 sources, however, are used only for treating blood for transfusions and to sterilize insects.

In addition, sealed radioactive sources and x-rays are used in industrial radiography as they have the potential to show flaws in metal castings or welded joints. The technique allows critical components to be inspected for internal defects without damage (non-destructive analysis).

Neutron activation analysis is an analytical technique based on the measurement of characteristic radiation emitted from radionuclides formed directly or indirectly by neutron irradiation of the irradiated material. Most commercial analysers use californium-252 neutron sources or Am-Be-241 sources.

Nuclear agents are radioactive materials generated from nuclear fission or fusion reactions. Therefore nuclear agents can be produced in nuclear power plants or by detonation of a nuclear weapon.

Nuclear fission is a process where heavy atomic nuclei materials (certain isotopes of uranium and plutonium) fragment into smaller nuclei, releasing energy and neutrons. Each fission reaction can initiate further fission reactions, leading to a selfsustaining nuclear chain reaction which can be controlled (as in nuclear power plants) or uncontrolled (as in the detonation of a nuclear bomb, where the chain reaction causes an immediate release of energy). Nuclear material is a term used to describe both fissile materials and non-fissile materials which are suitable for transformation into fissile materials. On the other hand, nuclear energy can also be released by fusion of elements with low atomic numbers. In a hydrogen bomb, two isotopes of hydrogen, deuterium and tritium are fused to form a nucleus of helium and a neutron. The process is called nuclear fission.

Fissile materials and nuclear materials have a wide range of applications, both civilian (nuclear power plants, research reactors) and military (nuclear submarines, and the construction of nuclear weapons). As a consequence, illicit trafficking nuclear materials has become a serious threat from the viewpoint of terrorist hazard.

3.1 Radiological Dispersion Devices

Radioactive material could be used in many instances to prepare weapons. Fission and fusion nuclear bombs are known by almost everybody worldwide, due to their use during the second world war. In that case the radioactive material is used as power supply for the deflagration and the consequent radioactive fall out is a secondary effect of the nuclear explosion. In a terroristic attack, radioactive materials could be used in many other ways. This kind of event is often called "radiological attack" and means the spreading of radioactive material with the intent to do harm. Radioactive materials are used every day in laboratories, medical centers, food irradiation plants, and for industrial uses; if stolen, or otherwise acquired, many of these materials could be used in a "radiological dispersal device" (RDD) [14]. A "dirty bomb" is one type of RDD that uses a conventional explosion to disperse radioactive material over a targeted area. RDDs could also include other means of dispersal such as placing a container of radioactive material in a public place, or using an airplane to disperse powdered or aerosolized forms of radioactive material.

Another kind of radiological device that could be used by terrorists is the so called RED (Radiation Emission/Exposure Device). This refers to a high-intensity radiation source placed in a public area to expose those individuals in close proximity. Prolonged exposure to a high intensity source may lead to acute radiation syndrome (ARS) or to cutaneous radiation syndrome (CRS), or to radiation burns.

The explosion of an RDD cannot cause mass casualties on the scale of a nuclear explosion. In a "dirty bomb" event all or most fatalities or injuries will probably due to explosion itself. While large numbers of people in a densely populated area around the detonation might become contaminated and require decontamination, few if any will be contaminated to a level that requires medical treatment. The need of medical intervention has to be investigated by the local health authorities that have to assess the persons who were very close to the point of release, or were in line with the path of the radioactive release. The other aspect to be considered for selecting people that have to be addressed to a medical triage is the assessment of the contamination level for each individual.

In general the health and environmental consequences from RDDs, will depend on

- Design of the device
- Type and quantity of radioactive material
- Characteristics of the area of the impact
- Pattern of dispersion following the release

Depending on the nature of the dispersion and the amount and type of radioactive material, RDDs may affect different kind of environments ranging from small and localized areas (e.g., a street, single building, or city block) to large areas, up to several square miles. Other hazards may also be present like fire, smoke, shock and shrapnel (from an explosion).

For the individuals living in the area of a RDD influence, the doses are expected to ranging from less than 1 mSv to some hundreds of millisievert. People staying closer to the strike point could be exposed to higher dose levels but these persons should be in a small number, not higher then few tenths of individuals.

On the other hand, it has to be noticed that the World Health Organization considers an individual dose of 500 mSv in 1 year acceptable for emergency work [15]. Therefore in general after an RDD impact it has to be expected that the health effects to the involved individuals should be restricted to the stochastic injuries for most of them, with minor deterministic tissue reactions for those exposed to doses closed to 500 mSv or higher.

In general there is no way of knowing how much warning time there will be before an attack by terrorists using a RDD and how high the maximum dose level due to the attack will be, so being prepared in advance and knowing what to do and when, is extremely important. Participation to International exercises like those of the INEX series (international nuclear emergency exercises), organized under the OECD Nuclear Energy Agency's (NEA) Working Party on Nuclear Emergency Matters (WPNEM) [15], is extremely important for testing, investigating and improving national and international response arrangements for nuclear accidents and radiological emergencies.

3.2   Nuclear Weapons

Nuclear weapons are, by definition, devices using fission and fusion reactions to produce large amount of power.

The explosive yield of a nuclear weapon is the amount of energy released when a nuclear weapon or device is detonated. Such quantity is usually expressed as a TNT[2] equivalent, either in kilotons (thousands of tons of TNT), in megatons (millions of tons of TNT), or sometimes in terajoules (TJ). The energy released during a nuclear explosion, depending on the size and type of the weapon, can range from tens of tons of kiloton to hundreds of megatons.

Fusion bombs are based on the fusion of light nuclei such H and require a quite sophisticated technology to be realized. Fission bombs are based on the fission of heavy nuclei and are of greater concern as far as the CBRNe point of view because they are based on a technology similar to that available in nuclear power plants. Improvised Explosive Devices could be realized if a sufficient quantity of fissile material is available (critical mass). The only naturally available isotope suitable for fission is U^{235}, but its concentration in natural uranium is so low that an enrichment process is necessary. A byproduct of this operation is depleted uranium (DU). In general, Special Nuclear Materials (SNM) are both fissile and fertile materials i.e. materials that can be easily made fissile for neutron absorption. A full list can be found in the IAEA documents, but the nuclides of most interest are U^{235}, U^{238}, Pu^{239}, Th^{232}, U^{233}.

Illicit smuggling of SNMs is tough to monitor because SNMs have high densities ($10–15$ Kg/cm^3) and hence dangerous masses can occupy tiny volumes. "Significant quantities" are: 4 kg of Pu^{239} and 25 Kg of highly enriched uranium (HEU) corresponding to spheres of few cm in diameters. Furthermore, passive detection is very difficult because such small items can easily be hidden under heavy shields.

Active inspection projects are at the moment under investigation; they use high energy beams of gamma rays to induce photo-fissions that can be detected with suitable equipment.

4   Emergency Response Planning

4.1   Equipment for Emergency Responders

As reported in par. 2 the hazard from ionizing radiation can derive from "*internal*" and "*external*" exposure, and so emergency responders must be equipped with suitable instruments to detect and assess the consequent risk arising from the spreading of radioactive material.

---

[2] The explosive yield of trinitrotoluene (TNT) is considered to be the standard measure of bombs and other explosives. The *ton* of TNT is a unit of energy defined to be 4.184 gigajoules. Similarly, the *kiloton* of TNT is a unit of energy equal to 4.184 terajoules, while the *megaton* of TNT is a unit of energy equal to 4.184 petajoules. These quantities are traditionally used to describe the energy output, and hence the destructive power, of a nuclear weapon.

The risk of *external exposure* can be assessed by measuring the equivalent dose in air taking into account the contribute coming from the background. The nine key radionuclides for RDDs (see References) are: Am^{241}, Cf^{252}, Cs^{137}, Co^{60}, Ir^{192}, Pu^{238}, Po^{210}, Ra^{226}, Sr^{90}; almost all of them (except Po^{210} and Sr^{90}) have gamma emissions that can easily be detected even with simple instruments.

For the *internal exposure*, some additional consideration has to be taken into account: radionuclides can penetrate into the body for ingestion, inhalation or through wounds from a contaminated surface or directly from the air. Contamination measurements can be carried out by instruments with a particular geometry and high sensitivity to alpha and beta radiation. Instead, air concentration measurements cannot be performed directly, and a sample of air volume is required for analysis. Airborne particulate can be collected with an aspiring pump and filters of suitable grid step (paper filters are easy to find and can be usefully used as a first assessment method). If the contaminant is a gas, the collection has to be performed by a different kind of sampler: carbon canisters or absorbing gels can be usefully employed.

In principle, any system able to produce a measurable quantity for the interaction between radiation and matter could be used to assess the exposure to radiological and nuclear agents, included chemical and calorimetric systems. In practice, most detectors are based upon two methods:

Detection technique	Description
Ionometric	Direct collection of ions generated by the passage of the radiation
Scintillation	Indirect inference of the exposure through the collection of the light generated by the radiation passing through special media

A very typical mounting for the first type of detectors is a wire stretched inside a gas-filled cylinder; the wire and the holder make up the armours of a capacitor to which a high voltage is applied.

The total charge produced by the passage of an ionizing radiation through the active volume is collected and measured as an electric signal that can be extracted and processed in different ways. The simplest system is an RC circuit to turn this charge pulse into an easily measurable voltage. The decay constant determines the integration time and can be made very short (less than a second) to obtain a quick but erratic response because of the intrinsic statistic fluctuations or very long (many tens of seconds) to get a precise but slower response. More sophisticated electronics can perform a pulse amplitude analysis through a multi-channel analyser (MCA). Different names are used for such devices based on the amount of voltage applied to the central electrode and the consequent nature of the ionizing events (Fig. 2).

If the voltage is large enough for the primary electron-ion pair to reach the wires but not high enough for secondary ionization, the device is called *ionization chamber*. The collected charge is proportional to the number of ionizing events, and hence to the total energy released to the active medium. Such devices are excellent radiation dosimeters, but the signal is very weak and can be confounded with the

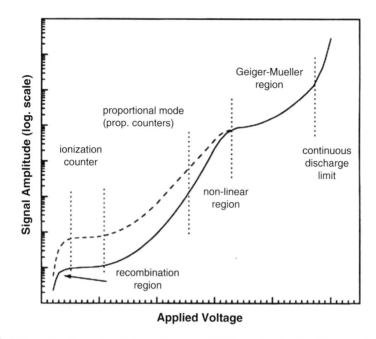

Fig. 2 I. Recombination region II. Ionization region III. Proportional region IV. Limited proportional region V. Geiger-Mueller region VI. Continuous discharge region (Image adapted from [16])

background if the dose rate is low. At a higher voltage, the number of ionizations associated with a particle detection rises steeply because of secondary ionizations: the device becomes a *proportional counter*. A single event can generate a voltage pulse proportional to the energy loss of the primary particle. At a still higher voltage, an avalanche pulse is produced by a single event in the devices called *Geiger counters*. These instruments are very robust, simple to operate and provide signals of high amplitude, but any correlation with the energy released is entirely lost, and so they are excellent instruments for source detection but return no information about the absorbed dose.

The gas can be substituted with a solid medium like a semiconductor, but the principle of operation remains the same when one swaps the capacitor for a junction and the ion-electron pair with an electron-hole pair. This mounting requires a cooling system and is therefore rarely found on the field, while is more common in high-resolution laboratory applications.

Scintillators are substances which emit light when struck by ionizing radiation. Old equipment was made up by simple phosphor screens counted by the naked eyes while modern scintillation counters use a large array of specially engineered crystals albeit the most common is by far NaI doped with Thallium. Because of a complex interaction of the radiation with the lattice and the presence of allowed states for the free electrons in the forbidden gap due to the doping, the emission of a flash of light follows quickly the passage of the IR.

Photomultiplier tubes are used to convert the light pulses into signals that can be processed by a suitable electronic chain. The decay times are very short, of the order of 200 ns, and the magnitude of the output pulse from the photomultiplier is proportional to the energy loss of the primary radiation. *Organic scintillators* such as a mixture of polystyrene and tetraphenyl butadiene have the advantage of faster decay time (about 1 ns), can be moulded into many useful configurations, and have an equivalent Z very close to that of the human body, which makes them useful for dosimetric purposes.

Personal dosimeters are a different type of device intended as a protection of the single operator, rather than for the public, or the environment. For this reason, this is not usually a concern of the responder but rather of the Radiation Protection Advisors (RPA) in charge of the safety of the team; just to present a complete set of useful information we, therefore, summarize in the following the main types of devices and their features:

Dosimeter	Characteristics
Film badge	Photographic films can react with visible light as well as with IR. The blackening is proportional to the absorbed dose and hence, with a suitable mounting and energy filters, the film badges make up excellent dosimeters
Thermo luminescence dosimeters	TLDs are based upon a physical process quite complicated and analogous to the scintillation described above. These dosimeters show high sensitivity and after reading can be used again
Electronic dosimeters	Many different types of ED exist. The simplest, dosimetric pens, are pre-charged capacitors, while the more sophisticated are miniaturized detectors. Although they don't require processing and can be read directly, the sensitivity is quite low and, worse, the response can be heavily influenced by nearby electromagnetic fields

5 Countermeasures

Broadly speaking, three time phases have been identified that are common to radiation accidents and to terrorist events. These are: early, intermediate and late phases. Of course, the transition from one time phase and the other is unavoidably gradual, therefore some actions may overlap. The phases are an essential tool to conceptualise the dominant pathways through which effects on people are manifested and to assess associated countermeasures after a radiological incident.

At its broadest level, the early phase starts at the beginning of the radiological incident. Therefore actions are essentially based on predictions of potential consequences. The major exposure pathways are inhalation and external irradiation from radioactive material suspended in the plume. The early phase may last from hours to days. The intermediate phase begins once the uncontrolled release of radiation is over and the radioactive plume has been dissipated. The pathways of concern are irradiation from contaminated ground, inhalation of resuspended material, ingestion of contains tend food. This phase may last from weeks to months. Depending on the radiation accident, the late phase may last from weeks to years and involve possible incorporation of long-lived radionuclides in the food chain. The late phase ends when restrictions are lifted and unrestricted access of previously involved areas is allowed.

In 2007 the ICRP published a revised System of Protection [6] that supersedes the 1991 recommendations. In the publication, three essential definitions are given:

- *Projected dose*: the dose that would be expected to be incurred if no protective measure(s) – were to be taken.
- *Residual dose*: the dose expected to be incurred after protective measure(s) have been fully implemented (or a decision has been taken not to implement any protective measures).
- *Averted dose*: the dose prevented or avoided by the application of a protective measure or set of protective measures, i.e., the difference between the projected dose if the protective measure(s) had not been applied and the expected residual dose

Regarding operational intervention doses, ICRP 1990 recommendations [17] provided intervention levels for public exposures, distribution of stable iodine, sheltering, temporary evacuation, and permanent relocation. On the other hand, in the recent ICRP 103 publication [6, 17] reference levels are presented as "bands" or ranges of doses for different types of situations. The bands provide flexibility for planners or decision makers to decide on an appropriate level of exposure, while taking into account other non-radiological considerations specific to each situation. In emergency situations the reference level will be expressed as the total residual dose to an individual as a result of the emergency that the regulator would plan not to exceed, either acute or, in case of protracted exposure, on an annual basis.

Framework for source-related dose constraints and reference levels according to ICRP 103 publication.

Bands of constraints and reference levels (mSv)	Type of situation
Greater than 20–100 mSv	Emergency situations, where events with uncertain consequences require urgent protective actions such a sheltering and evacuation to minimize the impacts of possible radiation exposures
Greater than 1–20 mSv	Existing situations, where radioactivity is already present in the environment at the time actions are taken to reduce radiation exposures. If doses are optimized below this reference level it is safe to live in the contaminated area
1 mSv or less	Exposures are usually controlled by action taken directly on the source for which radiological protection requirements can be planned in advance

As a general rule, no matter the specific time phase, the goal of any radiation countermeasure is to reduce the total dose to the most exposed members of the public and to minimize the total collective population dose. Especially during the late phase it may be necessary to consider countermeasures aimed to decrease the projected dose for a particular pathway. It is worth noticing that all countermeasures have an associated risk. Therefore the use of a specific countermeasure requires considering both risks and benefits, following the general principle of doing more good than harm. Furthermore, final judgments and decisions concerning the selection of any countermeasure should be considered in the overall context of the disaster.

5.1 Early Phase Countermeasures

The early phase last from hours to days depending on incident and require immediate decisions about responses. These are generally based on preliminary radiation measurement data, predictions of meteorological conditions and of possible exposure pathways. Of course, actions should be taken quickly and modified when additional data become available.

During the early phase of an incident, there are three main exposure pathways from airborne releases: (I) direct exposure to radioactive materials in an atmospheric plume, (II) inhalation of radionuclides from immersion in a radioactive atmospheric plume and inhalation of ground-deposited radionuclides that are resuspended into a breathing zone (III) external exposure due to the deposition of radioiodine and particulates from a radioactive plume that can continue to emit beta and gamma radiation as "groundshine" after the plume has passed.

The initial response force for any radiological event is emergency personnel arriving at the scene, among which fire department, police and first aid team. All personnel responding to a radiological incident should keep on mind three key principles in order to reduce exposure: (I) minimise the time spent on the scene, (II) maximise the distance from radiation sources and (III) use protective shielding (whenever possible). Furthermore, all first responders should be given a personal dosimeter and should wear appropriate clothing to prevent contamination.

Appropriate early countermeasures include evacuation of downwind areas, temporary shelter–in-place, administration of stable iodine and establishment of controlled areas.

Evacuation is the prompt removal of people from an area to avoid or reduce high-level, short-term radiation exposure from the plume or from deposition of radionuclides. Evacuation can be 100% effective in avoiding dose if completed before plume arrival. Evacuation is appropriate when its risks and secondary effects are less severe than the risk of the projected radiation dose. According to ICRP publication 63, the operational dose range for evacuation extends from 50 mSv to 0.5 Sv[3] in 1 week and is almost always indicated if the projected average effective dose is

---

[3]When a dose range is given, the lower value represents the lowest effective dose at which the countermeasure is *likely* to be justified. The larger value represents the effective dose at which the countermeasure is *almost always* justified [20].

likely to exceed 0.5 Sv within a day [18]. As a rule of thumb, evacuation will seldom be justified at less than 10 mSv over the first 4 days [19].

Sheltering-in-place is the a precautionary action of staying or going indoors immediately, possibly selecting a small, interior room, with no or few windows. Sheltering-in-place should be preferred to evacuation whenever it provides equal or greater protection. As a general rule, planners and decision makers may consider implementing sheltering-in-place when projected doses are in the range 5–50 mSv in 2 days [18].

Administration of stable iodine (e.g. potassium iodide) reduces or blocks the uptake of radioactive iodine in the thyroid of exposed individuals. The potential benefit of iodine prophylaxis will be greater in neonates, infants and small children, which represent the most sensitive groups for a number of reasons. Firstly, a higher radiation dose is accumulated per unit intake of radioactive iodine due to the small size of the thyroid gland. Secondly, the thyroid of the young infant has a higher yearly thyroid cancer risk per unit dose than the thyroid of an adult and. Ultimately, the young will have a longer time span for the expression of the increased cancer risk. According to ICRP publication 63, the operational dose range for stable iodine administration extends from 50 to 500 mSv [18].

5.2 Intermediate Phase Countermeasures

The intermediate phase is assumed to begin once the terrorist attack has occurred and the contaminated clouds have dissipated. Generally it is the time from weeks to months after the event is brought under control and information derived from environmental measurements can be used to assess the need of additional protective actions. Additional countermeasures include (I) relocation (II) personal decontamination (III) interdiction of food and water supplies.

Relocation involves moving people out of contaminated sites and settling them in safe areas for a long-term period. Relocation is generally initiated once extensive measurements are available. As a general rule, the decision to relocate a population should be made by local authorities based on projected effective dose levels and other factors that may have an impact on the safety of people. According to ICRP publication 103, the reference level for relocation is typically between 20 and 100 mSv/year, depending on the situation. It is worth noting that according to ICRP publication 63, a monthly projected effective dose of 10 mSv would justify relocation [18].

No matter the decision to evacuate or relocate, contaminated people should be decontaminated. External decontamination procedures are planned to minimize or prevent internal contamination, that is much a greater issue than external contamination. In fact, radionuclides deposited on the skin rarely deliver doses high enough to pose hazard to the patient. Generally, the removal of the other clothing of the contaminated individual will remove most of the surface contamination. After this preliminary action, if inhalation is suspected, a nasal sample from both nostrils should be taken. If needed other samples may be taken (smears of skin contamination, hair and nails). Finally, depending on the extent of contamination, a total body shower may be planned. It is worth noting that complete decontamination is almost impossible because part of radioactive material remains fixed to the skin. As a general rule, contamination values twice background are usually considered adequate [20].

In the case of foodstuff and water contamination, interdiction of food and water supplies can be a protective action. The interdiction of any single category of food is almost always justified if the yearly effective dose is likely to exceed 10 mSv [18].

5.3 Late Phase Countermeasures

The late phase may last from weeks to years. During this time actions may be undertaken to reduce levels of environmental contamination. Major countermeasures in the late phase of a radiological accident involving the dispersion of long-lived radionuclides are relocation/resettlement, foodstuff restrictions, agricultural countermeasures and clean-up of contaminated areas.

Temporary relocation and foodstuff restrictions have been described in Sect. 5.2. As opposite to temporary relocation, permanent resettlement elsewhere is when the period of temporary relocation from home exceeds a few years. According to ICRP publication 63, permanent resettlement is justified when the effective absorbed dose exceeds 1 Sv in lifetime (70 y) [18].

Following a radiological or nuclear accident, rural environments may be contaminated for many years. Therefore appropriate restoration strategies must be developed. Agricultural countermeasures means the use of technical measures to reduce the activity levels in foodstuffs. The main objective is to reduce radiation doses to humans. However, providing reassurance to consumers and people living in contaminated areas is also an essential objective. Many agricultural countermeasures aim to minimize the transfer of radionuclides from soils to crops, animals and animal products. Furthermore, remediation strategies may involve monitoring of food production, processing and cooking stage.

Finally, clean-up of large contaminated areas could be very costly and cause inconvenience to people living in the involved area. The aim of clean-up of contaminated areas is to minimize the external irradiation from deposited activity, reduce the transfer of radioactive material to humans, animals and foodstuffs, and reduce the potential for resuspension and spread of radioactive material.

6 New Threats and Emerging Technologies for Radiological and Nuclear Defense

It is generally believed that new potential threats may result from recent technologies. Emerging technological innovations might be used by terrorist groups seeking to develop cheaper and more powerful R/N weapons. Emerging technologies include nanotechnology, cyber-technology, drone technology and three dimensional printing. In addition, with the advent of the Internet confidential and sensitive scientific data found their way online. As an example, the diffusion of technical information through the Internet regarding the assembly of nuclear weapons has increased the risk that a terrorist organization with the right material could develop its own radiological or nuclear device. Therefore the ease of access to information is likely to contribute to a significant escalation in potential dangers and the consequent need for an increasingly rapid response.

Nanotechnology is emerging as a tool that will increasingly change the way that experts and authorities approach CBRN events. Nanotechnology encompasses a wide array of existing technologies offering the ability to manipulate individual atoms to effectuate more complex and efficient structures than otherwise possible. For example, advances in nanoparticles and nanorobotics are already altering conceptions of surgery, mechanization, and construction.

Recent research has pointed out the potentially catastrophic damage that nanotechnology could do if intentionally weaponized [21, 22]. In particular, the effects of a weapon engineered using actual nanotechnology could be dramatic. Nanotechnology could potentially be used to facilitate the dispersal and delivery process of explosives. For example, the Israeli army has developed and deployed the Dense Inert Metal Explosive (DIME), an explosive device that scatters microparticles of shrapnel at intense heat and speed. Of additional concern is the prospect of a DIME-style weapon incorporating nanoparticles which could directly enter the body and the bloodstream.

Cyber-attacks have been an increasing source of concern in recent years. Cyber challenge includes both inherent vulnerabilities in nuclear systems as well as the threat from hackers seeking to gain access to these systems in order to disable or damage them. In particular, nuclear facilities are a tantalizing target for digital sabotage because a meltdown could result in a major radiological event. Malicious codes, technically called *worms*, can be specifically formulated to target the systems that direct the inner operations of nuclear facilities. The issue is critically important for new generation nuclear plants as control rooms would employ digital systems to operate the facility. Those state-of-the-art instruments and systems make them ideal targets for hackers [23].

Computer-induced nuclear accidents have occurred in the past. Between 1979 and 1984 a number of cyber-accidents occurred at the US North American Aerospace Defense Command (NORAD). The first took place in October 1979 when computers at NORAD indicated a missile attack from a submarine in the waters off the West Coast. A low level state of nuclear war was declared and in minutes ten jet interceptors took off from American and Canadian bases. It was later discovered that the "sabotage" was caused by someone accidentally loading a war game onto a computer at the operations center of the NORAD [24]. On June 3, 1980, false attack indications were caused at NORAD by a faulty component in a communications processor computer. A few days later, on June 6, 1980, false attack indications were again caused by the faulty component during operational testing [25]. In 1984, a computer malfunction indicated that a US nuclear-armed missile was about to fire [26]. More recently, in October 2010, the US Air Force lost contact with 50 Intercontinental Ballistic Missiles due to a combination of equipment failure and human error: a circuit card, improperly installed in a weapons-system processor, had been dislodged by routine vibration and heat [27]. This risk is only likely to increase due to the growing reliance on software and computers for nuclear weapons management.

The best example of cyber-sabotage is the recent *Stuxnet* attack, the only one that has caused direct destruction of a nuclear facility. The *Stuxnet* was discovered by Belarusian company VirusBlockAda in June 2010. It was only one of a number of pieces of malware aimed at the Iranian nuclear enrichment programme at Natanz, which together caused significant damage to underground uranium centrifuge plant and delayed considerably any Iranian bomb. In conclusion the challenge of the cyber age is multifaceted and the *Stuxnet* virus demonstrated just how real the threat of sabotage and destruction is.

Recent years have seen a proliferation in the use of all types of unmanned aerial vehicles, also called drones. Model aircraft and CBRN have always had a close relationship, as model planes have the potential to spray weaponized CBRN agents in large metropolitan areas. As a matter of fact, the prospect of increasing numbers of drones filling the skies poses a number of problems for critical infrastructure, including for the nuclear industry. Flyovers could be used for reconnaissance by hostile actors or terrorist groups, for example in the collection of photos or videos of guard movements and the site layout. In principle drones could also provide air support in the event of an actual ground-based attack by dropping explosives or delivering weapons inside the plant. Ultimately, drones could also be used to bomb spent-fuel pools, which are generally less shielded and surveilled than reactor cores.

On the other side of the spectrum, drones equipped with gamma probes and chemical sensors have now been specifically developed for use in counter-CBRN missions with the aim to identify chemicals, nuclear weapons and drugs from a safe distance.

Ultimately, the availability of low-cost 3D printers could bring both dangers as well as opportunities to the area of CBRN weapons. Three-dimensional printing is a process of making a 3D solid object from a digital file using additive processes. New 3D printing techniques could soon be used in the creation of products without the need for complex and expensive tooling and the time required in traditional manufacturing. Key benefits of additive manufacturing are that it enables customization, complex shapes and parts on demand. In particular, 3D printing could be used to help manufacturing complex explosives or critical components for an RDD. On the other side 3D printing of drones may well offer new tactical opportunities for immediate reconnaissance. Engineers at the University of Sheffield have successfully printed a 1.5 m–wide drone using 3D technology [28]. The low costs involved with creating the drone will pave the way for the building of unmanned aerial vehicle that can be built and deployed in remote situations potentially within as little as 24 h. In future, even unmanned ground vehicles could be produced at a lower cost with the aim to help with decontamination procedures and to reduce the number of emergency personnel exposed to radiological hazard [22].

In the final analysis, the potential threat posed by emerging technologies portends to be a significant issue in the near future. Evolution of technology and its exploitation for malicious intents is something security and law enforcement officials need to include in 'what if' scenarios.

Acknowledgment   We are grateful to Chatham House, the Royal Institute of International Affairs, for permission to reproduce the table originally published in Unal, B and Aghlani, S, "Use of Chemical, Biological, Radiological and Nuclear Weapons by Non-State Actors.", 2016.

References

1. Nathwani, A., Down, J., Goldstone, J., Yassin, J., Dargan, P., Virchis, A., Gent, N., Lloyd, D., Harrison, J.: Polonium-210 poisoning: a first-hand account. The Lancet. (2016).
2. Unal B., Aghlani S.: Use of Chemical, Biological, Radiological and Nuclear Weapons by Non-State Actors – Emerging trends and risk factors. Lloyd's Emerging Risk Report, Chatham House – The Royal Institute of International Affairs (2016)
3. Broad, W.J.: Document Reveals 1987 Bomb Test by Iraq, The New York Times, April 29, 2001.; http://www.nytimes.com/2001/04/29/world/document-reveals-1987-bomb-test-by-iraq.htm
4. Begley S. Protecting America: The Top 10 Priorities. Newsweek. 2001 Nov 5;139(19):26–40
5. Sovacool, B.K.: Contesting the future of nuclear power: a critical global assessment of atomic energy. World Scientific. ISBN: 978-981-4322-75-1
6. ICRP, 2007. The 2007 Recommendations of the International Commission on Radiological Protection. ICRP Publication 103. Ann. ICRP 37 (2–4).
7. Harrison, J., Streffer, C.: The ICRP protection quantities, equivalent and effective dose: their basis and application. Radiation Protection Dosimetry. 127, 12–18 (2007).
8. International Commission on Radiation Units and Measurements. 'Determination of dose equivalent resulting from external radiation sources'. Bethesda, MD: 1985. ICRU Report 39.
9. International Commission on Radiation Units and Measurements. 'Quantities and units in radiation protection dosimetry'. Bethesda, MD: 1993. ICRU Report 51.
10. Mattsson, S., Marcus Söderberg M.: Dose Quantities and Units for Radiation Protection. In Mattsson, S. Hoeschen, C.: Radiation protection in nuclear medicine. Springer, Berlin (2013).
11. ICRP, 2005. Protecting People against Radiation Exposure in the Event of a Radiological Attack. ICRP Publication 96. Ann. ICRP 35 (1).
12. Durakovic, A.: Medical effects of internal contamination with actinides: further controversy on depleted uranium and radioactive warfare. Environmental Health and Preventive Medicine. 21, 111–117 (2016).
13. Vaughan, J., Bleaney, B., Taylor, D.M.: Distribution, excretion, and effects of plutonium as a bone seeker. In: Hodge HC, Stannard JN, Hursh JB, editors. Handbook of experimental pharmacology—uranium, plutonium, transplutonic elements. Berlin and NY: Springer-Verlag; 1973
14. RDD Handbook: Handbook for Responding to a Radiological Dispersal Device First Responder's Guide—The First 12 Hours, CRCPD, September 2006
15. World Health Organization, web: www.who.int/mediacentre/factsheets/fs371/en/INEX exercises, web: www.oecd-nea.org/rp/inex/
16. http://www.ebah.com.br/content/ABAAAfHM4AL/6938905-charged-particle-measurement?part=2
17. ICRP, 1991. 1990 Recommendations of the International Commission on Radiological Protection. ICRP Publication 60. Ann. ICRP 21 (1–3).
18. ICRP, 1992. Principles for Intervention for Protection of the Public in a Radiological Emergency. ICRP Publication 63. Ann. ICRP 22 (4).

19. PAG – Manual Protective Action Guides And Planning Guidance For Radiological Incidents, U.S. Environmental Protection Agency Draft for Interim Use and Public Comment March 2013
20. NCRP Report No. 138, Management of Terrorist Events Involving Radioactive Material (2001)
21. Bradley, L.D.: Regulating weaponized nanotechnology: how the international criminal court offers a way forward. Georgia Journal of International and Comparative Law, 41(3), 728–729 (2013)
22. Unal, B., Aghlani, S.: Use of Chemical, Biological, Radiological and Nuclear Weapons by Non-State Actors: Emerging trends and risk factors. Lloyd's Emerging Risk Report, Innovation Series (2016)
23. Matishak, M.: Nation's nuclear power plants prepare for cyber attacks. 27 August, NTI Global Security Newswire (2010) [online]. Available at: http://www.nti.org/gsn/article/nations-nuclearpower-plants-prepare-for-cyber-attacks/
24. Broad, W.: Computers and the U.S. Military Don't Mix. Science. 207, 1183–1187 (1980).
25. US General Accounting Office, "NORAD's missile warning system: what went wrong?", (15 May 1981), http://www.gao.gov/assets/140/133240.pdf
26. Gregory, S.: The hidden cost of nuclear deterrence: nuclear weapons accidents, United Kingdom: Brassey's (1990) p.97
27. Schlosser, E.: Neglecting our nukes, Politico, (16 September 2013), http://www.politico.com/story/2013/09/neglecting-our-nukes-96854.htm
28. https://www.sheffield.ac.uk/news/nr/3d-printing-trials-of-unmanned-aircraft-1.364084

Laser Based Standoff Techniques: A Review on Old and New Perspective for Chemical Detection and Identification

Pasqualino Gaudio

Abstract The active remote sensing standoff detection is a very interesting methodology that could be used with the aim to reduce the risk for the health, in the case of intentional (terrorism or war) or accidental (natural or incident event) diffusion in the air of chemical agents. At the present day, there are several laser-based methodologies that could be applied for this aim but the future developments seem to be the integration of two methodologies. The integration of two methodologies could guarantee the development of a network of low-cost laser based systems for chemical detection integrated with a more sophisticated layout able to identify the nature of a release that could be used only in the case that the anomalies are detected. The requirements for standoff detection and identification are discussed in this paper, including the technologies and some examples for chemical traces detection and identification. The paper will include novel techniques and tools not tested yet in operative environments and the preliminary results will be presented.

Keywords Remote sensing • Lidar • Standoff system • Detection chemical agent • Dial • Identification chemical agent

1 Introduction

The standoff detection and identification of chemical threats is a clever methodology to allow to detect the substances from a safe distance. In the common dialect detection and identification seem to be two similar actions. Instead in the technical jargon the two words shown a distinct meaning. The word detection means to "discover the existence of something", in the particular case the threat; identification means "a process of exact recognizing", in the specific case the substance. For

P. Gaudio (✉)
Department of Industrial Engineering, University of Rome Tor Vergata,
Via del Politecnico 1, 00133 Rome, Italy
e-mail: gaudio@ing.uniroma2.it

© Springer International Publishing AG 2017
M. Martellini, A. Malizia (eds.), *Cyber and Chemical, Biological, Radiological,*
Nuclear, Explosives Challenges, Terrorism, Security, and Computation,
DOI 10.1007/978-3-319-62108-1_8

155

several decades was developed standoff detection systems that operate in Mid-IR spectral region able to detect chemical threats but often unable to identify the particular substance dispersed in the atmosphere. The systems based on FTIR spectrometer methodologies, based on the collection of samples of gas in the atmosphere, are only able to identify the substances. These techniques are usually used with laboratory apparatus where is possible to collect the gas sample into the cell.

After a critical analysis of the existing devices for the detection and/or identification of chemical agents (CAs) by optical techniques, several commercial devices able to detect and identify the CAs (both chemical warfare agents (CWA) and toxic industrial chemicals/toxic industrial materials (TIC/TIM)) were identified; these technologies work however in very short range only or require a sample to be analysed in laboratory condition by an absorption cell.

The currently commercialized instruments are only able to detect the putative variation of the environmental chemical background in a restricted and confined area, providing an on/off response type (usually known as trigger systems).

Another interesting application of these techniques is the monitoring of confined environment in order to control the emission or presence of a chemical threat in safety and security applications. In this field over the chemical detection, a huge effort is doing on standoff detection of explosives laser based [1]. Wallin and colleagues present different laser-based physics phenomena potentially able to trace detection using spectroscopic technologies for explosives detection that could be used in the real application. These techniques should be tested in the real environment; however, currently they aren't yet sufficiently mature.

As for the detection of chemical substances in open field usually passive apparatus are proposed (for example FLIR thermocamera), they are able to detect the threat only but they are completely unable to identify the substances [2]. Passive MIR detection system, that uses ambient thermal radiation to detect the threat shows lower sensitivity and range if compared with the laser-based techniques.

With these premises, the absorption could constitute the most sensitive techniques for the detection of trace pollutants and chemical threats in MIR region if compared with other parts of the spectrum; it could be justified because this technique enables higher sensitivity detection and higher specificity chemical identification [3, 4] of a wide range of substances. In the past and in the present laser-based active remote sensing systems are used under a wider range of environmental conditions and for several civilian and military applications.

Currently, important researches involve the possibility to use laser-based standoff methods with different physical proprieties in order to identify exactly the threat. Raman spectroscopy, differential absorption or laser-induced breakdown spectroscopy-based techniques are currently applied in important standoff systems for indoor and outdoor applications [5, 6] with restricted requirements to use lasers that operate in the eye-safe region [7]. An alternative approach to the problem could be the creation of a network composed by a low-cost detection systems, LIDAR (LIght Detection And Ranging) based already tested in an urban [8], industrial and wooded [9] area to detect pollutants and emissions sources. These network systems could be used as a first alarm and can be coupled with DIAL (DIfferential Absorption

Lidar) based system, used as an identification system, able to investigate and to identify the nature of the threat. Two principal aspects are not completely performed in standoff detection/identification systems: first of all, a technological aspect that consists of the reduction of the apparatus dimension; second, a scientific aspect that consists in the capability of the system to fast and exactly identify and reduce false alarm. The first aspect could be solved using new technologies diode laser based for both detection [10, 11] and identification systems [12]; the second aspect could be solved by building a sufficient and complete database of fingerprints using the same standoff system developed for this specifics purpose.

In the present review the laser-based detection systems and the laser based identification systems will be presented and discussed coupled with the last methodologies for exact identifications for CWA. In conclusion, a new proposal for use both the methodologies for the detection and identification, in order to develop a "smart sensor", will be presented.

2 Laser-Based Standoff Systems for Chemical Detection

In the last decades and in particular after the first active remote sensing application starting with Fiocco and Smullin [13] by their pioneering studies of the upper region of the atmosphere, the lasers have played a fundamental role in environmental problems and in particular they opened new vistas with regard to remote sensing of the atmosphere. It is immediately appreciated that the lidar, as with radar, could provide spatially resolved measurements in real time. Lidar techniques were used in rapid succession from the invention of the laser. Lidar methodology is been plenty used to evaluate and detect profiles of pollutants emission. Fiorani et al. [14] developed a self-aligning detection applied to lidar system operating at 351 nm in order to analyze PBL (Planetary Boundary Layer) dynamics over the Naples city (Italy). The same application was carried out by Del Guasta [15]; he developed a lidar operating at two laser wavelength – the fundamental and second harmonic of Nd-YAG laser 1064–532 nm – in order to evaluate PBL dynamics in Florence (Italy). This last experiment represented the first use of the remote-sensing technique for long-term monitoring study of urban aerosols. About urban pollutants, an interesting review was proposed by Mazzoleni et al. [16]. In this work, the UV lidar measurements were used to quantify mass column content of exhaust emissions of vehicles. Particulate concentration measurements in rural area around Iowa City (USA) were carried out by Likar and Eichinger [17], while another aerosol distribution monitoring on urban area were proposed by He et al. [18] on Gorica city (Slovenia) and by Mei et al. [19] where a new approach –Scheimpflug lidar – for atmospheric aerosol monitoring at Lund (Sweden) was tested. In conclusion, two interesting utilization of lidar methodologies, the foundation of an aerosol lidar network, EARLINET [20], and the first utilization of lidar system in the space in LITE experiments [21] need to be reported.

It is very interesting to observe that any density variation along laser path can be observed as air irregularity on backscattered signals; the variation can be detected as anomalies due to emission in the atmosphere by pollutant sources, threats by terroristic event and so on. On the basis of this evidences, in recent time, the first use of plume automatic tracking was proposed by Andreucci and Arbolino [22].The authors showed in their numerical simulation that a lidar system allows the evaluation of the particulate concentration and/or density variation as a function of the range/altitude of the ground with the capability to estimate the plume shape and its form. In conclusion, their results showed that lidar systems are suitable for plume tracking. This scientific work opens the use of lidar system for environmental control in order to detect sources emission not only due to industrial release but the produced by the fire too. The authors developed a numerical model of smoke plume [23] produced by a forest fire in order to evaluate optical parameters with the aim to understand the limits of the application of lidar methodologies to detect the smoke [24]. The first measurement of forest fire by lidar was carried out by Vilar et al. [25]. They demonstrate the capability of a lidar system operating at 532 nm to detect a small forest fire. Certainly, the advantages in early fire detection respect the traditional methodologies (passive techniques) are not comparable. It is true that an automatic detection process requires a rapid identification of smoke plume, but backscattering signal profile can often present other peaks due to different causes don't attributable to fire events.

This specific problem was fixed by Bellecci et al. [9, 26] working on the reduction of the false alarm using both lidar and dial methodologies in order to detect the water vapor increment in smoke plume due to a combustion process of vegetable. This approach could be used adopting both remote sensing technologies dial and Raman in order to measure, after the peak detection, the amount of water vapor in the atmosphere due to a combustion process. This approach was used for early detection of forest fire by several authors at different laser wavelength [27] but the idea to use the identification of precursor of a forest fire to reduce false alarms was an innovative approach applied by Bellecci et al. [9]. The same approach can be used in threat detection in the atmosphere using a low-cost system; only after the anomalies are been measured, it activates the second standoff system, able to identify the threat if its fingerprint is contained in a very accurate absorption database. For this last purpose is very important to know the absorption coefficients with high efficiency. It is well know how the interpretation of backscattered signal and the atmospheric measurements require the high accuracy of absorption coefficients, including pressure and temperature dependence, that can be measured only by laboratory instruments. In case the identification of a release in the atmosphere or in a confined environment, using a standoff systems, is needed, the implementation of a database of absorption coefficients, using laboratory apparatus, is fundamental. In conclusion, there are several optical based techniques-tested or not in realistic environments – able to detect chemical threats. The creation and implementation of real databases is therefore essential for the application of the techniques to real cases.

2.1 Active Remote Sensing System for Detection

Laser-based remote sensing techniques, such as lidar and the powerful improvement techniques dial [14, 28, 29], are increasingly being useful and valuable options for a wide range of applications which can be divided into airborne and terrestrial types.

Lidar/Dial systems are currently used for precision agriculture, archaeology, geology and soil science, atmospheric remote sensing and meteorology, military, mining, physics and astronomy, robotics, spaceflight, surveying and transport.

As already stressed, lidar systems operate on similar principles that radar (radio detection and ranging) and sonar (sound navigation and ranging). In the case of lidar, the radio waves are substituted by a light pulse in the range of optical radiation (generally: ultraviolet, visible or near infrared) that are emitted into the atmosphere. Wavelengths used for lidar depend on the specifics application and range from about 250 nm to 11 μm [28]. Light from the beam is scattered in all directions from molecules and particulates in the atmosphere. A portion of the light is backscattered toward the lidar system. This light is collected by a receiving optics that measures the amount of backscattered light as a function of distance [29].

Lidar measurements have broad application in the characterization of the atmosphere, ranging from the determination of properties of cloud particles [30] or aerosols [31] to the profiling of trace gas concentrations [32], air temperature [33], or wind velocity [34]. A lidar system consists essentially of two principal blocks: a transmitter and a receiver equipment [14]. The first is a laser source of short, intense light pulses, whose characteristics are well known (high spatial/temporal coherence, low beam divergence, uniform polarization, brightness and monochromaticity of light, etc.). It may include additional components such as beam-expanding optics. The latter is primarily composed of a telescope with a detector placed at, or sometimes slightly offset from [14], its focal plane. They, respectively, collect the backscattered light and convert it into an electrical signal. Finally, a computer/recording system, which digitizes the electrical signal as a function of range and time of acquisition, is used for data acquisition, processing, evaluation, display, and storage as well as controlling the other basic functions of the system. The general scheme of a lidar system is reported in Fig. 1.

These systems can have additional components that differ considerably according to type and purpose of the lidar. Moreover, as shown in Fig. 2, it is possible to distinguish between bistatic (a) and monostatic (b) setup [14].

In the bistatic configuration, transmitter and receiver are separated by a distance. Conversely, when they are in the same place and/or part of the same system, lidar setup is called monostatic. Most modern lidars use monostatic configuration with either coaxial or biaxial arrangement. In a coaxial system, the axis of the laser beam is coincident with the axis of the receiver optics. Instead, in the biaxial arrangement, the laser beam only enters the field of view of the receiver optics beyond some predetermined range. It should be noted that the geometric arrangement of the emitter and receiver optics determines the degree of signal compression at distances close

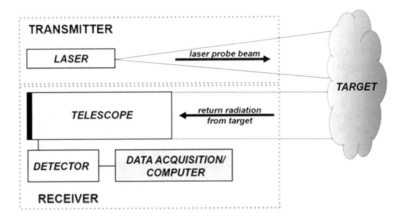

Fig. 1 A conceptual scheme of a biaxial lidar system

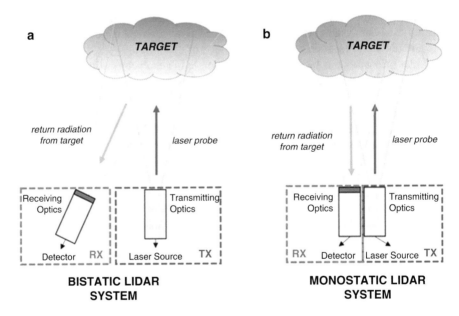

Fig. 2 Bistatic (**a**) and monostatic (**b**) lidar configurations

to the lidar and, therefore, it can affect the final performances of a lidar apparatus. In fact, at short distances, the laser beam cannot completely be imaged into the detector. As it will be shown in the following chapters, only a part of the actual lidar return signal is measured. Moreover, some practical adjustments can be used to suppress light outside the transmission band, such as for the background radiation (e.g. the use of interference filters placed in front of the detector) [14, 28].

A technical improvement of lidar system was developed by Quantum Electronic and Plasma Physics research group (QEP) (http://www.qepresearch.it) of the

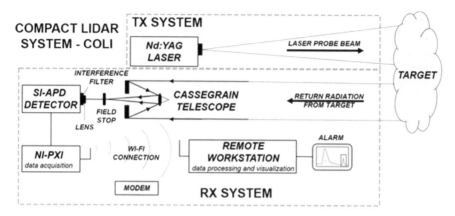

Fig. 3 Technical scheme of COLI in its final configuration

University of Rome Tor Vergata, where a miniLidar system "COLI" (Compact LIdar system) was developed. It consists of a compact, robust, stable, scanning mobile lidar system based on monostatic configuration (the laser and telescope are located in the same place [35] with a biaxial arrangement but the laser beam and the telescope axis are separated, and the laser beam only enters the field of view of the receiver optics beyond some predetermined range [36]). In Fig. 3 the final configuration scheme is reported.

The system is essentially composed of a transmitter and a receiver block. A commercial and relatively standard pulsed Nd:YAG Laser operating at 1064 nm in Q-switching mode is used as the transmitter. The output laser energy has been reduced in order to prevent saturation of the receiver in the near field. Backscattered light is collected by a commercial telescope in Cassegrain configuration. In order to suppress the background in daytime operation a 2 nm wide interference filter (Omega Optical Inc., USA) center wavelength that 1064 nm is used. Finally, the backscattered signal is collected and focalized by a telescope on detector area of IR-enhanced avalanche photodiode (APD) and data acquisition is performed using an NI-PXI digitizer.

In Table 1 the principal characteristics of COLI system are reported.

As reported in the previous table, the laser beam divergence is larger than the telescope field of view. As already reported by Kovalev [39], a consequence of the use of a narrow field of view is that the lidar system becomes increasingly difficult to align.

With the aim to overcome this last problem, the FOV of the telescope was corrected inserting a short focal lens in front of APD detector area in order to arrange the FOV of all receiver system and to make compatible with laser beam divergence avoiding the reduction of the lidar system performances.

The digitization parameters reported above yield a range resolution − ΔR of about 1.5 m ($\Delta R = c\Delta t/2$ [39], where c is the speed of light and Δt is the temporal

Table 1 Specifications of COLI in its final configuration

Laser – Big sky CFR 400 by Quantel, France	
Pulse frequency	10 Hz
Energy pulse	400 mJ (max.)
Pulse time width	8 ns
Beam divergence	< 4.5 mrad (full angle)
Beam diameter	< 7 mm
Telescope – MAHK 130 by Ziel, Italy	
Nominal focal length	1300 mm
Primary mirror diameter	102 mm
f-number	f/12.7
Field of View	1.2 mrad
Detector – EG&G C30954/5E, URS corporation, USA	
Responsivity	34 A/W (typ.)
Quantum efficiency	38% (typ.)
Diameter	1.5 mm
Response time	< 5 ns
Gain (M)	100 (typ.)
Digitizer – NI-PXI 5122, national instruments, USA	
Resolution	14 bit
Sampling rate	100 Ms./s
Bandwidth	100 Mhz

resolution between laser shots, equal to 10 ns). The maximum theoretical detection range has been found around 1.5 Km.

The laser-telescope system is able to scan the atmosphere in both vertical (elevation range from 0 to 90 degrees) and horizontal (azimuth range from 0 to 270 degrees) path, using computer-controlled motors incorporated into the telescope mount. Therefore, once deployed in the field, COLI is completely autonomous since it is remotely controlled by our own software package written in LabVIEW, developed for mechanical handling of laser-telescope block and data acquisition and processing procedures written using MATLAB software. The pictures of the COLI system is reported in Fig. 4.

The COLI acquisition system is able to acquire and elaborate the backscattered signals from particulate measurements in the atmosphere. The backscattered signals are described by elastic lidar equation [14] followed reported (1):

$$P_r\left(\lambda,R\right) = O\left(\lambda,R\right)\frac{A_r}{R^2}P_0\left(\lambda\right)\frac{c\tau}{2}\beta\left(\lambda,R\right)\exp\left(-2\int_0^R\alpha\left(\lambda,R'\right)dR'\right) \qquad (1)$$

where: $P_r(\lambda,R)$ is the backscattered power received at the specific laser operative wavelength λ from the distance R, $O(\lambda,R)$ is the overlap and transmission factor determined by the geometric considerations of the receiver optics, the transmitter and receiver efficiencies, and the overlap of the emitted laser beam with the field of

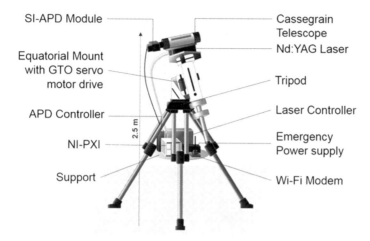

SI-APD Module

Equatorial Mount
with GTO servo
motor drive

APD Controller

NI-PXI

Support

Cassegrain
Telescope

Nd:YAG Laser

Tripod

Laser Controller

Emergency
Power supply

Wi-Fi Modem

2.5 m

Fig. 4 Picture of COLI system

view of the receiver. The term A_r/R^2 is the acceptance solid angle of the receiver optics with collecting area Ar. $P_0(\lambda)$ is the emitted power of a laser pulse, c is the speed of light and τ is the pulse duration. The last two terms describe the optical parameters of the laser atmosphere interactions. The first backscattering coefficient $- \beta(\lambda,R)$ is a measure of the scattering (in $m^{-1} \, sr^{-1}$) in the backwards direction (i.e. towards the incident direction, at a scattering angle of 180°) for the laser atmosphere interaction constituents as aerosol, particles and molecules. The second one is the extinction coefficient $- \alpha(\lambda,R)$ that consists in a measure of the attenuation of the laser pulse (in m^{-1}) that across through the atmosphere due to the scattering and absorption by aerosol particles and molecules.

If the IR wavelength is used, usually when lidar systems are Nd–YAG-based [25], the amount of molecular scattering due to particulate scattering can be neglect and the lidar signal is able to provide aerosol backscatter information [29]. To improve laser intensity, an average of sampled data points is usually performed. In the analysis methodologies, the backscattering signals are first normalized to the laser energy and the background noise is subtracted. Usually, in the case of lidar, the background noise is defined as the average of data points sampled at the far end of every signal trace. At the end, each returned lidar signal is corrected using range – squared dependence [14] and Savitzky – Golay filter [37] is applied. This last algorithm often applied in lidar data processing [38–41].

In Fig. 5, a backscattered lidar returned signal acquired by COLI system is shown. In particular, the maximum operative range vs experimental SNR (calculated as in Ref. [42]) provides the minimum detectable signal (SNR = 1) [43] at about 1.5 Km.

A test campaign of COLI system was carried out in a suburban area of Crotone in the south of Italy [8]. During this campaign, a backscattering value related to urban traffic, in order to measure particulate increments due to a vehicle, was

Fig. 5 An example of
average signal-to-noise
ratio retrieved by
COLI. The SNR remains
well above 100 (20 dB) up
to more than 1 Km
(Source [8])

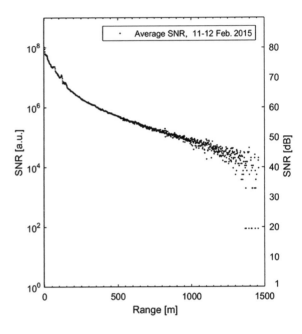

measured. A high linear correlation of PM concentration levels between lidar data
and conventional monitoring station was observed. In the same paper [8] the daily
variation of the particulate in urban and industrial areas is shown by reconstruction
of temporal maps; this innovative technique (real measurement) represents a strong
innovation if compared with the classical measuring technique (punctual system),
able to measure only few centimeter around the detector. The analysis of backscattered signals versus time was performed in order to investigate and detect potential
aerosol sources due to daily variation of vehicle emissions in high-speed congested
road. An interesting change in particulate concentration amount is strongly correlated to the human activity (Fig. 6).

Avery interesting experimental campaign as "case study" was carried out at
Lamezia Terme (a small town in the south of Italy) in order to demonstrate the capability of detection and topographic identification of pollutant sources. With this aim,
in an industrial area, a controlled fire was produced by the combustion of vegeta-
bles. Using a chimney in the same area, at a distance of about 400 m the COLI
system was located. In Fig. 7, the location of both COLI and chimney positions are
marked with color dots, yellow and black respectively.

In order to demonstrate the capability of detection of tenuous smoke, a two-
dimensional spatiotemporal distribution of horizontal scans was carried out covering an FOV of about 6.2°, where each scan took approximately 1 min to complete.

Meteorological data (temperature, relative humidity, wind speed and direction)
were collected at the same time of lidar measurements. These parameters were
acquired, approximately 5 m above the ground, by a conventional meteorological
station, assembled nearby the system location. Weather data were averaged every

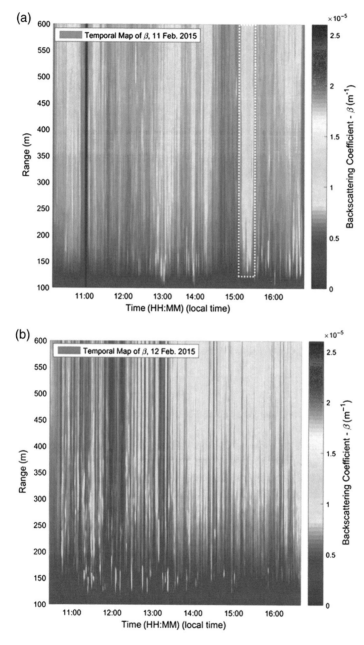

Fig. 6 In the plot is shown the temporal maps of backscattering coefficients. An increment of pollutants source is strictly related to human activity in the relation to closure of offices or schools (Source [8])

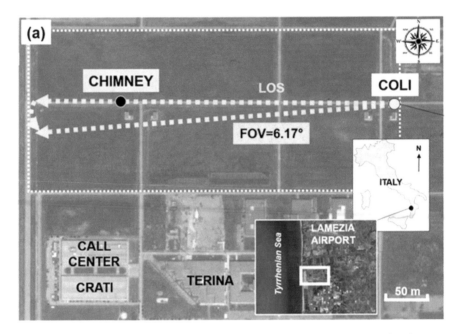

Fig. 7 The location of experimental area (*white dotted box*) of the "case study" during the campaign of Lamezia Terme (*South of Italy*). The lidar position, its FOV and the chimney are reported in (**a**)

quarter hour. Moreover, as you can see in Fig. 8, weather pattern disturbances was excluded.

As it is possible to observe in Fig. 8, both 2 days were characterized by the absence of precipitations; then attention was focused on the wind, the most sensitive parameter in air pollution monitoring.

In stable atmospheric conditions, as observed during the case study campaign, the wind direction was almost North-West during the measurement sessions, with an average value of about 3 m/s. The same result was obtained by the ENEA Agency; results reported in [44] perfectly fit with the region's climate trends. Therefore, meteorological conditions did not significantly affect lidar measurements, with the only exception of a moderate influence of the maritime air.

The backscattered lidar signal was analysis using the well known Klett method [45], obtaining the time evolution of the backscattered measured β versus range and time of acquisition. In this way, air pollutants dispersion maps of the investigated area for both measurements days were obtained (Fig. 9). An example of the temporal evolution of the smoke plume emission by an artificial source is shown in Fig. 9. Every plot report both Cartesian and polar representation. COLI system and chimney location are been highlighted with two different colors and the red arrow indicates the lidar scanning direction (counterclockwise). Figure 9 is referred to data acquired in case study measurements carried out on the 10th of June 2014.

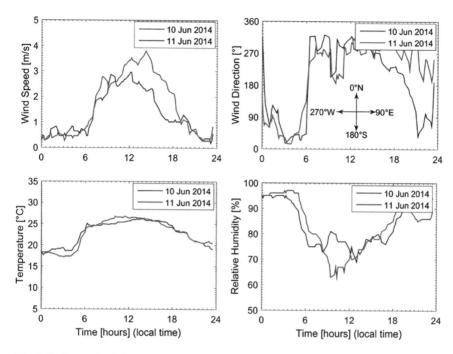

Fig. 8 Daily weather behavior of temperature, relative humidity, wind speed and direction measured at the same time of the experimental campaign. *Black* and *red* lines respectively reported the dates on 10 June 2014 and on 11 June 2014

In order to measure the background of the case study, the first plot (A) of Fig. 9 was obtained before of igniting the fire. The second and thirds plots represent the particles dispersion of a source in the experimental area. It is possible to observe that emissions in the second plot, source expands of about 200 m along laser direction; it can be attributed to the wind direction that was blowing in North-West direction in that period. The same consideration is been made in the third plot during measurements acquired in the afternoon session. The last plot (D) represents the same situation of the scan A, acquired at a different time; for this reason, a different background value is been measured. This last measurement was replicated the next day by maintaining the same system setup. Operating conditions are reported in Fig. 10. The principal driving of smoke dispersion is wind direction. In Fig. 10 it is possible to observe that the smoke produced during combustion of vegetal was diffused by a gust of northerly wind during the morning measurement. It can be justified by wind measurements reported in Fig. 8, while in the afternoon the fire smoke particles were scattered on about 200 m far respect COLI system location (Fig. 10b). Finally, the last scan (Fig. 10c) represent the background measurements.

The experimental campaign demonstrated the capability of lidar systems and in particular of miniLidar COLI developed to produce real measurements of particle diffusion in the free atmosphere. The COLI represents a smart compact and low-cost standoff detection system.

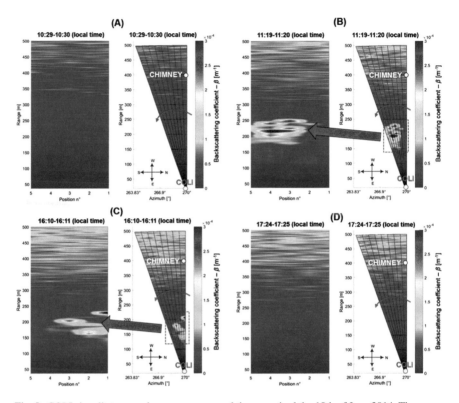

Fig. 9 COLI signal's temporal maps vs space and time acquired the 10th of June 2014. The scans (**a**) and (**d**) shown the situation in case no smoke, respectively before the lighting and after the extinguishing of the chimney. Whereas, (**b**) and (**c**) represents respectively the presence of scattered smoke due to ignition of the fire before and after midday. Each scan has been reported in both Cartesian (*left side*) and polar (*right side*) diagram. It is possible to notice that red spots rapidly change their position from the source of smoke even hundreds of meters. This was due to the presence of wind blowing from the North-West (**b**, **c**) (Source [67])

In this particular case study, the release in the atmosphere was simulated and the capability and sensitivity of lidar system to provide a fast and automatic detection of dispersion were tested.

2.2 Active Remote Sensing System for Identification

A very important goal for lidar application is to arrange standoff techniques able to identify chemical threats without contact. Several methods are used in order to identify trace gas in the atmosphere. In the case of standoff approaches, the best methodologies are optical based technologies. In the last years, a lot of attention was focused on the DIfferential Absorption Lidar (DIAL) methods; these techniques are

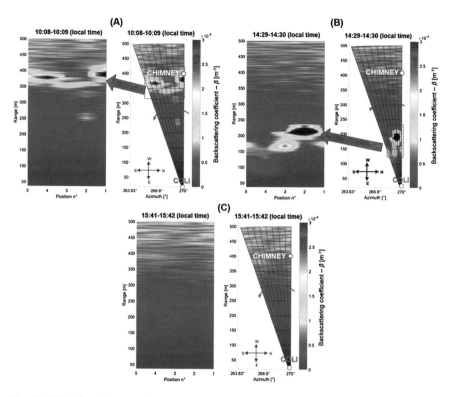

Fig. 10 COLI signal's temporal maps vs space and time acquired the 11th of June 2014. The scan (**c**) shows the situation of no smoke, after the extinguishing of the chimney. (**a**) and (**b**) are referred to the presence of scattered smoke. Each scan has been reported in both Cartesian (*left side*) and polar (*right side*) diagram. It is possible to notice that red spots rapidly change their position from the source of smoke even hundreds of meters. This was due to the presence of a gust of wind blowing, respectively, from North (**a**) and West (**b**) (Source [67])

able to give concentration maps of selected molecular species in the free atmosphere [13, 27, 29].

In the last 25 years, dial systems were used for identification and measurements of pollutants gases as SO_2, NO_2, NO, and ozone [29] showing an enough sensitivity to measure their air concentrations and distribution. Several dial systems were developed for atmospheric concentration evaluation from ground, airborne of space around the world, operating in both elastic [46–52] or inelastic (Raman) mode [47, 54, 55]. The use of dial system for the standoff identification of chemical can avoid the contamination of operators or equipment mitigating the decontamination issues. Active identification laser based as Raman [56] or laser induce breakdown spectroscopy [57] has been tested for explosive detection in the field with good results at the short range. However, for long range identification, the problem has been not resolved yet. In the case of chemical identification, technologies based on quantum cascade laser (QLC) were recently introduced; these techniques are based on very

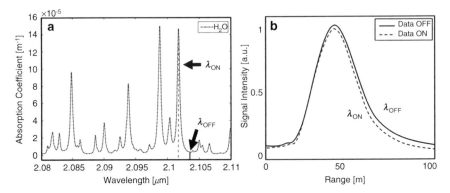

Fig. 11 Water absorption spectrum in the near infrared region (**a**) and typical DIAL signal as a function of range (**b**)

tunable compact laser sources but, due to a low output power, they can operate at the very short range [58]. The QLC seems to be able only to identify chemicals on the path of about 500 m using integrated measurements [11].A range resolved system based on tunable CO_2 laser have been tested (ADESIS- DEtection at DIStance) [59] showing a maximum range of 2 km for chemical profiles and about 16 km for topographical detection for some chemical vapor plumes (SF_6). Other CO_2 tunable laser were tested with frequency agile tuned [60] advanced detection algorithms [61] and for water vapor detection [62] and evolution [63]. Although dial CO_2 based system seems to be an "old" technology, its tunable capacity and reliability is still relevant for example in forest fire detection of water vapor [9, 26] or in case of finalizing to create a database of chemicals [64], fundamental knowledge for their subsequent identification in the atmosphere.

Before giving the last results on developing of miniDial system CO_2 based is necessary to consider how the fundamentals identification technologies dial based work.

Starting from lidar equation, the dial methodologies operate at minimum two wavelengths. If the analysis is carried out using two wavelengths only, ones of these, strongly absorbed by the species under investigation, is selected (λ_{ON}), whereas the second one, λ_{OFF}, is absorbed not at all or at least much less. The two wavelengths should be selected according to their absorption characteristics for the chemical under investigation; it is crucial that one of the two wavelengths is strongly adsorbed by the chemical respect to the other one; the absorption by the species of the second wavelength will be negligible and can be ignored [14, 29], as in the example reported in Fig. 11.

In Fig. 11a a water vapor absorption spectrum is plotted in the spectral range between 2 and 2.11 μm. If two laser lines signed by two arrows are sent in atmosphere, the two backscattered signals λ_{ON} (grey dashed line), and another off-line, λ_{OFF} (black solid line), the difference in the returns between the two wavelengths is then due only to absorption by water vapor molecules, as we can see in Fig. 11b. By this approach, it is possible to evaluate the vapor concentration profile. In particular, if two power backscattered lidar signals due to λ_{ON}, P_{ON},

$$P_r^{ON}\left(\lambda_{ON},R\right) = P_0 \frac{c\tau}{2} \frac{A_r}{R^2} \xi\left(\lambda_{ON}\right) \xi\left(R\right) \beta\left(\lambda_{ON},R\right) \exp\left(-2\int_0^R \alpha\left(\lambda_{ON},R'\right)dR'\right) \quad (2)$$

And λ_{OFF}, P_{OFF},

$$P_r^{OFF}\left(\lambda_{OFF},R\right) = P_0 \frac{c\tau}{2} \frac{A_r}{R^2} \xi\left(\lambda_{OFF}\right) \xi\left(R\right) \beta\left(\lambda_{OFF},R\right) \exp\left(-2\int_0^R \alpha\left(\lambda_{OFF},R'\right)dR'\right) \quad (3)$$

where $\xi(\lambda)$ represent the receiver's spectral transmission at respectively wavelengh and $\xi(R)$ represent the probability of radiation from position r in the target plane at range R reaching the detector [14]. It is possible to calculate the ratio of the two backscattered power return signals to obtain [14]:

$$\frac{P_r^{ON}\left(\lambda_{ON},R\right)}{P_r^{OFF}\left(\lambda_{OFF},R\right)} = \frac{\xi\left(\lambda_{ON}\right)\beta\left(\lambda_{ON},R\right)}{\xi\left(\lambda_{OFF}\right)\beta\left(\lambda_{OFF},R\right)} \exp\left\{-2\int_0^R [\alpha\left(\lambda_{ON},R'\right)-\alpha(\lambda_{OFF},R')]dR'\right\} \quad (4)$$

where the output power of the laser is usually normalized to laser power output. The atmospheric extinction coefficients defined by the following equation:

$$\alpha_{tot}\left(\lambda,R\right) = \alpha_a\left(\lambda,R\right) + \alpha_s\left(\lambda,R\right) + \sigma\left(\lambda,R\right)\cdot N\left(R\right) \quad (5)$$

where: α_a is the absorption coefficient due to others molecular species respect to the which of interest, α_s represents both Raylegh and Mie scattering contributions, N(R) is the density profile of the selected trace under study of gas of interest at range R and, finally, σ is the absorption cross-section at wavelength λ [14].

It is needed to underline that the choice of the pair $\lambda_{ON} - \lambda_{OFF}$ is the principal aspect in dial technique evaluation. The exact knowledge σ_{ON} and σ_{OFF} of absorption cross section on the species under investigations is primarily relates to accurately extraction of N(R) profile.

In consequence of some algebraic manipulations and definition of ΔR as the system's spatial resolution, it is possible to invert the Eq. (4), obtaining the density profiles [14]:

$$\bar{N} = \frac{1}{2\left[\sigma_{ON}\left(\lambda_{ON}\right)-\sigma_{OFF}\left(\lambda_{OFF}\right)\right]\Delta R} \left\{\ln\left[\frac{P_{OFF}\beta\left(\lambda_{ON},R\right)\xi\left(\lambda_{ON}\right)}{P_{ON}\beta\left(\lambda_{OFF},R\right)\xi\left(\lambda_{OFF}\right)}\right] - \delta S\right\} \quad (6)$$

with

$$\delta S = -2\int_0^{R_T} \left(\alpha_s\left(\lambda_{ON},R\right)-\alpha_s\left(\lambda_{OFF},R\right)\right)dR \quad (7)$$

In several cases, the difference between λ_{ON} and λ_{OFF} is extremely low and then the term δS is approximately equal to zero; under this hypothesis, it is possible to simplify the Eq. (7) allowing to rewrite the Eq. (6) in the following way [14]:

$$\bar{N} = \frac{1}{2\left[\sigma_{ON}\left(\lambda_{ON}\right) - \sigma_{OFF}\left(\lambda_{OFF}\right)\right]\Delta R}\left\{\ln\left[\frac{P_{OFF}\left(\lambda, R_{T}\right)}{P_{ON}\left(\lambda, R_{T}\right)}\right]\right\} \tag{8}$$

The last equation represents the capability of the dial-based identification system to provide the average density of gas traces in the interval ΔR to specify the average concentration between the system location and the final target (ΔR distance).

These systems can provide also concentration measurements – concentration profiles – of the chemicals under investigation. At this purpose it is needed to solve the Eq. (4) to obtain the spatially resolved trace gas concentration [29]:

$$N(R) = -\frac{1}{2\Delta\sigma}\frac{\partial}{\partial R}\ln\left[\frac{P_{ON}(R)}{P_{OFF}(R)}\right] \tag{9}$$

where

$$\Delta\sigma = \sigma_{ON}\left(\lambda_{ON}\right) - \sigma_{OFF}\left(\lambda_{OFF}\right) \tag{10}$$

is the differential absorption cross section of the measured gas.

The Eq. (9) permits to define the punctual concentration profiles of gas traces for all the points in the ΔR distance, without other calibrations except for wavelength calibration and detector linearity [53].

A very sensitive issue is the knowledge of absorption coefficients for several laser wavelength measured on different chemicals potentially dispersed in the atmosphere. It is needed the exact "finger prints" of several chemicals and it is imperative to measure the absorption coefficient in order to improve the absorption database of gases substances. It is important to underline that in the case of organic compounds as Chemical Warfare Agent (CWA), Toxic Industrial Chemical (TIC), Toxic Industrial Materials (TIM) or Volatile Organic Compounds (VOC) the absorption spectra in the IR region are not completely separated but can be shown some overlapping; therefore their analysis need at least standard multivariate data analysis technique as Principal Component Analysis (PCA) to develop a classification model aimed to identify organic chemical compound in atmosphere [65]. Preliminary results on the capability of PCA selection, realized using the preliminary database of absorption coefficient measured by mini TEA CO_2 laser is shown in Fig. 12.

Another important goal was the realization of a very compact and low-cost dial system. At this aim, the mini TEA CO_2 laser used for the absorption coefficient measurements was used for the development of a CO_2 based miniDial by the QEPM group of the University of Rome Tor Vergata (Fig. 13). Although the developed tool is not cheap, the dimensions of the transmission & receive system are extremely

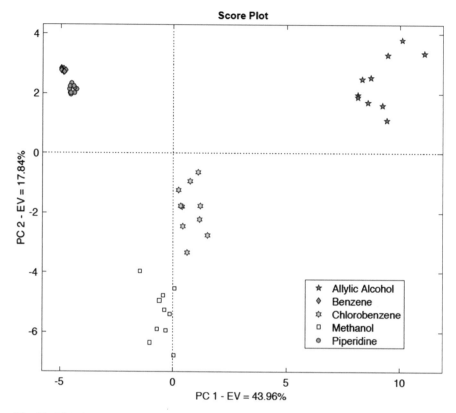

Fig. 12 PCA. Scores plot of the PCA model built with dataset related to the five measured compounds (Source: [67])

low. Probably the semiconductor-based laser technologies permit to dispose on the tunable source but today this approach is not able to produce high peak power. As a result, currently is not possible to develop miniDial systems suitable only for very short range or as gas sensing [11, 66].

3 Conclusions and Remarks

In this paper, the standoff optical based laser technologies are presented underlining the principal properties of detection by lidar and identification by dial methodologies.

The smart use of optical methodologies for detection and identification of CWA not only could be used together to create an innovative solution to the problem of public safety and security: it can result in the creation of a detection systems network. A low-cost lidar based system was tested in urban and industrial areas to

174   P. Gaudio

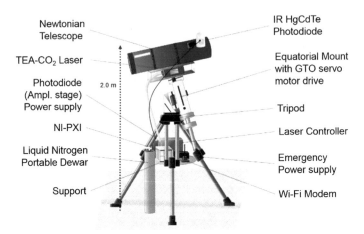

Fig. 13 The mini dial system developed at University of Rome Tor Vergata by Quantum Electronic and Plasma Physics Group

detect pollutants coming from urban traffic or industrial emissions. This approach shows the capability to provide a first alarm; coupled with an identification system (dial) it can result in the determination of the nature of the threat [67]. Which are the problems not completely resolved yet?

First of all, it is needed to create an absorption spectra database in order to consider it as fingerprints of chemicals that could be identify in atmosphere; moreover, it is important to optimize the system reducing the dimensions and the costs, improving at same time its performances in terms of reduction of false alarms, to increase the robustness, the acquisition speed and the reliability during the acquisition and elaboration data phases. Finally, it is important to improve the analysis for dial system in order to increase the accuracy and the stability of concentration profile adopting new methodologies (as an example, the multi-wavelength dial as proposed by Chengzhi Xiang to map atmospheric concentration compounds conducts inversion with one on-line and multiple off-line wavelengths [68]).

Acknowledgements The Author is very grateful to Dr. Stefano Parracino for the Figs. 1, 2, 3, 4, 7, and 8 provided by his PhD thesis.

References

1. Wallin, S., Pettersson, A., Ostmark, H., Hobro, A.: Laser Based Standoff Detection of Explosive: a Critical Review. Anal. Bioanal. Chem. 395, 259–274 (2009).
2. FLIR System, http://www.flir.it
3. Sasic, S., Ozaki, Y.: Raman, Infrared, and Near-Infrared Chemical Imaging. Wiley (2010).
4. Goyal, A.K., Myers, T.R.: Active Mid-Infrared Reflectometry and Hyper Spectral Imaging. Laser-Based Optical Detection of Explosives. CRC Press. 168–212(2015).

5. Deutsch, E.R., Kotidis, P., Zhu, N., Goyal, A.K., Ye, J., Mazurenko, A., Norman, M., Zafiriou, K., Baier, M., Connors, R.: Active and Passive infrared Spectroscopy for the Detection of Environmental Threats. Proc. SPIE. 9106, 91060A (2014).
6. Thériault, J.M., Puckrin, E., Hancock, J., Lecavalier, P., Jackson Lepage, C., Jensen, J.O.: Passive Standoff Detection of Chemical Warfare Agents on Surfaces. Appl. Opt. 43, 5870–5885 (2004)
7. Lewis, I.R., Daniel Jr., N.W., Chaffin, N.C., Griffiths, P.R., Tungol, M.W. Raman spectroscopic studies of explosive materials: towards a field able explosives detector, Spectrochim. Acta A Mol. Biomol. Spectrosc. 51, p. 1985–2000 (1995).
8. Parracino, S.,Richetta, M., Gelfusa, M., Malizia, A., Bellecci, C., De Leo, L., Perrimezzi, C., Fin, A., Forin, M., Giappicucci, F., Grion, M., Marchese, G., Gaudio, P.: Real-Time Vehicle Missions Monitoring Using a Compact LIDAR System and Conventional Instruments: First Results of an Experimental Campaign in a Suburban Area in Southern Italy. Opt. Eng. 55, 103107 (2016).
9. Bellecci, C., Francucci, M., Gaudio, P., Gelfusa, M., Martellucci, S., Richetta, M., LoFeudo, T.: Application of a CO2 Dial System for Infrared Detection of Forest Fire and Reduction of False Alarm.Appl. Phys. B. 87, 373–378 (2007).
10. Deutsch, E., Haibach, F., Mazurenko, A., Williams, B., Hulet, M., Miles, R., Goode, M.: Identification of CWAs Using Widely-Tunable Quantum Cascade Lasers. Proc. Chem. Biol. Defense Sci. Technol. Conf.(2011)
11. Goyal, A.K., Kotidis, P., Deutsch, E.R., Zhu, N., Norman, M., Ye, J., Zafiriou, K., Mazurenko, A.: Detection of chemical clouds using widely tunable quantum cascade lasers. Proc. SPIE. 9455, 94550L (2015).
12. Spuler, S.M., Repasky, K.S., Morley, B., Moen, D., Hayman, M., Nehrir, A.R.: Field-Deployable diode Laser-Based Differential Absorption Lidar (DIAL) for Profiling Water Vapor. Atmos. Meas. Tech. 8, 1073–1087 (2015).
13. Fiocco, G., Smullin, L.D.: Detection of Scattering Layers in the Upper Atmosphere (60–140km) by Optical Radar. Nature. 199, 1275–1276 (1963).
14. Measures, R.M.: Laser Remote Sensing – Fundamentals and Applications. Wiley (1984).
15. Del Guasta, M.: Daily Cycles in Urban Aerosol Observed in Florence by Means of an Automatic 532–1064 nm LIDAR. Atm. Env. 36, 2853–2865 (2002).
16. Mazzoleni, C., Kuhns, H.D., Moosmüller, H.: Monitoring Automotive Particulate Matter Emissions with LIDAR: A Review. Remote Sens. 2, 1077–1119 (2010)
17. Likar, K., Eichinger, V.: Monitoring of the Particles Above the Unpaved Road by Lidar Technique. GLOBAL NEST J. 13, 309–316 (2011)
18. He, T.Y., Stanič, S., Gao, F., Bergant, K., Veberič, D., Song, X.Q., Dolžan, A.: Tracking of Urban Aerosols Using Combined LIDAR-Based Remote Sensing and Ground-Based Measurements. Atmos. Meas. Tech. 5, 891–900 (2012).
19. Mei, L., Brydegaard, M.: Atmospheric Aerosol Monitoring by an Elastic Scheimpflug Lidar System. Opt. Exp. 23, A1613-A1628 (2015).
20. Bösenberg, J.: EARLINET: A European Aerosol Research Lidar Network. In Proc. 20th International Laser Radar Conference (IPSL) (2000).
21. Cuomo, V., Di Girolamo, P., Esposito, F., Pappalardo, G., Serio, C., Spinelli, N., ... Bellecci, C.: The LITE Correlative Measurements Campaign in Southern Italy: Preliminary Results. Appl. Phys. B. 64, 553–559 (1997).
22. Andreucci, F., Arbolino, M.V., Sozzi, R.: Lidar Utilisation for Industrial Stack Plume Automatic Tracking. Il Nuovo Cimento C. 14, 453–462(1991).
23. Andreucci, F., Arbolino, M.V.: A study on forest fire automatic detection systems 1. Il Nuovo Cimento C. 16, 35–50, (1993).
24. Andreucci, F., Arbolino, M.V.: A study on forest fire automatic detection systems 2. Il Nuovo Cimento C. 16, 51–65, (1993).
25. Utkin, A.B., Lavrov, A.V., Costa, L., Simões, F., Vilar, R.: Detection of Small Forest Fires by Lidar. Appl. Phys. B. 74, 77–83 (2002).

26. Bellecci, C., De Leo, L., Gaudio, P., Gelfusa, M., Lo Feudo, T., Martellucci, S., Richetta, M.: Reduction of False Alarms in Forest Fire Surveillance Using Water Vapour Concentration Measurements. Opt. Laser Technol. 41, 374–379 (2009).

27. Lavrov, A., Utkin, A.B., Vilar, R., Fernandez, A.: Application of Lidar in Ultraviolet, visible and Infrared Range for Early Forest Fire Detection. Appl. Phys. B. 76, 87–95 (2003)

28. Weitkamp, C.: Lidar Range-Resolved Optical Remote Sensing of the Atmosphere. Springer. (2005).

29. Kovalev, V., Eichenger, W.E.: Elastic Lidar, Theory, Practice and Analysis Methods. Wiley. (2004).

30. Reichardt, J., Reichardt, S.: Determination of Cloud Effective Particle Size From the Multiple Scattering Effect on Lidar Integration-Method Temperature Measurements. Appl. Opt. 45, 2796–2804 (2006).

31. Berkoff, T., Welton, E., Campbell, J., Valencia, S., Spinhirne, J., Tsay, S.C., Holben, B.: Observations of Aerosols Using the Micro-Pulse Lidar NETwork (MPLNET). Proceedings of IEEE International Geoscience and Remote Sensing Symposium. 3, 2208–2211 (2004).

32. Uchino, O., Tokunaga, M., Maeda, M., Miyazoe, Y.: Differential-Absorption-Lidar Measurement of Tropospheric Ozone with Excimer-Raman Hybrid Laser. Opt. Lett. 8, 347–349 (1983).

33. Vaughan, G., Wareing, D.P., Pepler, S.J., Thomas, L., Mitev, V.: Atmospheric Temperature Measurements Made by Rotational Raman Scattering. Appl. Opt. 32, 2758–2764 (1993)

34. Mikkelsen, T., Mann, J., Courtney, M., Sjholm, M.: Wand scanner: 3-D Wind and Turbulence Measurements From Three Steerable Doppler Lidars. IOP Conf. Ser. Earth Environ. Sci. 1, 012018 (2008)

35. McClung, F.J., Hellarth, R.W.: Giant Optical Pulsations From Ruby. J. Appl. Phys. 33, 828 (1962).

36. Reichardt, J., Wandinger, U., Serwazi, M., Weitkamp, C.: Combined Raman Lidar for Aerosol, Ozone, and Moisture Measurements. Opt. Eng. 35, 1457–1465 (1996)

37. Schafer, R.W.: What Is a Savitzky-Golay Filter? IEEE Signal Processing Magazine 28, 111–117 (2011).

38. Sumnicht, G.K., McGee, T.J., Twigg, L., Gross, M., Beyerle, G.: Savitzky-Golay Filtering for Ozone Retrieval and Vertical Resolution Algorithm Optimization. J. Opt. Soc. Am. (1999).

39. Poreh, D., Fiorani, L.: Software for Analyzing and Visualizing Laser Remote Sensing (LIDAR) Data. J. Optoelectron. Adv. M. 12, 1231–1236 (2010).

40. Azadbakht, M., Fraser, C.S., Zhang, C., Leach, J.: A Signal Denoising Method for Full-Wave Form LIDAR Data. Proceedings of the ISPRS Annals of Photogrammetry, Remote Sensing and Spatial Information Sciences, Antalya, Turkey, 11–13 (2013).

41. Iarlori, M., Madonna, F., Rizi, V., Trickl, T., Amodeo, A.: Effective Resolution Concepts for Lidar Observations. Atmos. Meas. Tech. 8, 5157–5176, (2015).

42. Gaudio, P., Gelfusa, M., Richetta, M.: Preliminary Results of a Lidar-Dial Integrated System for the Automatic Detection of Atmospheric Pollutants. Proc. SPIE. 8534, 853404 (2012).

43. Saleh, B.E.A., Teich, M.C.: Semiconductor Photodetectors. Fundamentals of Photonics. Wiley. (1991).

44. Enteper le Nuovetecnologiel'Energia e l'Ambiente – ENEA, http://www.enea.it/

45. Klett, J.D.: Stable Analytical Inversion Solution for Processing Lidar Returns. Appl. Opt. 20, 211–220 (1981).

46. Ancellet, G., Papayannis, A., Pelon, J., Megie, G.: DIAL Tropospheric Measurement Using a Nd:YAG Laser and the Raman Shifting Technique.J. Atmos. Ocean. Tech.6, 832–839 (1989).

47. Stefanutti, L., Castagnoli, F., Del Guasta, M., Morandi, M., Sacco, V.M., Zuccagnoli, L., Godin, S., Megie, G., Porteneuve, J.: The Antarctic Ozone LIDAR System. Appl. Phys. B. 55, 3–12 (1992).

48. Sunesson, J.A., Apituley, A., Swart, D.P.J.: Differential Absorption Lidar System for Routine Monitoring of Tropospheric Ozone. Appl. Opt.33, 7045–7058 (1994).

49. Fiorani, L., Calpini, B., Jaquet, L., Van Den Bergh, H., Durieux, E.: A Combined Determination of Wind Velocities and Ozone Concentrations for a First Measurement of Ozone Fluxes with a

DIAL Instrument During the MEDCAPHOT-TRACE Campaign. Atmos. Environ, 32, 2151–2159 (1998).

50. Zanzottera, E.: Differential Absorption Lidar Techniques in the Determination of Trace Pollutants and Physical Parameters of the Atmosphere. Crit. Rev. Anal. Chem. 21, 279–319 (1990).

51. Killinger D.K., Menyuk, N.: Remote Probing of the Atmosphere Using a CO2 Dial System. IEEE J. Quant. Electron. 17, (1981).

52. Menyuk, N., Killinger, D.K., DeFeo, W.E.: Laser Remote Sensing of Hydrazine, MMH, and UDMH Using a Differential-Absorption CO2 Lidar. Appl. Opt. 21, 2275–2286 (1982).

53. Amediek, A., Fix, A., Wirth, M., Ehret, G.: Development of an OPO System at 1.57 μm for Integrated Path DIAL Measurement of Atmospheric Carbon Dioxide. Appl. Phys. B. 92, 295–302 (2008).

54. Kempfer, U., Carnuth, W., Lotz, R., Trickl, T.: A Wide-Range Ultraviolet Lidar System for Tropospheric Ozone Measurements: Development and Application. Rev. Sci. Instrum. 65, 3145–3164 (1994).

55. Reichardt, J., Wandinger, U., Serwazi, M., Weitkamp, C.: Combined Raman Lidar for Aerosol, Ozone, and Moisture Measurements. Opt. Eng. 35, 1457–1465 (1996).

56. Moore, D.S., Scharff, R.J.: Portable Raman Explosives Detection. Anal. Bioanal. Chem. 393, 1571–1578 (2009).

57. Wallin, S., Pettersson, A., Östmark, H., Hobro, A.: Laser-Based Standoff Detection of Explosives: a Critical Review. Anal. Bioanal. Chem. 395, 259–274 (2009).

58. Deutsch, E. R., Kotidis, P., Zhu, N., Goyal, A. K., Ye, J., Mazurenko, A., ... Connors, R.: Active and Passive Infrared Spectroscopy for the Detection of Environmental Threats. SPIE. 91060A–91060A (2014).

59. Carlisle, C.B., Van Der Laan, J.E., Carr, L.W., Adam, P., Chiaroni, J.P.: CO2 Laser-Based Differential Absorption Lidar System for Range-Resolved and Long-Range Detection of Chemical Vapor Plumes. Appl. Opt. 34, 6187–6200 (1995).

60. Carr, L.W., Fletcher, L., Crittenden, M., Carlisle, C.B., Gotoff, S.W., Reyes, F., D'Amico, F.M.: Frequency-Agile CO2 DIAL for Environmental Monitoring. SPIE. 282–294(1994).

61. Carr, L.W., Warren, R.E., Carlisle, C.B., Carlisle, S.A., Cooper, D.E., Fletcher, L., ... Reyes, F.: Multiple Volatile Organic Compound Vapor Chamber Testing with a Frequency-Agile CO2 DIAL System: Field-Test Results. P. Soc. Photo-Opt. Ins. 291–302 (1995).

62. Bellecci, C., Caputi G.E., De Donato F., Gaudio, P., Valentini M.: CO2 Dial for Monitoring Atmospheric Pollutants at University of Calabria. Il Nuovo Cimento C. 18, 463–472 (1995).

63. Bellecci, C., Dalu, G.A., Aversa, P., Federico, S., Gaudio, P.: Evolution of Water Vapour Profile Sover Complex Terrain: Observation and Comparison with Model Simulation in the Valley of Cosenza. Proc. SPIE. 3865, 108–118 (1999).

64. Bellecci, C., Gaudio, P., Gelfusa, M., Martellucci, S., Richetta, M., Ventura, P., Antonucci, A., Pasquino, F., Ricci, V., Sassolini, A.: Database for Chemical Weapons Detection: First Results. Proc. SPIE. 7116, (2008).

65. Gaudio, P., Malizia, A., Gelfusa, M., Martinelli, E., DiNatale, C., Poggi, L.A., Bellecci, C.: Mini-DIAL System Measurements Coupled with Multivariate Data Analysis to Identify TIC and TIM Simulants: Preliminary Absorption Data Base Analysis. J. Phys. Conf. 778, 012004 (2017).

66. Frish, M.B., Wainner, R.T., Laderer, M.C., Green, B.D., Allen, M.G.: Standoff and Miniature Chemical Vapor Detectors Based on Tunable Diode Laser Absorption Spectroscopy. IEEE Sens. J. 10, 639–646 (2010).

67. Gaudio, P., Malizia, A., Gelfusa, M., Murari, A., Parracino, S., Poggi, L.A., Lungaroni, M., Ciparisse, J.F., Di Giovanni, D., Cenciarelli, O., Carestia, M., Peluso, E., Gabbarini, V., Talebzadeh, S., Bellecci, C.: Lidar and Dial Application for Detection and Identification: A Proposal to Improve Safety and Security. J. Instrum. 12 (2017).

68. Xiang, C., Ma, X., Liang, A., Han, G., Gong, W., Yan, F.: Feasibility Study of Multi-Wave Length Differential Absorption LIDAR for CO2 Monitoring. Atmosphere, 7, 89 (2016).

Implementing an Information Security Program

Cliff Glantz, Joseph Lenaeus, Guy Landine, Lori Ross O'Neil,
Rosalyn Leitch, Christopher Johnson, John Lewis, and Robert Rodger

Abstract The threats to information security have dramatically increased with the proliferation of information systems and the internet. Chemical, biological, radiological, nuclear, and explosives (CBRNe) facilities need to address these threats in order to protect themselves from the loss of intellectual property, theft of valuable or hazardous materials, and sabotage. Project 19 of the European Union CBRN Risk Mitigation Centres of Excellence Initiative is designed to help CBRN security managers, information technology/cybersecurity managers, and other decision-makers deal with these threats through the application of cost-effective information security programs. Project 19 has developed three guidance documents that are publically available to cover information security best practices, planning for an information security management system, and implementing security controls for information security.

Keywords Information security • Security controls • Information security management system • Risk-based security • Cybersecurity • Blended attack

1 Introduction

Threats to information security have increased dramatically with the proliferation of computer-based information systems and the connection of those systems to the internet. It appears that a wide range of adversaries are finding the theft or manipulation

C. Glantz (✉) • J. Lenaeus • G. Landine • L.R. O'Neil • R. Leitch
Pacific Northwest National Laboratory, Richland, WA, USA
e-mail: cliff.glantz@pnnl.gov; joseph.lenaeus@pnnl.gov; guy.landine@pnnl.gov; lro@pnnl.gov; rosalyn.leitch@pnnl.gov

C. Johnson
University of Glasgow, Glasgow, Scotland
e-mail: Christopher.Johnson@glasgow.ac.uk

J. Lewis • R. Rodger
National Nuclear Laboratory, Sellafield, UK
e-mail: john.g.lewis@sellafieldsites.com; robert.m.rodger@nnl.co.uk

© Springer International Publishing AG 2017
M. Martellini, A. Malizia (eds.), *Cyber and Chemical, Biological, Radiological,
Nuclear, Explosives Challenges*, Terrorism, Security, and Computation,
DOI 10.1007/978-3-319-62108-1_9

of information from chemical, biological, radiological, nuclear, and explosives (CBRNe) facilities to be easily achievable with a low probability of detection, attribution, and retribution. Adversaries may range from unsophisticated and inexperienced recreational hackers to well-funded, sophisticated actors working for organized crime or nation states. Unfortunately, the ability of attackers to achieve their malicious objectives has far surpassed the ability of many CBRNe facilities and associated institutions to protect information assets. One reason for this asymmetric result is that the attackers only need to compromise information assets to achieve their objectives. In contrast, CBRNe facilities have multiple business priorities (e.g., facility productivity, profitability, reliability) that divert attention from security issues. In some cases, the business priorities of CBRNe programs may conflict with the goal of having an effective information security program. Often the issue of information security is not recognized or adequately characterized until an information security event occurs and the facility and associated organization have to face adverse consequences. Even if CBRNe facilities identify their information security risks, they often lack the resources and expertise to effectively or completely address the underlying problems.

In recognition of this situation, the European Union CBRN Risk Mitigation Centres of Excellence Initiative[1] funded a project (Project 19) to assist CBRN facilities in addressing the challenge of information security.[2] Project 19 consists of multidisciplinary experts from the United States Department of Energy's Pacific Northwest National Laboratory, and the United Kingdom's National Nuclear Laboratory and the University of Glasgow.

The objective of Project 19 is to provide guidance on the security of information technology (IT) structures and data exchange mechanisms for CBRN facilities. This effort includes providing information on the managerial, operational, and technical security controls needed to address threats, characterize adversaries, identify vulnerabilities, and enhance defense and mitigation capabilities. A security control is a "safeguard or countermeasure… designed to protect the confidentiality, integrity, and availability" of an information asset or system and "meet a set of defined security requirements." (NIST 2013).

Project 19 is designed to help CBRN security managers, IT/cybersecurity managers, and other decision-makers typically involved in the information life cycle (e.g., development, acquisition, communication, use, disposal) to develop and implement appropriate and cost-effective information security programs. The guidance provided by Project 19 is based on international standards and best practices, the experience of the information security experts on the document writing team, and the many CBRN facility personnel and government officials from around the world who participated in stakeholder reviews.

The following guidance documents were developed by the project management team:

---

[1] The initiative is developed with the technical support of relevant international and regional organizations, the European Union Member States, and other stakeholders, through coherent and effective cooperation at the national, regional, and international level.

[2] The European Union's "CBRN" program does not include explosives in its scope of work. However, Project 19's information security work is applicable for all CBRNe facilities.

- Information Security Best Practices for CBRN Facilities [1]. This document provides recommendations on best practices for information security and high-value security controls;
- Information Security Management System Planning for CBRN Facilities [2]. This document provides guidance for developing information security planning documents that establish information security roles, responsibilities, and policies;
- How to Implement Information Security Management System (ISMS) Security Controls Using a Risk Based Approach at CBRN Facilities [3]. This document provides a guide for selecting security controls that can be used to implement information security plans and policies.

In addition to the three documents, a two-day train-the-trainer workshop on information security for CBRN facilities was developed. That workshop is designed to introduce the need for information security to CBRN facility decision-makers and others with oversight responsibilities for CBRN materials and facilities.

In the following sections, we provide summary information to provide some information security basics for CBRNe facilities and introduce the reader to these informative guidance products.

2 Information Security Best Practices

Decision-makers and workers at CBRNe facilities, organizational managers responsible for these facilities, and regulatory agencies (governmental and nongovernmental) are often in need of risk-based guidance on the best practices for protecting information security. Not all facilities can afford to purchase, install, operate, and maintain expensive security systems; therefore, decisions on information security have to balance security risks and resource constraints. When resources are limited, as they almost always are, information security investments should focus on the processes and practices that provide the greatest risk reduction. UNICRI [4] provides guidance on information security best practices to help decision-makers and those with information security responsibilities design an appropriate risk-based information security program.

A variety of international standards and national policy documents provide supporting guidance for developing an information security program. Some are based on laws and regulations and follow a compliance-based approach to information security (e.g., do what the law says). Others take a risk management approach that provides the freedom to evaluate information security threats, vulnerabilities, and consequences and then manage the risks so they fall within acceptable levels. Many information security guidance products have a lot in common. The International Organization for Standardization (ISO) 27,000 series ([5]) is one of the more popular sets of products because it is an international standard and is applicable for a wide range of infrastructures and industries. However, it is not customized to focus on a given type of industry; standards are not the most user-friendly of documents; and the cost to acquire the standards may deter its purchase and use by some personnel with information security responsibilities.

The ISO 27001 series presents a framework for an ISMS. It outlines seven performance areas:

1. Context of the Organization
2. Leadership
3. Planning
4. Support
5. Operation
6. Performance Evaluation
7. Improvement.

These performance areas can be divided into five stages – Governance, Plan, Do, Check, and Act. The relationship among the seven performance areas and their stages in the information security program are illustrated in Fig. 1.

Fig. 1 The information security life cycle [4, 6]

In an environment where information systems and associated processes are continually evolving, the information security threat and adversary capabilities are changing, and new information security technologies are emerging, a continuous process is needed to maintain the accuracy and performance of the ISMS. Within this life cycle, the application of industry best practices and high-value security controls is recommended for information security programs.

2.1   Best Practice Security Recommendations

UNICRI [4] presents a series of best practice recommendations for information security. These recommendations are listed in the following subsections.

Implement a Graded Approach to Protect Critical Systems CBRNe facilities contain numerous computer systems that contribute to the overall operation of the facility. Some of these are IT systems that are used to create, acquire, process, communicate, or store data. IT systems support management and decision-making. Operational technology (OT) systems are hardware and software that monitor and/ or control physical devices, processes, and events in a facility. The compromise of an IT or OT system could potentially affect the ability of a system or network to perform its intended function and result in a loss of confidentiality, integrity, or availability of data or system function.

It is important to recognize that all information or information systems within a CBRNe facility are not of equal importance. The relative significance of a system to the overall operation of a CBRNe facility is dictated by the functions it serves. While the loss of some types of information or systems may have negligible impact upon the facility, the loss of others may have more serious consequences. For example, the compromise of information related to CBRNe security systems or control systems that perform safety and security functions may pose substantially greater security risks than the compromise of information involving routine business functions. Correspondingly, the level of protection afforded to information or information systems within a CBRNe facility should be based on the importance of the system. The use of a graded approach to information security ensures that a disproportionate amount of security resources are not applied to systems that represent relatively low risk.

Integrate Physical and Cybersecurity Physical security and cybersecurity are often treated as separate problems. At many CBRNe facilities, physical security specialists and cybersecurity specialists have little or no interaction. This organizational disconnection can result in a failure to recognize and address significant crosscutting security vulnerabilities and underestimate the associated security risks. Crosscutting issues include cyber-enabled physical attacks and physical-enabled cyberattacks.

Strong lines of communication are needed between a facility's physical and cybersecurity programs to address blended attacks. In many cases, efficiency can be increased and costs reduced by combining the physical and cybersecurity programs into a single, integrated information security program that can more efficiently deal with a broad range of threats and attack pathways.

In a related issue, security evaluations and exercises at CBRNe facilities often only focus on traditional types of physical attacks. Because of the added complexity and the lack of experience with cyberattacks, cyber components are often excluded from traditional security training programs, evaluations, and exercise scenarios. By not training physical security forces to recognize and withstand a cyber-enabled physical attack, or evaluating their performance under such an attack, physical security forces may be inadequately prepared to deal with a cyber-enabled physical attack that can threaten sensitive information or information systems.

Secure the Data Center Both physical and cyber security protections are needed to safeguard information assets in facility data centers. An information security manager must also consider the potential threat of a natural disaster (e.g., earthquake, cyclone, flood) or the misuse of the facility's infrastructure (e.g., triggering fire sprinklers to activate without a fire being present) to damage information assets. Access control, both physical and digital, needs to limit access to information assets. The facility should establish requirements for with whom, when, and how information assets can be physically or digitally accessed.

Secure the Human Every authorized user is a potential threat to have an unintentional or malicious impact on information security. To mitigate this threat, an organization must train every user to understand the nature of information security threats, vulnerabilities, and consequences. Authorized users should understand the actions they should take to support information security under various circumstances.

Access to information assets will vary from position to position (or individual to individual), and therefore the threat to information security varies as well. An information security program should characterize its authorized users. Those users that have the greatest access to information assets should receive a greater level of background screening, training, and auditing.

When staff members are terminated or change departments within an organization, information security personnel should promptly address the change in access requirements. According to the Software Engineering Institute Computer Emergency Response Team (CERT) Insider Threat Center, 90% of the staff members that commit IT sabotage had some form of privileged access (CERT 2010). When users change departments, information security personnel should promptly reconsider access requirements and grant only the access permissions that are needed.

Secure Data Exchange In 2012, the United States established a National Strategy for Information Sharing and Safeguarding. This strategy focuses on providing guidance for more effective integration and implementation of policies, processes, standards, and technologies that promote secure and responsible information sharing [7]. The goal was to ensure the right information is shared with the right people and at the right time. Care must be taken when sharing sensitive data with other departments, particularly when using an untrusted medium like the internet. An information security manager is responsible for securing the data in transit and in storage while still making the data available for legitimate users. An important tool for information security during both storage and transport is the use of encryption.

For information shared with external organizations, agreements should be reached on how the data will be used and stored. The agreement should clearly document the scope of responsibility.

Adopt Information Security Policies, Standards, and Procedures Successful information security management relies upon the existence of properly developed policies and procedures. Effective information security policies establish the overall security objectives, goals, and practices. These policies provide additional detail on roles and responsibilities for information security within an organization. The policies developed should be specific to each CBRNe facility and clearly identify the objectives, requirements, and needed references to define the conduct of the program to secure and protect information assets from threats. Where possible, the developed site-specific information security policies should be designed to complement existing security policies that may exist in organizations with more than one facility. When site-specific policies conflict with those of existing organizational policies, or when adherence to the organizational policies would result in an unacceptable increase in risk to the systems existing within the CBRNe facility, the site-specific cybersecurity policies should take precedence.

CBRNe facilities should develop procedures, based on published technical standards where applicable, to implement their information security policies. The developed policies and procedures need to be periodically reviewed to ensure that they are sufficient to address the risks identified for the information and information systems they are intended to protect. In addition, issues resulting from evolving threats, vulnerabilities, and security controls should be addressed. It is recommended that CBRNe information security personnel review ISO 27002, Information Technology – Security Techniques – Code of Practice for Information Security Management (ISO/IEC 2013) and NIST SP 800–12, An Introduction to Computer Security: The NIST Handbook, (NIST 1995). These publications discuss the areas of applicability when developing site-specific security policies to protect information systems and other critical systems.

Identify Credible Threats and Vectors of Attack The magnitude of a threat is dependent on an adversary's motivations, capabilities, and opportunities to carry out an attack. Attackers use tools, tactics, and techniques to carry out their threats. "Motivations" may include political, ideological, religious, financial, or emotional drivers. "Capabilities" refers to the knowledge, skills, and tools necessary to conduct an attack. "Opportunities" represents the situational circumstances that would support the initiation of an attack. The relative magnitude of the threat directly corresponds to the proportional quantities of each of these elements. An attacker with a clearly stated or well-known intent, a high degree of capability, and knowledge could represent a serious threat. In contrast, an attacker that has weak motivations, minimal capabilities, and limited opportunities would represent a negligible threat. It is important to recognize that threat levels can change rapidly. For example, the availability of new attack tools and techniques, or the discovery of new vulnerabilities, can dramatically alter motivations, capabilities, and opportunities. Threats may involve physical or cyber attacks, be targeted at a specific facility or more general in nature, and rely on non-malicious as well as malicious actions.

Attack trees [1] are diagrams that show all of the key steps needed to conduct an attack against a target. Attack trees are a powerful technique because, unlike other forms of analyses, they require the analyst to adopt the mindset or perspective of the attacker. Attack trees are used to identify stages in the attack where the application of appropriate and cost-effective security controls may achieve the greatest amount of risk reduction.

Whether using attack trees or another methodology, the systematic assessment of credible threats and attack vectors can provide valuable information to help inform the development and application of an effective program of information security.

Identify, Eliminate, and Mitigate Vulnerabilities For an attack to succeed, it must exploit some inherent weakness or vulnerability. The term "vulnerability" is defined as a weakness in the physical or electronic configuration of an asset that could allow an adversary to compromise the security of the asset. All operational systems have vulnerabilities in their technology, implementation, or human components. The concept of total and complete security is an illusion. Generally, the more complex the system, the more likely it is to contain vulnerabilities. To make a given target less susceptible to an attack, it is necessary to harden the target. Hardening of a target can be accomplished through vulnerability elimination or mitigation.

Vulnerability elimination requires the configuration or redesign of a system to ensure that the identified weakness no longer exists. In the case of digital systems, this involves addressing vulnerabilities identified in system hardware, firmware, operating system software, and application software. The engineering costs involved with vulnerability elimination can be significant. A robust design process will take potential future costs into consideration, attempting to identify and eliminate security issues early in the design process rather than later, where the expense is considerably greater.

Vulnerability mitigation seeks to address identified weaknesses either through reconfiguration of the system or through the application of security controls. A few examples of types of vulnerability mitigation are:

- Removal of unnecessary user accounts and software
- Elimination of unneeded operating system services and communication ports
- Implementation of enhanced access controls
- Prompt application of security patches for operating systems
- Implementation of system security monitoring
- Enhancement of the defensive architecture through network segmentation and communication restrictions
- Implementation of encryption mechanisms.

Implement Defense-in-Depth Principles The more complex an information system, the more likely it is to contain weaknesses or vulnerabilities. To address vulnerabilities that are difficult to eliminate or mitigate, protective strategies that embrace the concept of defense-in-depth or resilience should be implemented.

Defense-in-depth represents an approach to system security where multiple layers of security are implemented to protect against failure of a single component or layer. The concept of defense-in-depth originates from a military strategy that seeks to delay, rather than prevent, the advance of an attacker by yielding space in order to buy time. The time gained during the attack allows more time for informed decision-making to counter the assault. Defense-in-depth can be visualized as a series of concentric layers of security in which the vulnerabilities that exist for a given layer are prohibited from existing within the adjacent layers. A threat agent seeking to attack such a system would be forced to identify and exploit non-identical vulnerabilities existing at each successive security layer.

A CBRNe information security program should implement a diverse set of protective strategies capable of detecting, isolating, and neutralizing unauthorized activities in a timely manner. These protective strategies should exhibit defense-in-depth characteristics, ensuring that the failure of any single element of a strategy does not result in successful compromise of an information asset. The detection of unauthorized activity should trigger preplanned security response mechanisms that seek to delay or prevent the advance of an attack.

Implement Life Cycle Security The security of an information system is dependent on the level of security present throughout an asset's life cycle. A security lapse in any phase of the life cycle can affect information security. To create secure information systems, security issues should be addressed during the design, procurement, implementation, testing, operations, maintenance, and decommissioning phases of all systems. Institute of Electrical and Electronics Engineers (IEEE) Standard 1074–1995, *IEEE Standard for Developing Software Life Cycle Processes* [8], is a useful resource for addressing the software component of life cycle issues. A variety of resources involving personnel security are available to support the human aspects of life cycle security.

Information Security Awareness Training The information security program implemented by CBRNe facilities should ensure that facility personnel, including contractors, are aware of the cyber and physical security requirements associated with their assigned duties and responsibilities. The training program should help personnel:

- Identify information security scope, goals, and objectives
- Identify training staff and other resources to support awareness training
- Identify target audiences and their appropriate levels of training
- Provide training updates and enhancements to support a continued quality improvement objective
- Periodically evaluate program effectiveness against program goals and objectives

The following levels of information cybersecurity awareness and training should be provided:

- General information security awareness training for all staff members and contractors
- More in-depth training on applicable information security concepts and practices for personnel involved in the development, use, or storage of sensitive information assets
- Advanced cybersecurity training for individuals designated as information security specialists with responsibilities to protect information systems, related assets, and networks from information security threats

Analyze and Select Security Controls Securing information assets involves assessing the sensitivity of information assets, the capabilities of potential adversaries, the identification of credible threats and attack vectors, and the characterization of the existing security posture of the system and its operational environment. Strategic security objectives should be defined and appropriate risk-based and cost-effective security controls (sometimes called countermeasures) should be identified and implemented to meet security objectives.

Many potential security controls are available for use in an information security program. In Sect. 2.2, we dig deeper into this topic by reviewing security controls that are generally considered to have high value from a risk-benefit standpoint. More detailed guidance on selecting security controls for information systems is provided in [3].

Perform Security Monitoring While security controls may be deployed with the intent of deterring, delaying, detecting, or denying an attack, no set of security controls can eliminate all vulnerabilities. Therefore, CBRNe information networks should be adequately monitored for anomalous behavior or activity that would suggest a potential breech in security. A robust information security program will include capabilities for the collection and analysis of indication and warning data in order to detect and respond to intrusions.

While system log management solutions are extremely helpful in identifying potential problems, other resources are also needed to identify and prevent potential problems. These resources include network intrusion detection and prevention systems, host intrusion detection and prevention systems, antivirus systems, and file integrity systems. To provide more comprehensive security, a Security Incident and Event Manager (SIEM) application is often employed. A SIEM is an integrated suite of security tools used to manage multiple security applications and devices. Most SIEM applications will have the capability to interpret or normalize the information coming from various security devices and provide the ability to present a unified view of security across a network.

Implement an Information Security Incident Forensic Response and Recovery Capability An incident response and recovery process is an important element for information security. If a security incident involving computer systems and networks at a CBRNe facility occurs, an immediate goal is to identify the problem, implement corrective actions, and return the facility to normal operations in a timely manner. It is also important to safeguard the forensic data necessary to assess the event and be able to counter similar types of attacks in the future.

Incident response for an information security event should be integrated with the incident response programs for cyber and physical security events; these programs should meet any national and international reporting requirements. If there is a cyber component to the compromise of information, cybersecurity incident response will come into play. If there is a physical or personnel component to the compromise, physical security incident response will be involved. For many potential events, both the cyber and physical incident response teams will need to coordinate operations.

An information security incident response and recovery process should include preparation, identification, containment, eradication, recovery, post-incident analysis, forensics activities, and information security awareness training.

2.2   High-Value Security Controls

In this section, we examine several types of security controls that are considered high value. If resource constraints prevent the implementation of a comprehensive set of security controls, then those controls highlighted here are perceived as providing a high benefit-to-cost ratio.

CBRNe facilities can procure and install a broad range of managerial, operational, and technical security controls to enhance their information security. The Council on CyberSecurity reports that there is an enormous set of "good things" that a facility organization can use to improve its security, "but [it does] not always [have] clarity on what to prioritize" [9].

Australian Signals Directorate Top Four Strategies The Australian Signals Directorate (ASD) has a developed a set of Top Four Strategies to mitigate the effects of cyberattacks, including those that seek to compromise information security [10]. ASD is widely cited by cybersecurity experts and organizations as providing leadership in the development of simple strategies that can provide reasonable protection from many, but not all, threats to information security. Their top four strategies are:

1. Application Whitelisting
2. Application Patching
3. Patching Operating Systems
4. Minimizing Administrative Privileges

The top four strategies are designed to mitigate about 85% of the cyber intrusion techniques that are detected by ASD's Cyber Security Operations Centre [10].

Council on CyberSecurity Top 20 The "The Critical Security Controls for Effective Cyber Defense" are a set of 20 security controls that were developed and are maintained by the Council on CyberSecurity (Table 1). These controls use

Table 1 The top 20 security controls from the council on CyberSecurity (2014)

1.	Inventory of authorized and unauthorized devices
2.	Inventory of authorized and unauthorized software
3.	Secure configurations for hardware and software on mobile devices, laptops, workstations, and servers
4.	Continuous vulnerability assessment and remediation
5.	Malware defenses
6.	Application software security
7.	Wireless access control
8.	Data recovery capability
9.	Security skills assessment and appropriate training to fill gaps
10.	Secure configurations for network devices such as firewalls, routers, and switches
11.	Limitation and control of network ports, protocols, and services
12.	Controlled use of administrative privileges
13.	Boundary defense
14.	Maintenance, monitoring, and analysis of audit logs
15.	Controlled access based on the need to know
16.	Account monitoring and control
17.	Data protection
18.	Incident response and management
19.	Secure network engineering
20.	Penetration tests and red team exercises

approaches that have been shown to work in real-world applications [9]. They represent a powerful tool for prioritizing information security activities by concentrating on a manageable number of key security controls that have high benefit-to-cost ratios. Use of these controls within a comprehensive information security program is endorsed by a wide range of organizations, including governmental agencies and private consultants. Five critical tenets for effective cybersecurity are captured in the top 20 controls:

1. Offense informs defense. Use information about real-world attacks to create effective defenses.
2. Prioritization. Focus on implementing security controls that provide the greatest risk reduction and are readily implementable.
3. Metrics. Establish easy-to-understand measures of your security controls so that their effectiveness can be readily understood by decision-makers and other members of the workforce.
4. Continuous monitoring. Conduct continuous security monitoring to test and validate the effectiveness of your security controls.
5. Automation. Automate defenses to allow for a rapid assessment and response to potential attacks [9].

The Council provides guidance, based on real-world feedback from users, on the steps that should be followed to implement their top 20 controls:

- Step 1. Perform an initial security gap assessment to determine what security controls are in place and where gaps remain.
- Step 2. Develop an implementation roadmap by selecting the specific controls (and sub-controls) that need to be implemented and schedule the various implementation phases based on risk considerations.
- Step 3. Implement the initial or next phase of controls by identifying existing tools and their current utilization, new tools that are needed, processes that need to be implemented or improved, and training needs.
- Step 4. Integrate security controls into operations by focusing on continuous monitoring, attack mitigation, and lifecycle security.
- Step 5. Report and manage progress against the implementation roadmap developed in Step 2. Repeat Steps 3–5 for the next phases outlined in the implementation roadmap to maintain an ongoing process of information security assessment and improvement. (Council on CyberSecurity 2014).

The Critical Security Controls [9] complement ongoing work in security standards and best practices such as the following:

- The Security Content Automation Program (SCAP) (NIST 2014c)
- NIST Special Publication 800–53, *Recommended Security Controls for Federal Information Systems and Organizations* (NIST 2013b)
- ASD's "Top 35 Strategies to Mitigate Targeted Cyber Intrusions" (ASD 2014)
- ISO/International Electrotechnical Commission (IEC) 27,002:2013 Information technology – Security techniques – Code of Practice for Information Security Controls (ISO/IEC).

2.3 Additional Information

The *Information Security Best Practices* document UNICRI [3] goes into more detail on the topics described in this section. It also presents information on common deficiencies and problems observed in existing information security programs. The document also provides a description of useful additional sources of information.

3 Information Security Management System Planning

The systematic protection of information requires a comprehensive information security program that is implemented through an ISMS. An ISMS is a set of policies, procedures, practices, technologies, and responsibilities that protect the security of information. The protection of information involves cybersecurity, physical security, and personnel security. Attacks on information security may involve an attempt to exploit just one type of security or it may involve blended attacks (i.e., cyber-enabled physical attacks, and physically-enabled cyber attacks). Information security also involves protecting infrastructure resources upon which information security systems rely (e.g., electrical power, telecommunications, and environmental controls) [2].

Two major types of plans typically control an organization's ISMS – its risk management plan and security plan. The risk management plan identifies and tracks threats to and vulnerabilities in the CBRNe information systems. It also identifies how the facility will respond to these threats and documents the facility's monitoring approach. The risk management plan should cover risk assessment, response, and management.

The security plan provides the framework for implementing the facility's information security strategies and covers:

- Business environment
- Asset management
- Security control implementation
- Configuration management
- Contingency planning and disaster recovery
- Incident response
- Governance
- Monitoring and auditing
- Awareness and training

Information security programs should be risk-based. While the same general outline for information security planning documents is followed regardless of the security risk, the level of detail will vary according to the risk level. It is up to the CBRNe facility, in consultation with its competent authorities and any parent organization under which the facility operates, to determine the appropriate level of detail and length of the information security planning documents needed to support its information security plans.

For example, a CBRNe facility with a low level of information security risk (e.g., it possesses little or no sensitive information and would experience minimal consequences if its information security were compromised) may prepare short and simple risk management and security plans. These may be quite easy and inexpensive to develop and implement. In contrast, an organization with a high level of information security risk (e.g., it creates, processes, or stores sensitive or classified information and would experience substantial consequences if its information security were compromised) may divide its risk management and security plans into multiple planning documents, some of which may be lengthy and detailed. For example, the risk management plan may be broken up into separate risk assessment, risk response, and risk monitoring planning documents and each section of the security plan may be a standalone planning document.

The ISMS planning document [2] provides detailed guidance on each of the above components of the risk management plan and the security plan that form the framework of the ISMS.

4   How to Implement Security Controls

Security controls cover managerial, operational, and technical actions that are designed to deter, delay, detect, deny, or mitigate malicious attacks and other threats to information systems. The protection of information involves the application of a comprehensive set of security controls that address cyber, physical, and personnel security. It also involves protecting infrastructure resources upon which information security systems rely (e.g., electrical power, telecommunications, and environmental controls). The application of security controls is at the heart of an ISMS. The selection and application of specific security controls should be directed by a facility's information security plans and policies. However, many CBRNe facilities lack the expertise to identify and implement a comprehensive, risk-based, and cost effective set of security controls. UNICRI [3] provides detailed guidance to assist facilities in performing this activity.

The guidance provided in UNICRI [3] for information security controls is presented from a risk management perspective. This document introduces risk-based security controls that are associated with an ISMS. These include the suite of security controls that are needed for the risk management plan and the security plan (as outlined in Sect. 3).

It is often helpful to illustrate the concept of security controls for information systems by using a comparable physical security example. Think of a current-day information security program as being analogous to the security program of a castle in the Middle Ages. Like modern information security programs, castles relied on managerial, operational, and technical security controls to defend themselves from attackers.

Examples of management and operational security controls within the castle included assigning roles and responsibilities for various security activities, hiring and training soldiers, providing assignments and schedules for members of the castle defense force, purchasing armaments, developing and implementing procedures for allowing visitors to enter the castle, conducting security inspections of castle defenses to identify potential vulnerabilities, characterizing the attack capabilities of potential adversaries, inspecting defenses for vulnerabilities, and maintaining communication with neighboring castles to provide each other with timely notification of attacks.

At the Middle Ages castle, technical security controls were designed and implemented. Examples include the design, construction, and operation of a water-filled moat; a drawbridge over the moat to restrict access to the castle gate and walls; high and thick castle walls; and a massive iron gate to block attackers from entering the interior of the castle. These and other technical security controls form the defensive architecture of the castle.

Many security controls for information systems serve analogous functions to those employed for castle defense. For example, management and operational security controls govern the assignment of information security roles and responsibilities; the hiring and training of staff to address information security issues; assigning staff to perform information security monitoring activities; conducting information security vulnerability assessments; procuring information system hardware and software; developing and implementing access control and authentication procedures; characterizing the attack capabilities of potential adversaries; identifying security vulnerabilities; and maintaining communication with applicable government, organization, and CBRNe industry colleagues to provide timely notification of attacks.

Similarly, modern information security programs design and implement technical security controls. Examples include: the design, implementation, and operation of multiple defensive layers within the facility's IT structure; firewalls to stop unauthorized access and communications; "demilitarized zones" or perimeter networks to securely exchange data with other systems; intrusion detection systems to detect unauthorized access; automated monitoring to detect attacks; and the elimination of backdoor entrances into the information system.

UNICRI [3] should be used concurrently with its companion ISMS planning document [2] to identify general facility security strategies and then to choose specific security controls that meet the facility's needs and goals. In UNICRI [3], potential security controls are presented to cover a range of information security program topics. The potential list of security controls is derived from the controls presented in the National Rural Electric Cooperative Association's (NRECA) Guide to Developing a Cyber Security and Risk Mitigation Plan and Cyber Security Plan [11, 12].

Each topic area includes a checklist (in table form) summarizing the various security best practices and controls that a facility should consider for implementation, based on the risk level of the facility and the available resources. After reviewing the various security control options presented in the checklist, including consulting the references provided to gather more information, the facility should select and implement an appropriate set of security controls based on risk levels and resource constraints. These security controls should then be tracked to ensure they are appropriately used and maintained and that associated responsibilities, assignments, deliverables, and deadlines are documented.

Once security controls are in place, they should undergo a thorough, periodic review (e.g., every 1–3 years) as part of the information security program review. As risks evolve and available resources for information security change, the checklists in this document should be reevaluated to determine if more rigorous security controls are required. In some cases, where risks or available resources decrease, a less rigorous set of controls might be appropriate to maintain an adequate level of information security.

5   Workshop on Enhancing Information Security

The Project 19 team conducted the workshop "Enhancing Information Security Management Systems Defending CBRN-related Information, Materials, and Facilities from Evolving Threats" in Zagreb, Croatia on 11–12 April, 2015. The workshop was a featured element in the World Congress on CBRNe Science & Consequence Management. The aim of the Information Security workshop was to enhance the awareness of regional, governmental, and academic institutions on the importance of information security for CBRNe facilities and the steps they could take to enhance information security in their CBRNe facilities.

The workshop was designed as a train-the-trainer workshop. The first day of the two-day workshop focused on the practicalities of information security breaches, including threats, vulnerabilities, and consequences of such breaches. The second day focused on best practices and guidelines for optimal information security. The workshop consisted of a series of lecture presentations and interactive class exercises. Table 2 presents the agenda for the workshop. Workshop materials with supporting notes are accessible at http://unicri.it/news/article/2015-05-18_Project_19_Workshop.

Table 2 Agenda for the two-day CBRN information security workshop

Day 1
1. Welcome, introductions, and objectives of workshop
2. Information security assets
3. Information security threats
    Exercise 1: introduce Plant Alpha, our example facility. Identify its adversaries and their capabilities
4. Consequences of a loss of information security
    Exercise 2: identify potential consequences from an information security breach
5. Present examples of information security compromises
6. Information security vulnerabilities
    Exercise 3: identify potential information security vulnerabilities at Plant Alpha
7. Performance areas for information security programs
    Exercise 4: putting together an information security plan for Plant Alpha
8. Review key sources of information and guidance on information security
9. Best practice recommendations – part I
    Exercise 5: enhancing personnel security at plant alpha
10. Day 1 wrap up

Day 2
11. Day 2 introduction
12. Best practice recommendations – part II
    Exercise 6: implementing attack vectors analysis at plant alpha
13. Best practice recommendations – part III
    Exercise 7: performing security monitoring at plant alpha
14. Quick improvements to upgraded information security
    Exercise 8: selecting quick improvements for plant alpha
15. How to get started?
16. Incident investigation and forensics
17. Sources of information security assistance
18. Closing remarks and feedback

Acknowledgments This chapter was produced in connection with Project 19 of the European Union Chemical Biological Radiological and Nuclear Risk Mitigation Centres of Excellence Initiative. The initiative is implemented in cooperation with the United Nations Interregional Crime and Justice Research Institute and the European Commission Joint Research Center. The initiative is developed with the technical support of relevant international and regional organizations, the European Union Member States, and other stakeholders, through coherent and effective cooperation at the national, regional, and international level. Special thanks go to Odhran McCarthy and the staff at the United Nations Interregional Crime and Justice Research Institute for their support, patience, and technical guidance during this project.

References

1. Schneier B.: Modeling Security Threats. In: Dr. Dobb's Journal (1999). Accessed March 20, 2017 at https://www.schneier.com/paper-attacktrees-ddj-ft.html
2. United Nations Interregional Criminal Justice Research Institute: How to Implement Security Controls for an Information Security Program at CBRN Facilities. UNICRI, Turin, Italy (2015b). Accessed March 20, 2017 at http://www.pnnl.gov/main/publications/external/technical_reports/PNNL-25112.pdf
3. United Nations Interregional Criminal Justice Research Institute: Information Security Management System Planning for CBRN Facilities. UNICRI, Turin, Italy (2015c). Accessed March 20, 2017 at http://www.pnnl.gov/main/publications/external/technical_reports/PNNL-24874.pdf
4. United Nations Interregional Criminal Justice Research Institute: Information Security Best Practices for CBRN Facilities. UNICRI, Turin, Italy (2015a). Accessed March 20, 2017 at http://www.pnnl.gov/main/publications/external/technical_reports/PNNL-25112.pdf
5. International Organization for Standardization: ISO/IEC 27000:2014 Information Technology—Security Techniques—Information Security Management Systems—Overview and Vocabulary (2013)
6. International Organization for Standardization: ISO 27001 Controls and Objectives Annex A (2014)
7. The White House Plan: National Strategy for Information Sharing and Safeguarding (2012). Accessed March 20, 2017 at https://obamawhitehouse.archives.gov/sites/default/files/docs/2012sharingstrategy_1.pdf
8. IEEE Standard for Developing Software Life Cycle Processes in: Institute of Electrical and Electronics Engineers IEEE Standard 1074-1995 (1997). Accessed March 20, 2017 at http://arantxa.ii.uam.es/~sacuna/is1/normas/IEEE_Std_1074_1997.pdf
9. Council on Cyber Security: Critical Security Controls for Effective Cyber Defense, Version 5 (2015). Accessed March 20, 2017 at https://www.cisecurity.org/critical-controls/Library.cfm
10. Australian Signals Directorate: 'Top 4' Strategies to Mitigate Targeted Cyber Intrusions: Mandatory Requirement Explained. Australian Signals Directorate/Defence Signals Directorate, Australian Government, Department of Defence, Intelligence and Security (2013). Accessed March 20, 2017 at http://www.asd.gov.au/publications/Top_4_Strategies_Explained.pdf
11. National Rural Electric Cooperative Association: Guide to Developing a Cyber Security and Risk Mitigation Plan. NRECA/Cooperative Research Network Smart Grid Demonstration Project. Arlington, Virginia (2014a). Available by using the download tool at https://groups.cooperative.com/smartgriddemo/public/CyberSecurity/Pages/default.aspx. Accessed March 20, 2017
12. National Rural Electric Cooperative Association: Cyber Security Plan Template. NRECA/Cooperative Research Network Smart Grid Demonstration Project. Arlington, Virginia (2014b). Available by using the download tool at https://groups.cooperative.com/smartgrid-demo/public/CyberSecurity/Pages/default.aspx. Accessed November 23, 2015.

13. NIST - National Institute of Standards and Technology. 1995. An Introduction to Computer Security: The NIST Handbook. NIST Special Publication 800-12. National Institute of Standards and Technology, Gaithersburg, Maryland. Accessed July 26, 2017 at http://csrc.nist.gov/publications/nistpubs/800-12/handbook.pdf.
14. NIST - National Institute of Standards and Technology. 2013a. Guide to Industrial Control Systems (ICS) Security. NIST Special Publication 800-82, revision 1, National Institute of Standards and Technology, Gaithersburg, Maryland. Accessed July 26, 2017 at http://nvlpubs.nist.gov/nistpubs/SpecialPublications/NIST.SP.800-82r1.pdf.
15. NIST - National Institute of Standards and Technology. 2013b. Security and Privacy Controls for Federal Information Systems and Organizations. NIST Special Publication 800-53, revision 4. Accessed July 26, 2017 http://dx.doi.org/10.6028/NIST.SP.800-53r4.
16. NIST - National Institute of Standards and Technology. 2014. "The Security Content Automation Protocol (SCAP)." Accessed July 26, 2017 at http://scap.nist.gov/.
17. CERT. 2010. "Insider Threat Deep Dive: IT Sabotage." Published September 22, 2010. Accessed July 26, 2017 at http://www.cert.org/blogs/insider-threat/post.cfm?EntryID=57.
18. Council on CyberSecurity. 2014. "The Critical Security Controls for Effective Cyber Defense Version 5.1" Accessed July 26, 2017 at http://www.counciloncybersecurity.org/critical-controls/.
19. ASD – Australian Signals Directorate. 2014. "Top 35 Strategies to Mitigate Targeted Cyber Intrusions." Accessed July 26, 2017 at https://www.checkpoint.com/asd-top-35-mitigation-strategies/.
20. ISO/IEC – International Standards Organization/International Electrotechnical Commission. 2013. Information Technology—Security Techniques—Code of Practice for Information Security Controls. ISO/IEC 27002:2013. Accessed July 26, 2017 at http://www.iso.org/iso/catalogue_detail?csnumber=54533.

Bio-risk Management Culture: Concept, Model, Assessment

Igor Khripunov, Nikita Smidovich, and Danielle Megan Williams

Abstract Biorisk Management Culture (BRMC) is a subset of an organizational culture that emphasizes responsible conduct in life sciences, biosafety, and biosecurity. BRMC is further defined as an assembly of beliefs, attitudes, and patterns of behavior of individuals and organizations that can support, complement or enhance operating procedures, rules, and practices as well as professional standards and ethics designed to prevent the loss, theft, misuse, and diversion of biological agents, related materials, technology or equipment, and the unintentional or intentional exposure to (or release from biocontainment of) biological agents. Effective BRMC could also significantly contribute to preventing proliferation of biological weapons as an integral part of a comprehensive WMD non-proliferation strategy including culture. Given the complexity of biosafety/biosecurity oversight systems, the need for evidence-based decision-making (e.g. on staffing, areas for improvement, choice of training programs), and the ability to detect behavioral changes associated with a particular intervention, it is important to periodically assess the strengths and weaknesses of BRMC. The purpose of this paper is to apply the experience in culture assessment and enhancement accumulated in other domains to biorisk management with due regard for its special features. This methodology is not prescriptive and leaves much latitude to its users. With appropriate modifications, the model can be applicable to a wide range of institutions including biological research and public health laboratories, diagnostic facilities, and bioproduction facilities. The BRMC and its systematic assessment (conducted periodically) are critical to understanding inter alia, the role of the human

I. Khripunov (✉) • D.M. Williams
Center for International Trade and Security, University of Georgia, Athens, GA, USA
e-mail: i.khripunov@cits.uga.edu; dmw04298@uga.edu

N. Smidovich
Saltzman Institute of War & Peace Studies, School of International & Public Affairs, Columbia University, New York, USA
e-mail: nsmidovich@hotmail.com

© Springer International Publishing AG 2017
M. Martellini, A. Malizia (eds.), *Cyber and Chemical, Biological, Radiological, Nuclear, Explosives Challenges*, Terrorism, Security, and Computation,
DOI 10.1007/978-3-319-62108-1_10

factor, the strengths and weaknesses of the bio-risk management framework, causality of system breakdowns or analysis of incidents, sources of human error or breaches of biosafety/biosecurity, and the effectiveness of training.

Keywords Biosecurity • Security culture • Biorisk management culture • Biosafety

1 Introduction

The biomedical enterprise and the supporting laboratory infrastructure are instrumental in developing national capabilities (including medical countermeasures) to prevent and mitigate the risks of infectious diseases and environmental risks, whether naturally occurring, deliberate, or accidental. Work with biological agents and toxins in life sciences research and public health laboratories, diagnostic facilities, and bioproduction facilities is not without risks. Qualified and motivated personnel is critical to countering such risks.

As an example, in 2015, according to the annual report for the US Federal Select Agent Program, there were 233 potential releases of select biological agents reported. Of the 233 potential releases, 199 reports were determined to represent potential occupational exposure to laboratory workers. Of these 199, two reports were of seroconversions involving three laboratory staff with no known incident that were identified by routine annual screening of laboratory staff. Incidents involving the shipment of live pathogens, thought to be inactivated, and inappropriate inventory and accountability of pathogens have recently occurred [1, 2]. While none of these incidents resulted in illness, death, or transmissions to the surrounding environment, they highlight the potential of adverse effects on animal and public health due to inadvertent or deliberate release of biological agents inside or outside biocontainment.

These incidents prompted a comprehensive review of biosafety and biosecurity practices and oversight of federally-funded activities involving (but not limited to) biological select agents and toxins, consistent with the need to realize such activities' public health and security benefits, while minimizing the risks of misuse. Upon review, the US Federal Experts Security Advisory Panel's (FESAP) recommendations and implementation plan [3] seem to focus less on adding new rules and regulations (one notable exception being the Occupational Health and Safety Administration's proposed Infectious Diseases Standard) and more on guidance, incident reporting, training and education [4]. This is consistent with the constant grappling at the policy level with the challenge of striking the right balance between burdensome regulations and preventing unnecessary restrictions on scientific freedom and technological progress. Notably, FESAP's first recommendation addressed the need to strengthen and sustain the culture of biosafety, biosecurity, and the responsible conduct of science and in this context direct the development of semi-quantitative methods to evaluate the human factor by focusing on the efficacy of training, education, codes of conduct, and other interventions to reduce risk. The purpose of this paper is to suggest methods for assessing bio-risk management culture drawing on the practical experience in other domains, particularly nuclear and radiological.

2 Bio-risk Management: Objectives and Scope

Bio-risk management is a process designed to assess and control risks associated with the handling or storage and disposal of biological agents and toxins in laboratories and facilities. The objective of establishing such bio-risk management standards is to: (1) introduce and maintain a bio-risk management system capable to control and minimize risk to acceptable levels for employees, the community and others as well as the environment; (2) guarantee that the standards are in place and implemented consistently; (3) establish guidelines for biosafety and biosecurity training, awareness raising and best practice promotion; (4) accomplish certification and verification of bio-risk management compliance by a third party.

Risk assessment is key to an effective bio-risk management and should be conducted periodically or in the face of changing circumstances. For instance, risk assessments are initiated by management when work practices are changed including the introduction of new biological agents or alternations to work flow or volume. Other factors may trigger risk assessments: incidents or near misses, modifications in standard operating procedures (SOPs), new security risks reported by relevant government agencies, modifications in the national legal and regulatory framework, revised emergency and contingency requirements, and others.

Risk assessment lays the ground work for developing bio-risk management policy which should clearly state the overall bio-risk management priorities and a commitment to continuously improving the performance by personnel. Among such priorities is reducing the risk of unintentional and intentional release of, or exposure to biological agents and toxins; compliance with all legal and regulatory requirements; personal accountability; bio-risk information sharing with all stakeholders, and more.

Managing bio-risks in a laboratory or facility relies on a hierarchy of tools including the elimination of risk (i.e. by inactivation of biomaterials), substituting the hazard for one with less risk (i.e. by using surrogate BA or attenuated strains), risk isolation (via access restrictions for instance), engineering (i.e. biosafety cabinets) or administrative (supervision, training, SOPs) controls, and use of personnel protective equipment (PPE).

As a particular case of bio-risk management, assessing the risks associated with dual use research of concern (DURC) is a provision of the US Policy for Institutional DURC Oversight which requires consideration and judgment of the potential risks associated with conducting the research in question or communicating its results. Such consideration includes: the ways in which knowledge, information, technologies, or products from the research could be misused to harm public health and safety, agriculture, plants, animals, the environment, material, or national security;

the ease with which the knowledge, information, technologies, or products might be misused and the feasibility of such misuse; and the magnitude, nature, and scope of the potential consequences of misuse [5].

People are at the core of a bio-risk management program. Human fallibility (expressed for instance as human errors or intentional misuse of biological materials, equipment, and technology, or unprofessional attitudes toward biosafety and biosecurity) arguably constitute the largest source of biosafety and biosecurity risks in laboratories and other facilities working with biological agents and toxins. However, effective bio-risk management culture can be the critical element of any strategy for preventing and mitigating such risks. Therefore, a prevailing organizational culture in a laboratory or a facility emphasizing such human traits as professionalism and security awareness, personal accountability, compliance, mutual respect and cooperation, vigilance, and reporting are an important trigger of periodically held assessments and effective policy implementation. As emphasized by the World Health Organization, "One of the goals of the bio-risk management approach is to develop a comprehensive laboratory biosafety and biosecurity culture, allowing biosafety and biosecurity to become part of the daily routine of a laboratory, improving the overall level of working conditions and pushing for expected good laboratory management." [6].

3   Human Performance

Human performance refers to the actual behavior and results of people's actions, as opposed to an idealized or abstracted view of what they are supposed to do. The actual results are a complex combination of people's risk perception, motivation, ability to perform the task, leadership expectations, quality of work procedures, site conditions and many others. When things go wrong, it is common for event reports to describe the gap between the expected outcome and what may actually happen. By systematically evaluating the capabilities and limitations of personnel, effective human performance management can identify a prevailing organizational culture and whether it contributes to required task performance in a safe and secure way [7].

The International Atomic Energy Agency (IAEA) published a Nuclear Security Culture Implementing Guide in 2008 [8] emphasizing the role of the "human factor" when implementing programs aimed at enhancing systems of nuclear material protection, control, and accounting. The effectiveness of such programs is believed to be increased when influencing the attitudes, behaviors, and beliefs of personnel, encouraging personnel reliability, and addressing human performance.

Laboratory and biomedical facilities' leadership is responsible for providing a management system (through policies, practice, and adequate resources) that ensures safe and secure handling, storage, and transport of biological materials (a biological risk management system). However, most policy and management efforts to ensure the effectiveness of bio-risk management programs are focusing more on compliance with rules, regulations, and implementation of training and education programs, and less on cultural factors; e.g. on improving people's beliefs, attitudes and motivation in the support of biosafety, biosecurity and behavior of personnel handling biological agents and toxins. Despite a systematic and evidence-based approach developed in other domains, there is a lack of tools to assess the cultural element of individuals and organizations participating in bio-risk management. Generally, this means having a congruent set of behaviors, attitudes, and policies that enable a person or an organization to work in a safe and secure manner with biological agents and toxins and, in addition, having a process by which individuals and organizations respond appropriately and effectively to biological hazards. Bioethics are an important driving force in this process. They are the moral principles or values governing life sciences research and societal implications of certain biological research procedures, technologies, or treatments.

Effective culture does include compliance with rules and regulations as its major characteristic, while complementing and reinforcing the knowledge and skills acquired through biosafety, biosecurity, and bioethics training and work experience. A useful tool may be the ISO 19600 which was developed as a guide for compliance management rather than as a specification that provides requirements. ISO 19600 follows a risk-based approach to compliance management that is aligned with ISO 31000 (the ISO standard for risk management). As emphasized by ISO 19600, compliance is an outcome of an organization meeting its obligations, and is made sustainable by embedding its obligations in the organizational culture and integrating them with the organization's financial risk, quality, environmental, health and safety management process, and its operational requirements and procedures [9].

An organization's approach to compliance is ideally shaped by the leadership when applying core values and generally accepted governance, ethical and community standards. Embedding compliance in employee behavior depends, above all, on the leadership and the clear values of an organization, as well an acknowledgement and implementation of measures to promote compliant behavior. If the leadership of a laboratory or biomedical facility does not institutionalize biosafety and biosecurity as top priorities, then the personnel within those organizations will find it difficult to monitor, adapt, and improve their behavior, as they may not recognize their critical role in the bio-risk management process, or they may not feel motivated to provide critical feedback if leadership fails to act upon that information.

In this regard, the US National Science Advisory Board for Biosecurity (NSABB) stated, "Above all, good management practices are the foundation that underpins the development of a culture of responsibility, integrity, trust, and effective biosecurity. In addition, strong institutional and laboratory leadership, clear articulation of priorities and expectations, and an institutional framework that provides relevant education, training, performance review, and employee support will facilitate responsible practices, personnel reliability, safety, and security, while allowing research on biological select agents and toxins (BSAT) to flourish" [10].

Education and training should promote responsible and safe work practices that allow scientists to discuss, analyze, and resolve in an open atmosphere the potential dilemmas they may face in their research; including the risks and benefits of dual use research of concern (DURC), and the possibility of accidental or misuse of the life sciences. Cultural competence depends on effective risk communication based on a desire and responsibility to protect the health and safety of people and the environment while maintaining the public trust in the biomedical enterprise.

Any tool to assess cultural competence and thus the ability to detect changes associated with a particular intervention (such as training) should be based on defining the BRMC at the organizational level including its characteristics and indicators. A specific list of indicators will allow organizations to effectively identify, monitor, and control the cultural aspects of its bio-risk management activities [11]. In practice, a BRMC assessment and monitoring tool will set performance-based criteria for a bio-risk management system that could be used by an organization to do self- or external assessments against these criteria. The Voluntary Protection Program of the Occupational Health and Safety Administration [4] provides a good example of how such external assessments could take place by considering the required elements of a particular site's safety and health management programs.

4   Bio-risk Management Culture

The BRMC design is based on the organizational-culture model developed by Professor Edgar Schein from the Massachusetts Institute of Technology (MIT). Schein proposes that culture in organizations can exist in layers comprised of underlying assumptions, espoused values and artifacts. Some of the layers are directly observable while others are invisible and have to be deduced from what can be observed in the organization [12].

Cultures are formed by underlying assumptions about reality. In practical terms, this means that an organization will display observable artifacts and behaviors that relate to what it assumes about a variety of phenomena, such as vulnerability to an event. All of these assumptions or beliefs ultimately manifest themselves in observable forms such as policies, procedures, and behaviors.

The next layer of culture in organizations is espoused values- the principles which leadership claims to believe in and requires the organization to display in their actions. Culture predominantly manifests itself through its artifacts which is the third and observable layer. Thus, the protection equipment, people's behaviors, written documents, and work processes are all artifacts of the culture.

Schein's model was successfully used in the 1990s to develop nuclear safety culture following the Chernobyl accident in 1986, which amply demonstrated serious gaps in safety compliance and a disastrous failure of the human factor. Given the many synergies between safety and security as part of overall organizational culture, Schein's model provided a ready-made analytical framework for exploring and modeling nuclear security culture. Schein defined culture as "a pattern of shared basic assumptions that the group learned as it solved its problems of external adaptation and internal integration, that has worked well enough to be considered valid and, therefore, to be taught to new members as the correct way to perceive, think, and feel in relation to those problems" [12].

Applied to security as a subset of organizational culture, its essence is jointly learned relevant values, beliefs, and assumptions that become shared and taken for granted as a facility continues to successfully operate at an acceptable risk and compliance level. To paraphrase Edgar Schein, they became shared, sustainable and taken for granted as the new members of the organization realize that the beliefs, values, and assumptions prevailing among the leaders and the staff led to organizational success and so must be "right" [13].

The BRMC concept integrates both safety and security as well as bioethical and established norms of social responsibility. Reinforcing the norms of responsible conduct in life sciences is critical to: (1) counteracting the diversion of biological materials, equipment, or technologies for harmful purposes, and (2) fostering the long term health security and wellness of the public, animals, plants, the environment, and the economy. BRMC is defined as an assembly of beliefs, attitudes, and patterns of behavior of individuals and organizations that can support, complement or enhance operating procedures, rules, and practices as well as professional standards and ethics designed to prevent the loss, theft, misuse, and diversion of biological agents, related materials, technology or equipment, and the unintentional or intentional exposure to (or release from biocontainment of) biological agents. This definition has been adopted by the FESAP working group tasked with strengthening the culture of biosafety, biosecurity, and responsible conduct in the United States.

The BRMC is a combination of top down, bottom up approaches. Within the BRMC, practices are introduced from the top, while attitudes are pioneered from the bottom to contribute to a cultural build-up process. Also relevant to understanding common mechanisms of BRMC and overall organizational culture are their roles in detecting, interpreting, and managing departures from norms and expectations. What differentiates one organization from another is the extent to which people agree on what is appropriate and how strongly they feel about the appropriateness of the attitude or behavior.

Bioethics, research excellence, biosafety, and biosecurity represent the foundational value of BRMC, consistent with the World Health Organization guidance on responsible life sciences research for global health security [6]. Training in bioethics (defined in this paper as the moral principles or values governing life sciences research and the moral or societal implications of certain biological procedures, technologies, or treatments) provides the foundation for building common values and beliefs in an organization. However, in order to be relevant to BRMC, such training or education should address a broad range of issues including ethical theories, ethical concerns embodied in life sciences' practice, emerging technologies, dual use dilemmas, ethical research, and research integrity. Such training should also include considerations of how bioethical dilemmas are shaped by life science professionals' cultural values and beliefs about the concepts of biosafety, biosecurity, and responsible conduct. Additional topics for discussion may include potential differences between common morality and professional ethics, and between applied ethics and professional ethics, respectively.

Analyzing and providing examples of ethical and decision making frameworks (decision theory, the precautionary principle, rights-based approaches, deontological ethics, principle-based frameworks, etc.) will serve to promote dialogue and raise awareness on approaches for weighing the ethical and societal responsibility regarding issues pertaining to emerging technologies. Also, awareness and commitment to relevant codes of conduct (a formal statement of values and professional practices of a group of individuals with a common focus) for life sciences occupations or academic fields serve to define the expectations and are a persistent reminder of moral and ethical responsibilities of scientists, the equivalent of a Hippocratic oath to "do no harm". The element of professional ethics plays a much more important role in developing a comprehensive bio-risk management culture than in most other domains including nuclear.

Laboratory errors may lead to significant costs in time, personnel efforts, patient outcomes, and may endanger the health security of individuals, communities, animals, plants, and the environment. A vigorous quality management system does not guarantee an error-free organization but it can ensure a prompt detection of errors when they occur, while also contributing to a strong BRMC. Strengthening BRMC implies devoting training resources and consideration to: management systems; leadership and personnel behavior; principles for guiding decisions and behavior; and beliefs and attitudes on biosafety, biosecurity and responsible conduct. Performing assigned duties and responsibilities in a safe, secure, and responsible manner is key and must be emphasized during the culture assessment (which affirms that training is effective and personnel is motivated to competently apply knowledge, skills, and abilities to meet the standards of their professional practice).

As one of several distinct subsets of organizational culture, BRMC is first rooted in a country's national culture. Numerous constituent factors, among them history, traditions, geography, religion, and demography, contribute to national culture making it distinctly different from one country to another. National cultural values are learned early, held deeply, and change slowly over the course of generations. Organizational culture, on the other hand, is comprised of broad guidelines that are rooted in organizational practices learned on the job. More importantly, organizational culture is unlikely to trump national culture [14]. However, organizational culture has more common international traits due to globalized trade, communications, universal standards, and best practice sharing.

A culture emphasizing biosafety, biosecurity, and responsible conduct could also significantly contribute to strengthening barriers against proliferation of biological weapons and prevention of bioterrorism. This is consistent with the international norms embodied by the Biological Weapons Convention (BWC), which bans the possession and use of biological weapon, and UN Security Council Resolution 1540 designed to prevent the proliferation of WMD to non-state actors. At the 8th Review Conference of BWC in 2016, member states noted the value of national implementation measures in accordance with the constitutional process of each state party to implement voluntary management standards on biosafety and biosecurity, and to encourage the promotion of a culture of responsibility amongst relevant national professionals and the voluntary development, adoption and promulgation of codes of conduct [15].

While the examples above emphasize the importance of culture, there are currently no tools for specifically assessing and monitoring for the purpose of taking corrective actions to strengthen the organizational culture of life sciences institutions on the whole (including biological research and public health laboratories, diagnostic facilities, and bioproduction facilities) and the bio-risk management process in particular.

5   Model of Bio-risk Management Culture

Using Edgar Schein's three layers of culture (artifacts, espoused values, underlying assumptions) the BRMC model divides the first layer into three parts, resulting in a total of five elements. They are: (1) intangible beliefs and attitudes as drivers of human behavior (corresponding to what Schein calls "underlying assumptions"); (2) principles for guiding decisions and behavior (what Schein calls "espoused values"); (3) observable leadership behavior (specific patterns of behavior and actions which are designed to foster more effective BRMC); (4) observable management systems (the processes, procedures and programs in the organization which prioritize biorisk management and have an important impact on its functions); and (5) observable personnel behavior (the desired outcomes of the leadership efforts and the operation of the management systems). See Figs. 1 and 2.

Beliefs and attitudes are the foundation of BRMC and are drivers of people's behavior, including individuals who regulate, manage or operate biofacilities or activities in addition to those who could be affected by such activities. These beliefs

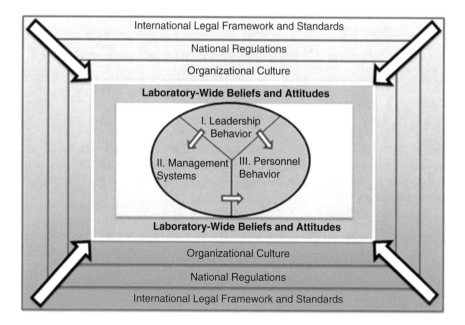

Fig. 1 Model of biorisk management culture

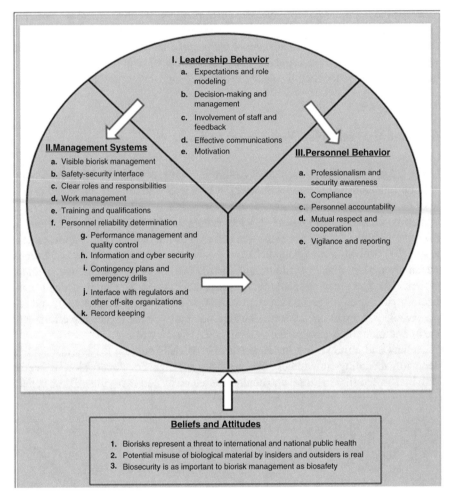

Fig. 2 BRMC characteristics

and attitudes are: (a) bio-risks represent a credible threat to international and national public health; (b) the potential misuse of biological materials by insiders and outsiders is real; and (c) biosecurity is as equally important to bio-risk management as biosafety.

Beliefs and attitudes can be reinforced in multiple ways:

(a) Raise awareness on the risks associated with working in a laboratory with biological materials (e.g., accidental exposure, infection or release; intentional theft and/or misuse; others such as radiological/chemical/physical safety),

(b) Refer to potential ramifications and impact on public health, animal health, or the environment should such risk events were to occur;

(c) Raise awareness and increase understanding of the ethical, legal, and societal issues and consequences concerning life sciences research, development, and associated technologies;

(d) Raise awareness and place emphasis on the importance of quality systems and practices in laboratory biosafety and biosecurity training and research design;

(e) Review codes of ethics and social responsibility guidelines in life sciences research;

(f) Review biosafety, biosecurity, and dual use research of concern regulations, guidelines, policies and procedures, and any other specified training requirements.

The core elements of biorisk management culture are a product of international legal framework (e.g. Biological Weapons Convention- BWC) and standards as well as national regulations, which are developed with due regard for the international practices. BRMC is a sublet of the organizational culture, which can be sustained only by being internalized by the workforce if it is reasonably consistent with the prevailing organizational culture.

Laboratory-wide beliefs and attitudes are primary drivers of managers' and staff members' behavior in support and recognition of the need for biorisk management. These beliefs and attitudes include: (1) The idea that biorisks represent a threat to international and national public health; (2) Potential misuse of biological material by insiders and outsiders is real; and (3) Biosecurity is as important to biorisk management as biosafety.

Figure 2 lists the BRMC characteristics as applicable to leadership behavior, management systems and personnel behavior.

Twenty-one characteristics of Biorisk Management Culture are applicable to the three observable segments in the model. Leadership behavior as the main source of effective BRMC includes actions designed to shape relevant management systems and influence personnel behavior. Management systems cover the processes, procedures, and programs in the laboratory, which prioritize biorisk management and contribute to its efficient implementation. Personnel behavior is the desired outcome of leadership efforts and the operation of the management systems. The meaning of each characteristic is illustrated by cultural indicators associated with these characteristics and are listed in the Appendix.

An effective bio-risk management culture requires a set of principles (Schein's "espoused values") that leaders can instill in the organization to guide policies, decision-making, management systems, and the behavior of people at all levels. Individuals should be familiar with these principles, and they should be applied clearly and consistently across the organization. The main principles include motivation, leadership, commitment, responsibility, professionalism, competence, and learning and improvement. They are all equally important, but education and training are integral to the implementation of other principles. Depending on the profile of the organization and its specific needs, these principles may include a wide variety of training modules comprising initial training, periodic training, ongoing programs, assessments, and quality assurance on training and trainers. Management systems ensure that bio-risk management requirements are integrated holistically with other health, environmental, compliance, and economic requirements. The overall objective is to achieve a risk-based balanced approach.

The BRMC model has 21 characteristics in the management systems and behavior (both leadership and personnel) segments; the proper assembly of which are expected to lead to more effective bio-risk management and culture improvement. The meanings of the characteristics are illustrated by culture indicators assigned to each characteristic (see Appendix), which also serve as a tool for their evaluation. The purpose of the characteristics-indicators package is to stimulate further thought and continuous learning. In other words, culture indicators are benchmarks which help evaluators take a culture measure, identify practical ways to improve it, and follow the dynamics of culture evolution in a longer-term perspective. They constitute a framework under which to facilitate BRMC change and development, promoting desirable and discouraging undesirable culture adjustments.

Leadership helps improve and sustain culture at all levels. Leaders play a vital role in dealing with malicious capabilities, unintentional personnel errors, inadequate organizational procedures, and management failures. They can promote new and different assumptions and patterns of thinking, and establish new patterns of behavior, as well as change the physical environment, the mentality and the guiding principles. Culture therefore tends to mirror the real intention, specific actions and priorities of the management. Given the diversity of laboratory types and other organizations, management includes individuals and groups who direct, control, and appraise the organization. Managers develop individual and institutional values as well as behavioral expectations for the organization to support the implementation of the safety and security management systems and act as role models in the promulgation of these values and expectations. Characteristics of management behavior include explicitly demonstrated expectations, effective decision-making process and management oversight, involvement of staff and feedback, effective communication, and motivation tools.

The management systems integrate characteristics that either relate directly to bio-risk management or are a part of the managerial framework without which safety and security cannot be ensured and maintained. These systems are designed and shaped by senior management, consistent with their vision of an effective culture and the need for appropriate management tools to facilitate and support this process. At the same time, management systems ensure that health, environment, safety, quality, compliance and economic requirements are not considered separately from bio-risk management requirements.

Characteristics of management systems include: visible and effective safety and security policy; the safety-security interface; clear definition of roles and responsibilities; work management; determination of personnel reliability; training and qualification; performance measurement and quality assurance; information security; feedback process; contingency plans and drills; interface with regulators and other off-site organizations; record keeping.

The ultimate objective of bio-risk management culture is to develop a set of desired standards of personnel behavior. Bio-risk awareness and culture are driven by personnel beliefs and attitudes. In the absence of an adequate regime, bio-risks may affect international and national public health and lead to serious social, security, economic, financial, environmental and other consequences. The behavior of bio-risk culture conscious personnel includes several key characteristics designed to improve the bio-risk management regime: professionalism and security awareness; compliance; personal accountability; mutual respect and cooperation; vigilance and reporting.

It may take some time for the biorisk management culture model to be validated, refined and implemented as a tool for human capacity building in support of effective bio-risk management. Though not a panacea, it can enhance and sustain the bio-risk management regime and contribute to its major objectives.

6 Assessment of Bio-risk Management Culture

The purpose of a BRMC assessment is to provide a clear picture the influence of the human factor has on an organization's bio-risk management process. Determining the baseline and charting trends over time can provide the management an early warning to investigate the causes of most problems revealed in order to take corrective actions. A prerequisite for successful assessment is ensuring confidentiality for participants throughout the entire process.

There are at least three assessment options for evaluating BRMC: basic, intermediate, and comprehensive. Their selection depends on many factors and circumstances including risk estimates, the size of the organization and workforce, and the record of previous incidents or near-misses.

Basic Based on statistical methods and information derived from a variety of sources, evaluated against selected criteria, this method would allow management to form a judgment of whether adequate measures are enforced to promote risk awareness and culture. While this audit-type assessment will not provide any insights into the drivers of personnel behavior, it may send a signal about potentially negative trends in the evolution of the bio-risk management regime and the need to take corrective action, including the launch of a more in-depth assessment.

(a) Percentage of employees who have received relevant refresher training during the previous quarter or year.
(b) Percentage of improvement proposals submitted, considered, or implemented during previous quarter or year.
(c) Percentage of communication briefs for employees that included relevant bio-risk management information.
(d) Number of inspections and work station visits conducted by senior managers, managers, or supervisors during the previous quarter or year.
(e) Number of employee suggestions relating to risk management improvements during the previous quarter or year.
(f) Percentage of routine organizational meetings with bio-risks as an agenda item.
(g) Number of received reports on noncompliance and irregular conduct.

(h) Number of actions (restriction, suspension, disqualification) administered under Human Reliability Program (HRP).
(i) Number of direct interactions between certifying officials and staff members.
(j) Frequency and nature of contacts with regulators.
(k) The amount of time to take corrective action.

Intermediate This type of assessment is based on managers' own "yes" or "no" judgment regarding the evolving structure and functionality of the risk component in the organization's management systems. Because they are non-interactive, these security management indexes have limited utility but can pinpoint the functional areas where major deficiencies or gaps are most likely to exist as a result of inadequate human performance. Compared to the basic method, the intermediate approach can stimulate managers' consideration of specific problems and justify a more comprehensive method. Risk management indexes requiring a "yes" or "no" response include:

(a) A risk management mechanism is established and posted.
(b) Processes are in place to identify mandatory requirements related to risk management.
(c) Staff are required to report laboratory accidents, incidents and near-misses.
(d) Significant security items are addressed at regularly held management meetings.
(e) Professional rewards and recognition is associated with the achievement of risk management goals.
(f) Measures exist to protect staff who report unlawful or irregular conduct.
(g) Roles and responsibilities for all professional positions are clearly defined in relevant documents.
(h) Staff are regularly tested to ensure competence in the standard operating procedures (SOP).
(i) Relevant performance results are compared to targets and regularly communicated to staff.
(j) Researchers have a designated person who will provide competent advice if they have safety or security questions or concerns relating to their professional activity.
(k) Feedback from staff is requested and analyzed.
(l) Periodic evaluation of relevant training programs is conducted and revisions are incorporated.
(m) Biosafety and biosecurity training is provided to all those working in laboratories.
(n) Education and/or training is offered on dual use issues.
(o) Education and/or training is offered on research ethics.
(p) Contingency plans are established to address possible unforeseen events.
(q) Processes and protocols exist for handling sensitive information.
(r) Checklists/detailed procedures for maintenance of safety and security systems exist.
(s) Training is provided to appropriate personnel to guide in identifying high-risk behavioral symptoms.

(t) An insider threat mitigation program is in place.

(u) The potential for misuse of the research is considered at all stages and appropriate action taken if necessary.

(v) Appropriate ethical research guidelines and practices have been posted and implemented.

(w) Adequate mechanisms exist for investigating and responding to non-adherence to ethical standards.

(x) Management processes are in place for changes that could affect the relevant function.

(y) Contingency plans are in place.

(z) Management level communication with local and national organizations involved in biorisk prevention and remediation is regularly performed.

(aa) A record of research projects exists and is maintained at institutional level.

(bb) A record of valuable biological materials exists and is maintained at institutional level.

An alternative to this method would be for a management team to review culture indicators in the Appendix and self-reflect on the state of risk management to identify human factor-related gaps. Such a quick look, however, would not preclude a more labor-intensive assessment should it become necessary to check whether the original findings were correct, the measures adopted by the management really worked, and if the organization is on the right track.

For a more complete picture, the results of the BRMC checklist survey and the answers to the questions above may be combined with quantitative data (for the previous quarter or year) concerning numbers of employees trained, biosafety and biosecurity incidents or near misses, cybersecurity breaches, inventory discrepancies, quality management errors, negative inspection results from reports of regulatory agencies, and other failures of compliance with policies and regulations. Data on the number of recommendations and suggestions from employees on bio-risk management improvements or specific organizational programs recognizing employees with an outstanding biosafety and biosecurity record may also be valuable for consideration.

Comprehensive[1] This is a multi-stage process comprising both non-interactive and interactive assessment tools. The assessment focuses on management and behavioral characteristics of the BRMC model. These characteristics are evaluated by comparing what the culture is at the present moment to their optimal parameters specified by culture indicators assigned to each characteristic as benchmarks. Due to the heavy focus on perceptions, views and behavior, sequential comprehensive assessments help to understand and explain the reasons for an organization's patterns of behavior in certain circumstances, devise optimal risk management

---

[1] Based on the assessment methodology suggested in the Draft Self-Assessment of Nuclear Security Culture in Nuclear Facilities and Transport, NST026, 2 July 2016. This technical guidance has been validated so far in Indonesia for research reactors, in Bulgaria for nuclear power plants, and in Malaysia for radioactive sources. Its release by the IAEA is expected before the end of 2017.

arrangements, and predict how the workforce may react to a wide range of risks. Due to the cost and time required for a comprehensive assessment, it may be reasonable to limit them to groups of individuals in the organization who are directly or indirectly associated with BA of concern.

An important initial step is establishing a self-assessment team which has a representative cross-section of relevant expertise tasked with drafting an assessment plan. At this initial stage, it is important to minimize the cost and avoid organizational disruption. Methods included in the plan are typically broken into two categories: (1) non-interactive methods (BRMC surveys, document review, and observations) and (2) interactive methods (individual interviews and focus-group discussions). As all of these methods have their strengths and weaknesses, a reasonable approach would be to combine a non-interactive method with an interactive method (e.g. carrying out a survey followed by a set of onsite interviews to fill possible gaps and clarify ambiguities). Other options are possible, but the choice should be made at the management's discretion. An external evaluators team is preferable for conducting these interviews and leading focus group discussions.

The comprehensive option contrasts sharply with two previous audit-type assessments (basic and intermediate), which accentuate technical and statistical issues more than less tangible human elements. Launching a comprehensive self-assessment requires conscious efforts to think in terms of how individuals and teams perceive bio-risks and, as a result, interact with one another, with the physical surroundings, the laboratory, and with the external environment. The results of a BRMC assessment can shed light on why different bio-risk management related issues emerge, what the root causes of problems may be, and how risk management can be improved.

Surveys are important for self-assessment because they establish a baseline for tracking changes over time. Survey statements are derived from culture indicators, but must be shortened and personalized to facilitate responses. A wide range of personnel at managerial and non-managerial levels are invited to participate in surveys, including scientific and security staff and specialists, bio-risk management advisors, veterinary and animal care staff, occupational health professionals, human resources representatives and other categories. In addition, when appropriate, contractors and suppliers should be invited to take part.

This method provides clear and straightforward data because anonymous respondents can express critical views without fear of adverse consequences. Once a baseline is established, and measures for improvement have been instituted, a follow-up survey can assess progress. It is up to management to determine the scoring scheme for the survey. The present paper suggests a scoring system employing a 7-point scale from 1 ('Strongly Disagree') to 7 ('Strongly Agree'). 'Neither Agree Nor Disagree' indicates that a respondent feels unable to pass judgment on a particular point and is requested to provide a reason in the comment space. This box is particularly important because it provides data subject to wide range of interpretations. This scheme indicates that a particular indicator is either fully observed and present (Strongly Agree – 7), completely unobserved and absent (Strongly Disagree – 1), or somewhere in between.

Strongly Disagree	Disagree	Somewhat Disagree	Neither Agree Nor Disagree	Somewhat Agree	Agree	Strongly Agree
1	2	3	4	5	6	7
Survey Statement						
If you have a comment, please leave it here						

Fig. 3 Survey format

Respondents to a survey are requested to offer comments if they have something else to say.

To calculate the results of the survey for each statement, all scores should be summed up and divided by the number of respondents. For example, if after summing up the total is 135 and the number of respondents is 30, the average score would be 4.5 (Somewhat Agree). Survey results are easier to manage, analyse, and store for future use if the score averaging for each statement is graphically represented as a histogram (Fig. 3).

The next step is to develop subgroups based on convergent or conflicting views amongst respondents. Each subgroup demands special scrutiny regardless of whether they represent predominantly negative, positive or conflicting views. The latter send a message that the workforce is split on an important issue of risk management. As evaluators identify convergent or conflicting views, and review comments from respondents, they should formulate themes that need to be explored further, with the help of qualitative data from other sources (e.g. interviews and focus groups).

Interviews play a significant role in the BRMC assessment because they allow for flexible questioning and follow-up clarifications from interviewees. This eases the task for getting at the deeper tenets of an organization's culture. Interviewees – who need to be carefully selected by experience, work positions, and skills – can give specific examples of past practices that they have observed or heard about and even supply explanations that would provide insight into people's beliefs and attitudes.

Limited to a smaller group of people, face-to-face interviews must include a senior manager designated with operational responsibility for overseeing the system for managing bio-risk. Members of the Bio-risk Management Committee established as an independent review group and/or Safety Committee are also valuable candidates.

Interviewers should consider preparing an informal "interview guide" listing groupings of topics and questions derived from survey results and other sources that can be asked in different ways to different participants. This helps the interviewer focus on the topic at hand while tailoring questions to the assessment goals.

Interviewers should be trained and briefed to ensure that they behave respectfully during the interview, in addition to showing empathy and open-mindedness. A major challenge is to establish trust and provide credible assurances of anonymity.

Efficient, structured, and focused note taking is a vital skill for each interviewer to master before launching the assessment process. Compared to individual face-to-face interviews, focus-group sessions have the advantage that interactions within a group setting often prompt and sustain discussions with minimal input from the interviewer. Group members share their experiences, views, and attitudes about the topic in question, thus eliciting responses from one another in a relatively short time period. The interviewer's role is to facilitate discussion while recording key points that emerge.

Document reviews and observations can take place prior to assessment to familiarize evaluators with past biosafety and biosecurity incidents, their root causes, and corrective measures taken. They can be a useful tool during the process of assessment. A major document for review is the bio-risk management policy, which states the overall objectives and a commitment to improve performance in this area. It may be integrated into the organization's HSE (health, safety, and environment) policies or adopted as a freestanding document. Other documented sources may include risk assessment and hazard identification, roles and responsibilities assignments, allocation of resources, procedures for handling sensitive information, investigation of incidents and accidents, recruitment procedures, training schedules, and internal audits and reviews.

Document review can supply insight into how management sets its priorities and how it intends for its policies, programs, and processes to operate in practice. Combined with surveys and interviews, a document review helps evaluators appraise differences between stated policies and procedures and actual behavior. A document review is, however, a labor-intensive process with administrative implications due to the sensitive nature of some documents.

The purpose of conducting observations is to record actual performance and behavior in real time while under different circumstances, especially at general meetings, training sessions and emergency drills. Observations are a well-established, time-tested, commonplace tool. To provide relevant input into cultural analysis, evaluators need access to records of recent observations. Previously recorded observations are often more reliable than observations conducted in the midst of a well-publicized assessment campaign when staff members are aware of the program and its purpose.

As appropriate, observations may focus on compliance with microbiological techniques such as animal handling, centrifugation, the correct use of vacuum pumps, pipetting, sonication, the use of biological safety cabinets, and others. In addition to routine work activity, important objectives of observation is to determine personnel's behavior during emergency drills, significant alterations to standard operating procedures (SOPs), construction of and modification to laboratories, plant and equipment or its operation as well as in other circumstances which may require both compliance and improvisation. Observational information comes mainly from observational notes. Effective use of observations depends on the ability to adequately develop notes as well as analyze and store them. Following each observation event, data collectors need to expand their notes into rich descriptions of what they have observed, making it possible for others to familiarize themselves with these data.

The analysis step is critical for comparing and integrating the findings of assessment tools. Without conducting an analysis, evaluators are at risk of merely reporting what they have learned and presenting a factual summary. Assessment starts as a fact-based process but must go well beyond the facts. The significant value that evaluators can bring is their interpretation of the findings, their analysis of underlying root causes, and their informed opinion about what problems might exist and how they can best be addressed.

Assessment reports may focus on specific BRMC related issues in the organization such as overconfidence and complacency; lack of a systemic approach toward risks; dysfunctional communication channels; subcultures inside the organization based on different perceptions of the role of biosafety and biosecurity; excessive dependence on technology while underestimating people's input; apathy or ignorance toward potential risks; or indifference to the experience of others. Assessment reports serve as a basis for senior managers to develop and implement corrective action measures to enhance BRMC as a component of overall organizational culture.

7 Self-Assessment of BRMC as a Supplementary Training Tool

In addition to evaluating culture, comprehensive self-assessments can perform a valuable education and training function similar to on-the-job training. Below are examples of why and how self-assessment can help to move organizations along and supplement conventional classroom practices in support of bio-risk management.

- Preparation for the self-assessment process highlights the importance of biorisk management throughout the organization and leadership's commitment. Usually, the head of the organization releases a directive stating the assessment's purpose, outlining the procedure for carrying it out, and explaining how the results will be used.
- Surveys and interviews proactively involve a major portion for the workforce who are no longer in the position of passive observers in a classroom. Those selected for surveys are supposed to grade survey statements (which are based on culture indicators) from 'strongly agree' to 'strongly disagree'. This assignment gives participants a chance for self-reflection and encourages them to understand how the culture indicators relate to their laboratories and facilities.
- Though some culture indicators are mostly generic, BRMC assessment teams at each organization are encouraged to either adjust them to their needs or develop

their own indicators consistent with their profile. As a result, survey respondents can clearly see how they fit into their organization's bio-risk management processes and structures. Moreover, since most indicators apply to the entire organization, it is recommended to personalize them and focus strictly on individual attitudes. Expressions of personal views from respondents not only facilitate the search for cultural root causes, but also reinforce and expand their knowledge.

- Interviews allow for personal contact between interviewer and a respondent, ideally for starting an unconstrained flow of information and providing a chance for respondents to review their own role in the bio-risk management process. An experienced interviewer may be in a position to identify gaps in cultural competencies or policies and procedures and will later recommend adjustments and updates to strengthen the relevant areas.
- The BRMC assessment process culminates in a final document summarizing the findings, setting the foundation for communicating key messages, and providing the baseline for subsequent self-assessment(s). A major purpose of the report submitted to the management for sharing with the organization is to foster a sense of shared responsibility among the staff. This makes each person within the organization a joint custodian of BRMC. The final communication phase acclimates senior management and the entire organization to the role of the human factor in bio-risk management, which, in turn, helps them learn lessons and encourages them to improve their professional skills. It is important to stress that the self-assessment findings should be debated rather than simply published in a report.
- The challenge for practitioners is how to combine BRMC assessments and traditional training in a way to raise the effectiveness of both in a mutually complementary manner. For example, the BRMC assessment can be preceded by a classroom training session to explain and discuss, among other topics, the meaning of culture indicators used for surveys, interviews or focus group discussion, and complement the classroom experience with practical insight.

BRMC and its assessment must be seen in their broad context as both an evaluation tool and as an individual's skills enhancer. Once instituted, this methodology will not only facilitate identification of vulnerabilities, but also help transform the human factor from a problem to be overcome into an asset to strengthen the bio-risk management process and effectively facilitate the mitigation of current and emerging bio-risks.

8 Conclusions

Life sciences research and other biotechnology and biomedical activities are essential to the development of new treatments and therapeutics, strengthening of health research systems, and promotion of public, animal, and environmental health surveillance and response activities. Yet these activities are not without risk to the health security and wellness of people, animals, and the environment, if biological agents are misused or otherwise released from biocontainment, either accidentally or intentionally.

History showed that focusing primarily on compliance with rules and regulations, and implementation of training and education programs without addressing metrics and measures for evaluating the baseline and progress in strengthening and sustaining an organizational culture is doomed to be ineffective in managing biological risks and threats. While a challenging task, a periodic assessment of BRMC will help the organization understand the efficiency/effectiveness of a bio-risk management framework, causality of system breakdowns or analysis of incidents, sources of human error or breaches of biosafety/biosecurity, and the efficiency/effectiveness of training.

A strong BRMC includes willingness to report concerns, response to incidents, and communication of risks. It is also enabled by the identification, collection, analysis, and dissemination of lessons learned and best practices.

High standards of professionalism are also intrinsic to BRMC as applied to the imperatives of preventing the proliferation and use of weapons of mass destruction (WMD). The states parties to the BWC declared in the Final Document of the Eighth Review Conference (7–25 November 2016) "their grave concerns at the threat posed by bioterrorism and the determination that terrorists must be prevented from developing, producing, stockpiling, or otherwise acquiring or retaining, and using under any circumstances biological agents and toxins, equipment, or means of delivery of agents and toxins for non-peaceful purposes" [15]. The concept of bio-risk management culture and its assessment as outlined in this paper can become an effective tool to address this threat on global and national scales.

Appendix: Bio-risk Management Culture Model and Indicators

The objective of this Appendix is to illustrate the characteristics of BRMC using culture indicators as benchmarks for actual characteristic's performance. The Appendix groups indicators around the relevant characteristics of the model. Some of them are generic by nature and should be treated as illustrations that can help each organization tailor a self-assessment project to its individual needs. Of particular importance is the development of additional indicators reflecting the profile of the organization and its activities.

As most characteristics overlap, so do some of their indicators.

Leadership Behavior

Expectations and Role Modeling

Leaders must establish performance expectations for BRMC to guide staff in carrying out their responsibilities as well as act as the role model.

- Culture Indicators:

    1. Senior management develops individual values, institutional values and behavioral expectations regarding bio-risk management and act as role models in the promulgation of these values and expectations.
    2. Managers are seen to comply with policies, procedures and processes.
    3. Management recognizes distinctively challenging bio-risk requirements in use, storage, and transport and consistently addresses them.
    4. Senior management demonstrates a sense of urgency to correct significant risk management weaknesses or vulnerabilities.
    5. Senior management personally inspects performance in the field by conducting walk-around, listening to staff and observing work being conducted, and then taking action to correct deficiencies.
    6. Senior management provides on-going reviews of performance as well as appraises role and responsibilities to reinforce expectations and ensure that key bio-risk management responsibilities are met.
    7. Managers visibly promote bioethical considerations such as transparency of decision-making, public participation, confidence and trust, accountability and vigilance in protecting society.
    8. Managers systematically identify relevant compliance obligations and their implications for the research, products and services.
    9. Facility leadership conveys the importance of bio-risk management and personnel reliability through communicating their own personal commitment to those goals.
    10. Management has a system of rewards for new, innovative ideas for improving bio-risk management.
    11. Senior management demonstrates its commitment by ensuring the availability of resources to establish, implement, maintain and improve the bio-risk management system.

Decision-Making and Management

The process through which an organization makes decisions is an important part of the bio-risk management culture. Adherence to formal and inclusive decision-making processes demonstrates to staff the significance that management places on risk management decisions, and improves the quality of decisions. Management oversight is essential to support bio-risk management related decisions and make them sustainable.

- Culture Indicators:

  1. Senior management develops the goals, strategies, and plans in an integrated manner, so that all personnel understand their collective impact on bio-risk management.
  2. Management ensures that a bio-risk-conscious environment permeates through organization, involving both bio-risk and non-bio-risk management personnel.
  3. Managers are ultimately held responsible for breaches of biosafety and biosecurity.
  4. Management determines the causes of bio-risk management breaches and near misses and takes remedial action to prevent their reoccurrence.
  5. Management takes diverse actions to avoid complacency among personnel and continuously challenges existing conditions to identify discrepancies that might endanger bio-risk management.
  6. Managers re-evaluate the effectiveness of the protective and preventive measures against insider threat particularly when there are changes in the operating processes and conditions.
  7. Regularly held management meetings adequately cover significant bio-risk management items as priority matters.
  8. Management plans, executes, and evaluates periodic bio-risk management exercises.
  9. Management periodically does a review of the overall site layout to ensure that VBMs are stored and used in their most secure location.
  10. Managers of laboratories address the dual use nature of the agents they are holding and take responsibility for deciding on the adoption of appropriate bio-risk management measures.
  11. Bio-risk managers conduct audits (assessments), provide remedial strategies for identified vulnerabilities and gaps, and ensure that the facility's risk assessment is regularly reviewed and updated.
  12. Management establishes the controls and puts in place documented procedures for monitoring the effectiveness of the controls applied to the risk management process.
  13. Senior management ensures that roles, responsibilities and authorities related to bio-risk management are defined, documented and communicated.
  14. Managers and scientists directly responsible for VBM safekeeping assess their relative scientific importance to ascertain whether these materials need to be maintained, protected and accounted for.
  15. Management explains, as appropriate, the necessity and significance of each decision regarding bio-risk management.

Involvement of Staff and Feedback

Management encourages raising bio-risk management concerns without fear of retaliation, intimidation, harassment, or discrimination. The value of feedback and its use must be clearly demonstrated to the entire workforce.

- Culture Indicators:

  1. Staff members are encouraged to make suggestions for improving bio-risk management and are properly recognized for their contributions.
  2. Management encourages staff members and contractors to raise bio-risk management-related concerns without fear of retaliation, intimidation, harassment, or discrimination.
  3. Staff members are involved, as appropriate, in identification, planning, and improvement of bio-risk management-related work and work practices.
  4. Senior management supports and improves mechanisms which staff members can use to contribute their insights and ideas to addressing bio-risk management-related problems.
  5. Laboratory bio-risk management activities include inputs from scientific directors, principle investigators, biosafety officers, laboratory scientific staff, maintenance staff, administrators, information technology staff, law-enforcement agencies and security staff.
  6. Managers play a critical role in establishing the tone of mutual trust and open communication that strengthens morale, communication, and laboratory cohesiveness.

Effective Communications

An important part of an effective bio-risk management culture is to encourage and maintain the flow of information, both upward and downward, within the organization.

- Culture Indicators:

  1. Commitments to compliance are communicated widely in clear and convincing statements supported by action.
  2. Management welcomes input from staff members and takes action, or explains why no action was taken.
  3. Competent individuals are designated to provide advice and guidance on bio-risk management issues (bio-risk management advisor) and are authorized to stop work in the event that it is considered necessary to do so.
  4. Management ensures that relevant bio-risk information is communicated to and from employees and other relevant parties.
  5. Senior managers communicate their vision of bio-risk management often, consistently, and in a variety of ways.
  6. Management adopts appropriate methods of communication to ensure that all messages are heard and understood by all staff on an ongoing basis.
  7. The results of self-assessments are evaluated by the management and its decisions and reasons for them are communicated to staff members.
  8. The systems of communication is regularly tested to check that messages are being both received and understood by the workforce at all levels.
  9. Management is using diverse methods of external communication including websites and email, press releases, advertisements and periodic newsletters,

annual (or other periodic) reports, informal discussion, open days, focus groups, community dialogue, involvement in community events and telephone hotlines.
10. Processes are in place to ensure that the experience of the senior staff is shared with new and junior staff members at the organization.

Motivation

The satisfactory behavior of individuals depends upon motivation and attitudes. Both personal and group motivational systems are important in improving the effectiveness of bio-risk management.

- Culture Indicators:

1. Managers encourage, recognize, and reward commendable attitudes and behaviour leading to bio-risk management improvements.
2. Rewards and sanctions relating to bio-risk management are known to the entire workforce.
3. The principles used to reward good performance in bio-risk management mirror those used to reward good performance in research and operations.
4. When applying disciplinary measures in the event of violations, the sanctions for self-reported violations are tempered to encourage the reporting of future infractions.
5. Senior management takes action to make career paths in bio-risk management career-enhancing.

Management Systems

Visible Bio-risk Management

An organization needs a bio-risk management, which states the commitment of the organization to manage bio-risks. The plan should describe the overall system in place and include measures to address an increased risk level, respond to relevant events, and protect sensitive information. This document should establish the highest expectations for decision-making and conduct and be supported by an atmosphere of professionalism and teamwork.

- Culture Indicators:

1. A bio-risk management plan which describes the overall system in place is implemented and its content is shared with staff on a need-to-know basis.
2. The implementation of bio-risk management plans is regularly reviewed against the evolving risk environment and actions are taken where necessary to address deviations from the plans.

3. The bio-risk management has respected status within the organization as a whole.
4. Procedures are in place to detect human errors, which may jeopardize bio-risk management, as well as to correct or compensate for them.
5. Bio-risk management plan defines how technical and administrative measures are implemented to counter insider threat.
6. Effective control procedures are established for valuable biological materials to track and document the inventory, its use, transfer and destruction.
7. Sound ergonomic principles in deploying equipment and operating procedures are followed in order to minimize the contribution of human errors to bio-risk management incidents.
8. An appropriate waste management policy for biological agents and toxins is established and explained to all staff.
9. The controls for the physical security of cultures, specimens, samples and contaminated materials or waste are implemented and maintained.

Biosafety-Biosecurity Interface

Biosafety and biosecurity culture address different risks (unintentional versus intentional) but they share a common goal of contributing to an effective bio-risk management. There are also challenges in their promotion, management, and coordination related to differences in approach and risk perception. This means that an optimal decision-making process requires an integrated concept that ensures the involvement of experts in each discipline on a continuous basis. Safety and security culture issues should be promoted and evaluated on mutually supporting and reinforcing terms.

- Culture Indicators:

  1. Policy documents recognize that good biosafety practices reinforce and strengthen laboratory biosecurity.
  2. Problems concerning the safety and security are promptly identified and corrected in a manner consistent with their importance and with due regard for their similarities and differences.
  3. Major decisions regarding safety and security are taken with the participation of experts on safety and security on a continuous basis.
  4. Organizational arrangements and communication links are established that result in an appropriate flow of information on the safety and security at various management and staff levels, as well as between them.
  5. While handling dangerous pathogens and toxins, the organization ensures that all emergency response personnel, including law enforcement, are aware of the safety issues on-site and the procedures to be followed if a security incident occurs.
  6. Research with biological agents and toxins (BSAT) involve not only addressing the scientific questions underpinning the research but also ensuring that the research is conducted in a safe manner and in a security environment.

7. Facilities, equipment, and processes are designed and run in both a safe and secure way with respect to bio-risk management.
8. Procedures for the safe and secure transport of cultures, specimens, samples and contaminated materials are established and implemented.

Clear Roles and Responsibilities

Members of all organizations need a clear understanding of "who is responsible for what" in order to achieve the desired results. A significant part of establishing an effective bio-risk management is the clear definition of roles and responsibilities. It is particularly important to review and update this system when organizational changes are being planned and executed.

- Culture Indicators:

1. Procedures are in place to define the roles, responsibilities and authorities of laboratory personnel who handle, use, store, transfer and/or transport VBM.
2. Staff members understand potential bio-risks well enough to accept their roles and responsibilities.
3. Staff members know why they are assigned bio-risk management-related functions, how these functions fit into the broader picture, and what impact their noncompliance may have on the organization.
4. A senior manager is designated with operational responsibility for overseeing roles and responsibilities for management of bio-risk.
5. The bio-risk management documentation describing roles and responsibilities is understandable to those who use it.

Work Management

All work must be suitable planned and managed to ensure that bio-risk management is not compromised.

- Culture Indicators:

1. The approach to risk assessment and management is defined with respect to its scope, nature and timing so that it is proactive rather than reactive.
2. Bio-risk management activities are integrated into the overall policies and administrative procedures of the facility.
3. Processes are in place to identify new and changed laws, regulations, codes and other compliance obligations to ensure ongoing compliance.
4. Measurable objectives for implementing the bio-risk management-related goals, strategies, and plans are established through appropriate processes.
5. Resources are allocated to establishing, developing, implementing, evaluating, maintaining and improving a robust compliance culture through awareness-raising activities and training.

6. Systems and controls are in place to avoid illegitimate or unethical research.
7. Provisions describing personnel management also address procedures and training for visitors, contractors, suppliers, and maintenance staff.
8. Internal transport security includes reasonable documentation, accountability and control over VBM in transit between secured areas of a facility as well as internal delivery associated with shipping and receiving processes.
9. Maintenance programs include the capacity to rapidly repair operational or other vital systems and to rapidly replace parts that have been damaged.
10. Operations and activities that are associated with possible biological risk and where control measures are applied are clearly identified.
11. An accurate and up-to-date biological agents and toxin inventory is established and maintained.
12. Criteria for work that requires prior approval are established.
13. Needs for personal protective equipment (PPE) are identified and suitable equipment is specified, made available, used and maintained appropriately within the facility.
14. Documented procedures to define, record, analyse and learn from accidents and incidents involving biological agents and toxins are established and maintained.

Training and Qualification

An effective BRMC depends upon staff having the necessary knowledge and skills to perform their functions to the desired standards. International standards and domestic regulations must be adequately covered by the training program. A systemic approach to training and qualifications is required for an effective BRMC at the user facility.

- Culture Indicators:

   1. A training program on bio-risk management exists with requirements and qualification standards established, documented, and communicated to the personnel.
   2. Training programs establish an environment in which all staff are mindful of bio-risk management policies and procedures, so that they can aid in detecting and reporting inappropriate behaviour or acts.
   3. Participation in bio-risk management training is given a high priority and is not disrupted by non-urgent activities.
   4. Training ensures that individuals are aware of the relevance and importance of their bio-risk management-related activities and how their activities contribute to the overall bio-risk management of their organization.
   5. Professional and bioethical eligibility and suitability for working with VBM of all personnel who have regular authorized access to sensitive materials is central to effective bio-risk training.

6. Leadership skills and best practices in bio-risk management are included in training programs for managers and supervisors.
7. Systems are in place to ensure that procedures and practices learned in training are applied in practice.
8. Staff are regularly tested to ensure competence in the Standard Operating Procedures (SOP).
9. Refresher training is readily provided if any gaps or need for reinforcement in the individual's skills or knowledge are identified by management, peers, or the individuals.
10. Training provides guidance on the implementation of codes of conduct and help laboratory workers understand ethical issues.
11. Training includes the development of communication skills among staff and partners as well as improvement of mutual collaboration.

Personnel Reliability Determination

Any barrier or procedure can be defeated with insider action. Therefore, effective processes for the determination of reliability and for mitigation of insider threat must be in place at sites, particularly those open to outside visitors and the public. The formal process should serve to assist in reducing the risk of authorized personnel engaging in illegal activities. Relevant elements of the security are important for the reliability.

- Culture Indicators:

    1. Measures are taken to determine the reliability of individuals involved in the use, storage, and management of VBM.
    2. Reliability measures are based on a graded approach and range from confirmation of identity to a comprehensive background check by the legitimate national authority, including a verification of references, as required by states' national practice.
    3. Appropriate background checks and psychological examinations of staff members are regularly performed by certified or reputable institutions or individuals.
    4. Persons whose reliability has not been certified are escorted by, or kept under continuous surveillance of a person who is authorized and qualified to perform such escort services.
    5. Staff members are aware and understand the importance of reliability determination.
    6. Training is provided to management and other appropriate personnel to guide them in identifying apparent high-risk behavioral symptoms and in applying other observational and analytical skills.
    7. Measures and procedures are in place to ensue reliability of personnel is regularly validated.

Performance Measurement and Quality Control

Quantified measures of performance, with associated goals, are essential in establishing management expectations and in involving staff to achieve desired results. Standard quality management practices should be applied to bio-risk management functions. Documented evidence of the benefits of quality management initiatives can convince personnel that quality service helps gain trust and support for the organization and the people in it.

- Culture Indicators:
    1. Compliance responsibilities are included into job descriptions and employee performance measurement processes.
    2. Quality assurance programs ensure that bio-risk management systems designed to a performance-based approach have adequate supporting documentation for effectiveness.
    3. Action is taken when bio-risk management performance does not fully match the goal.
    4. Performance results compared to targets are regularly communicated to staff.
    5. Potential for misuse of the research is considered and monitored at all stages and appropriate action is taken when necessary.
    6. Performance evaluations provide a regular and consistent venue to communicate expectations regarding security and safety and to convey an institutional commitment to bio-risk management and personnel reliability.
    7. Mechanisms exist for investigating and responding to non-adherence to ethical standards.

Information and Cyber Security

Controlling access to sensitive information is a vital part of the risk management function. Accordingly, the organization must implement control measures for the protection of sensitive information and prevention of cyber-attacks.

- Cultural Indicators:
    1. An information and cyber security function is established, funded, staffed, and visible.
    2. A policy and procedure is in place to identify sensitive information.
    3. Classification and control requirements are clearly documented and well understood by staff.
    4. Information and computer security are a component in the overall security plan which, among others, addresses the insider threat posed by adversaries with the ability to conduct cyber-attacks.
    5. Access to information assets is restricted.
    6. Protection of information is consistent with the level of risk it poses in terms of potentially compromising a VBM.

7. Regulations are in place that govern the markings and handling of information and how that information is gathered, maintained, distributed, documented, accessed, shared and stored within the facility and with counterparts.
8. Comprehensive bioethical reviews are carried out and documented before final decisions are reached on the publication of data.

Contingency Plans and Emergency Drills

The bio-risk management system must be in a continuous state of readiness to handle events at any time. An important element of the system is the set of contingency plans used to respond to attempted or successful malicious acts or address the safety and security breaches. Appropriate and realistic drills and exercises must be conducted periodically.

- Cultural Indicators:

    1. All contaminated or potentially contaminated waste items are identified and documented, and procedures are put in place to devise effective decontamination and other appropriate treatments.
    2. Plans and procedures to identify the potential for incidents and emergency situations involving biological agents, toxins and materials are established and maintained.
    3. Structured and realistic emergency exercises and simulations, including security drills are conducted at regular intervals.
    4. Emergency plans are effectively communicated to all employees and relevant third parties, and tested with the intention that everyone is aware of their obligations.
    5. Contingency plans are tested periodically through drills and other means to ensure that they are effective and current.
    6. Bio-risk management plans ensure that the laboratory personnel and external partners (police, fire brigade, and medical emergency personnel) participate in laboratory bio-risk management drills and exercises.
    7. Staff members are trained to effectively deal with novel and unexpected situations for which no procedures have been devised and when no management supervisor is available.
    8. Provisions are in place to ensure that bio-risk management readiness can be temporarily tightened during times of increased threat.
    9. Emergency scenarios take into consideration significant vulnerabilities to insider threat as a result of facility evacuation.
    10. Mechanisms are established that allow entry by emergency respondents but ensure uninterrupted and constant laboratory biosecurity, control, accountability and traceability of VBM.
    11. Vaccines, other preventative measures and treatments are available to minimize the consequences of natural or intentional releases of biological materials.

Interface with Regulators and Other Off-Site Organizations

Effective bio-risk management often involves several regulatory and law enforcement bodies. A constructive working relationship with each regulatory or law enforcement body is therefore important to ensure that information is exchanged. Bio-risk management matters involve not only the relationship between regulatory authority and regulated organizations but also policy making and other legal bodies.

- Culture Indicators:

    1. Procedures are established with local law enforcement regarding intelligence information and use of appropriately reliable and secure communications as well as reactions to an increased risk.
    2. Information on abnormal conditions and events significant to bio-risk management is made available to the regulatory authority and other relevant bodies including, where appropriate, other users.
    3. Involvement, roles and responsibilities of public health and other authorities in the event of a safety or security breach are clearly defined.
    4. Procedures are in place to reassure the public that the bio-risks inherent to laboratory work are controlled with appropriate safety and security tools to meet their expectations.
    5. External peer advice and review are valued as a source which can assist to establish, implement, systemize, integrate, and improve the bio-risk management regime.
    6. Staff members fully understand the regulatory body's role in the context of bio-risk management.

Record Keeping

Efficient record keeping and protection of sensitive information are vital to effective biosafety and biosecurity as well as accurate audits and inspections.

- Culture Indicators:

    1. Transfers of biological agents and toxins between laboratories at the facility or into and out the facility are recorded and controlled in line with the level of risk associated.
    2. Records relating the inventory of biological agents and toxins are current, complete and stored securely with adequate backup provisions.
    3. Inventories are regularly updated at storage sites.
    4. A record of research projects exist and are maintained at the institutional level.
    5. A record is kept of all persons who have access to, or monitor the use of keys granting access to storage containing sensitive material.
    6. Records, documents and data are established, controlled, and maintained to provide evidence of conformity to the requirements and they remain legible, readily identifiable and retrievable.

Personnel Behavior

Professionalism and Security Awareness

All organizations using VBM need their personnel to adhere to high standards of professionalism. Awareness is a key driving force for staff members to stay committed to robust biosafety and biosecurity. It is vital to reinforce this awareness and combat complacency.

- Culture Indicators:

    1. Bio-risk awareness programs are developed in a coordinated manner with biosafety and biosecurity awareness programs.
    2. Personnel working with BSAT demonstrate the ability to work well as a team, follow instructions, and adhere to standard operating procedures (SOP).
    3. Bio-risk awareness programs enable entire personnel to address the risks of blackmail, coercion, extortion, or other threats to staff members and their families.
    4. Staff members are prepared to face the unknown and improvise should it become necessary.
    5. Staff members notify their co-workers and managers when these co-workers are doing something that may downgrade bio-risk management.

Compliance

Regulations and procedures represent accumulated knowledge and experience. It is important that they are followed to avoid repeating errors which have already been identified and corrected. It is also important that procedures are clear, up to date, readily available, and user friendly so that personnel do not resort to departing from the approved process.

- Culture Indicators:

    1. Staff members are aware that commitment to compliance is maintained and that noncompliance and noncompliant behavior will be dealt with accordingly.
    2. Scientists adhere to codes of conduct as a model of their professional behavior and guidance to address and prevent the unethical use of biological research.
    3. The organization's instructions on bio-risk management are easy to follow because they are readily available, clear, and user friendly.
    4. Staff members who discover discrepancies in the implementation of bio-risk management procedures promptly report them to their supervisors.
    5. Staff members show trust in and acceptance of bio-risk management procedures.

6. Staff members avoid shortcuts in implementing biosafety and biosecurity procedures.
7. There is a well-developed practice of reminding staff members about the importance of following procedures.

Personal Accountability

Accountable behavior means that all workers know their specific assigned tasks related to bio-risk management (i.e. what they have to accomplish by when and what results should be achieved) and that they either execute these tasks as expected or report their inability to do so to their supervisor.

- Culture Indicators:

    1. Staff members consider themselves accountable for maintaining an adequate level of bio-risk management.
    2. Staff members believe that their personal accountability is clearly defined in appropriate policies and procedures.
    3. Staff members understand how their specific tasks support biosafety and biosecurity in their organization.
    4. Commitments are achieved or staff members give prior notification of their non-attainment to management.
    5. Evidence can be cited that staff members solicit advice or seek more information when they have doubt about implementing their safety or security related tasks.
    6. Goal-setting-principles-based approaches enable staff to deal with unpredicted and unfamiliar events in the most prudent and safest manner until an expert opinion can be obtained.
    7. Procedures and processes ensure clear single-point accountability before execution.

Mutual Respect and Cooperation

Mutual respect and teamwork is essential in biorisk management. An effective BRMC can best be found in an organization where there is extensive interpersonal interaction and where relationships between various groups are generally positive and professional.

- Culture Indicators:

    1. Teams are recognized and rewarded for their contribution to biosafety and biosecurity.
    2. Teamwork and cooperation are encouraged at all levels and across organizational and bureaucratic boundaries.

3. Researchers, laboratory workers as well as biosafety and biosecurity managers communicate and collaborate in their efforts to find the ethical balance for the activities performed.
4. Problems are solved by multilevel and multidisciplinary teams.
5. Team members support one another through awareness of each other's actions and by supplying constructive feedback when necessary.
6. Professional groups appreciate each other's competence and roles when interacting on biorisk management issues.
7. Training provides opportunities for discussions among all staff members while also strengthening their team spirit.
8. Team members are periodically reassigned to improve communications between teams.
9. Cross-training among different professional areas and groups is conducted to facilitate team work and cooperation.

Vigilance and Reporting

Biorisk management depends on the attentiveness and observational skills of staff. Prompt identification of potential vulnerabilities and reporting to superiors permit proactive corrective action. An appropriate questioning attitude is encouraged throughout the organization.

- Culture Indicators:

  1. Staff members notice and question unusual indications and occurrences and report them to management, as soon as possible, using the established process.
  2. Staff are aware of procedures and measures designed to protect those who report unlawful and irregular conduct.
  3. Staff members seek guidance when unsure of the safety and security significance of unusual events, observations, or occurrences.
  4. Staff members are trained in observational skills to identify irregularities in the implementation of bio-risk management.
  5. Staff members are aware of potential insider threat and its consequences.
  6. Staff members avoid complacency and can recognize its manifestations.
  7. Staff members accept and understand the requirement for a watchful and alert attitude at all times.
  8. A policy prohibiting harassment and retaliation for raising safety and security concerns is enforced.

Acknowledgements The co-authors would like to acknowledge a valuable contribution by Dana Perkins, PhD. Co-Chair, Federal Experts Security Advisory Panel Working Group on Strengthening the Culture of Biosafety, Biosecurity, and Responsible Conduct in the Life Sciences for the U.S. Department of Health and Human Services.

References

1. United States Government Accountability Office: High Containment Laboratories: Improved Oversight of Dangerous Pathogens Needed to Mitigate Risk, (GAO-16642) (2016).
2. United States Government Accountability Office: High Containment Laboratories: Comprehensive and Up-to-Date Policies and Stronger Oversight Mechanisms Needed to Improve Safety, (GAO-16-305) (2016).
3. Public Health Emergency (PHE): Implementation of Recommendations of the Federal Experts Security Advisory Panel (FESAP) and the Fast Track Action Committee on Select Agent Regulations (FTAC-SAR) (2015).
4. Occupational Safety and Health Administration (OSHA): Infectious Diseases Rulemaking. Retrieved from United States Department of Labor website: https://www.osha.gov/dsg/id/ (2016).
5. NIH Tools for the Identification, Assessment, Management, and Responsible Communication of Dual Use Research of Concern: A Companion Guide to the United States Government Policies for Oversight of Life Sciences Dual Use Research of Concern (2014).
6. World Health Organization (WHO): Biorisk Management: Laboratory Biosecurity Guidance, 12 (2006)
7. Canadian Nuclear Safety Commission: Human Performance Discussion Paper. DIS-16-05, 2–3 (2016)
8. IAEA: Nuclear Security Culture Implementing Guide (2008).
9. International Organization for Standardization: Compliance Management Systems-Guidelines. ISO 19600 (2014).
10. NSABB: Guidance for Enhancing Personnel Reliability and Strengthening the Culture of Responsibility (2011).
11. CWA 15793 CEN Workshop Agreement Laboratory Biorisk Management Standard, https://www.absa.org/pdf/CWA15793_Feb2008.pdf (2008).
12. Schein, E.: The Corporate Culture: Survival Guide. San Francisco, CA: Jossey-Bass, 16–20 (1999).
13. Schein, E. (3rd ed.): The Corporate Culture and Leadership. San Francisco, CA: Jossey-Bass, 17 (2004).
14. Hofstede, G., Hofstede, G.J., Minkov, M.: Cultural and Organizations: Software of the Mind. New York: McGraw-Hill, USA, 12–20 (2010).
15. BWC 8th Review Conference Final Document, http://www.unog.ch/80256EDD 006B8954/(httpAssets)/F277FA6A2B96BA98C125807A005B2F59/$file/2016-1129+Final+report+adv+vers.pdf (2016).

Preventing Nuclear Terrorism: Addressing Policy Gaps and Challenges

Kenneth Luongo

Abstract Since the collapse of the Soviet Union, there has been an intensified global effort to keep nuclear materials and weapons out of the hands of terrorists and non-state actors. This began with the creation and subsequent significant expansion of the Nunn-Lugar Cooperative Threat Reduction Program, further intensifying after the terrorist attacks of September 11, 2001. President Obama provided a special focus on this issue through the creation of the Nuclear Security Summit (NSS) process, resulting in four heads-of-state meetings. However, the NSS process did not address many of the difficult questions that could significantly strengthen nuclear security and constructed a weak bridge to future improvements. With the conclusion of the summits, the political momentum created by them is rapidly decreasing and the issue is settling back into the bureaucratic channels. The challenge now is how best to address the nuclear policy gaps and new threat challenges in a post-summit environment.

Keywords Nuclear security • Nuclear security governance • Nuclear and radiological terrorism • Nuclear security summits • Cooperative threat reduction efforts • HEU and plutonium minimization and removal • CPPNM • INFCIRC/869 • 5 Priorities

1 Introduction

Since the collapse of the Soviet Union, there has been an intensified global effort to keep nuclear weapons and materials out of the hands of terrorists and non-state actors. The possibility of a nuclear terrorist attack has been called a black swan occurrence, an unlikely but possible event that is the national security nightmare that keeps leaders around the world up at night.

There are three primary nuclear terrorism threats. The one considered most likely is the use of a radiological device that would spew radiation but not create a nuclear

K. Luongo (✉)
Partnership for Global Security, Washington, DC, USA
e-mail: kluongo@partnershipforglobalsecurity.org

© Springer International Publishing AG 2017
M. Martellini, A. Malizia (eds.), *Cyber and Chemical, Biological, Radiological,
Nuclear, Explosives Challenges*, Terrorism, Security, and Computation,
DOI 10.1007/978-3-319-62108-1_11

explosion. It is considered a higher risk because there are thousands of highly radio-active sources used in commercial and medical applications around the world. They regularly go missing. The other two threats come from the use of nuclear weapons materials in an improvised nuclear device or a illicitly procured nuclear weapon. Highly-enriched uranium (HEU) is of particular concern because it is used outside of military programs for civilian purposes, such as reactor fuel. A crude HEU gun-type device is considered to be the easiest nuclear weapon for terrorists to make [1, p. 16]. It would take an estimated 50–60 kg of HEU to make such a device [2]. The weapon would be large and heavy, but terrorists would need only basic infrastruc-ture support, such as a machining capability, and no advanced knowledge, to create it. A plutonium device would be much harder to develop without a more sophisti-cated technology.

A global effort to address the potential for nuclear terrorism began with the cre-ation and subsequent significant expansion of the Nunn-Lugar Cooperative Threat Reduction (CTR) program and further intensified after the terrorist attacks of September 11, 2001. These attacks left little doubt that if unscrupulous terrorist organizations obtain weapons of mass destruction (WMD) in the future, there will be little barrier to their use. A detonation of a nuclear weapon or the dispersal of radiological materials by terrorists would not only create deaths, casualties and psy-chological horror but significantly disrupt the geopolitical and global economic sys-tem. Currently many nations possess nuclear materials and eight countries possess nuclear weapons, so leakages could come from many locations. Amongst these, Russia possesses the largest stockpile of nuclear weapon material and weapons, and the security of its stockpiles is inadequate.

According to the International Atomic Energy Agency (IAEA)'s Incident and Trafficking Database, as of December 2015, there had been a total of 2889 con-firmed incidents reported. Amongst these, 454 involved unauthorized possession and related criminal activities, 762 involved reported theft or loss, and 1622 involved other unauthorized activities. Of these incidents, 188 occurred in 26 different coun-tries, and the majority were reported in North America. The United States had the highest number of reported cases with 59.4%, followed by France, Canada, Ukraine, and Russia. Between 1993 and 2015, incidents included HEU and plutonium. The number of reported incidents reported reached a peak in the early 1990s [3, 4].

President Obama provided a special focus on this issue through the creation of the Nuclear Security Summit (NSS) process that resulted in four heads-of-state meetings on the issue. However, the NSS process did not address many of the dif-ficult questions that could significantly strengthen nuclear security, and constructed a weak bridge to future improvements. Its primary achievements were accelerating activities that had already been planned and adding a few useful procedural and substantive additions to the agenda. Now that the summit process has ended the political momentum created by them is rapidly decreasing and the issue is settling back into the bureaucratic channels from which it was lifted beginning in 2010. The challenge now is how best to address the nuclear policy gaps and new threat chal-lenges in a post-summit environment.

1.1 Defining the Scope of Nuclear Security

To identify a useful end state set of goals for the international nuclear security system, it is important to clarify the definition of "nuclear security." The IAEA has been assisting countries with their nuclear security since the 1970s and is widely considered to be the foremost international authority on nuclear issues in many countries. At present, the IAEA defines nuclear security as "*the prevention and detection of, and response to, theft, sabotage, unauthorized access, illegal transfer, or other malicious acts involving nuclear material, other radioactive substances or their associated facilities*" [5]. This definition was adopted in late 2003, however, the definition of nuclear security has evolved in several ways since then, particularly through the NSS process.

2 Nuclear Security Panorama After the Collapse of the USSR

The collapse of the Soviet Union in 1991 created an unstable environment for nuclear and radioactive materials in Russia and the former Soviet states. The United States understood the importance of securing nuclear weapons and nuclear materials from falling into hostile hands. The escalating economic crisis and lax security in Russia worsened the situation. With a Cold-War sized nuclear complex, nuclear stockpiles and weapons-usable nuclear materials were stored at over 100 buildings located in over 50 different sites throughout Russia and Former Soviet Union (FSU). Furthermore, the post-Soviet economic crisis caused nuclear scientists, engineers, technicians, guards, and other personnel to face dire financial circumstances. It was feared that illicit trafficking and smuggling of nuclear and radioactive material could spike and that knowledgeable scientists could exchange their services to other governments or non-state actors.

These concerns were based on facts. Between 1992 and 1995, 15 kg of illicit nuclear material was intercepted. For example, in October 1994, 1.5 kg of HEU was seized in Podolsk, Russia, from Yuri Smirnov, a chemical engineer and employee of the State Research Institute of the Luch Scientific Production Association, who was suspected of stealing equipment from the Luch facility. Deteriorating economic conditions were a factor that influenced Mr. Smirnov to smuggle HEU in small quantities – 25–30 g – between May and October of 1992, without being noticed [6]. Furthermore, in 1993, Lithuanian authorities seized 4.4 t of Beryllium, a radioactive material that originated at the Institute of Physics and Power Engineering in Russia, 141 kg of which was contaminated with HEU [7].

Faced with this security situation the United States entered into numerous cooperative non-proliferation efforts. These activities were initiated under the CTR Program and the Department of Energy's (DOE) laboratory-to-laboratory efforts in the early-to-mid 1990s, and aimed at increasing U.S.-Russian partnership and decreasing the threats posed by the collapse.

2.1 Nuclear Security Bilateral Cooperative Efforts

The Nunn-Lugar Cooperative Threat Reduction policy began as a congressional initiative. Senators Sam Nunn and Richard G. Lugar built a bipartisan plan authorizing the use of the Defense Department funds to assist the FSU through what became known as the Cooperative Threat Reduction Program. In their Washington Post article, both Senators provided detailed and clear pictures of the economic and security situation in the FSU to the American citizens. They also highlighted the principal role the U.S. would be taking in the dismantlement of a massive military industrial complex, hoping to receive support from Congress and the American citizens [8]. Soon after, their amendment, the "Soviet Nuclear Threat Reduction Act of 1991" or Nunn-Lugar Act was enacted.

CTR was designed to address the potential leakage of WMD from the collapse of the Soviet Union. CTR and related programs focus on protecting and eliminating nuclear, chemical, and biological stockpiles; securing nuclear weapons-usable materials; and eliminating delivery systems.

Some of the program's first important successes came in 1992, when Ukraine, Belarus, and Kazakhstan agreed to return the nuclear weapons once inherited to Russia, and accede to the Non-Proliferation Treaty (NPT) as non-nuclear-weapon states. The same year, the U.S. helped Russia establish several science centers designed to provide alternative employment for scientists and technicians who had lost their jobs, and in some cases, had become economically desperate, as weapons work in Russia was significantly reduced. Since then, the Material Protection, Control & Accountability (MPC&A) Program evolved and expanded.

In 1993, the United States signed bilateral government-to-government MPC&A agreements with Russia, Belarus, Kazakhstan, Lithuania, Latvia, Ukraine, and Uzbekistan. The goal for these agreements was to strengthen MPC&A systems in FSU states and neighboring states, by providing them with the "*capability to deter, detect, delay, and respond to possible adversarial acts or other unauthorized use of nuclear material and, if necessary, aid in recovering nuclear material*" [9, p. 22]. That same year Russia agreed to blend down its HEU to 4–5% enrichment, which then would be purchased by the U.S. and used as a power reactor fuel.

The CTR Program also created several spinoff cooperative non-proliferation efforts. One set was focused on scientist redirection and included the creation of the International Science and Technology Center (ISTC), which focused on supporting weapon scientists by providing funds for them to pursue research projects; and, the Industrial Partnering Program (known as the Initiatives for Proliferation Prevention Program) (IPP), which aimed at creating U.S.-Russian business partnerships revolving around Russian-designed technologies.

Another was the Material Protection, Cooperation, and Accountability program that was designed to address what many considered to be the most dangerous proliferation danger in Russia, the loss or theft of HEU and plutonium from the vast Russian stockpile. It provided for continued installation of security, control, and accountancy upgrades and equipment to safeguard weapons-usable nuclear materials that were stockpiled in former Soviet states. It also financed the consolidation of nuclear materials into fewer sites and buildings. In 1994, MPC&A cooperation intensified as U.S. and Russian laboratories started to work directly with each other to improve the security of weapons-usable nuclear materials.

In 1996, the Defense Department relinquished administrative and funding responsibilities of these Nunn-Lugar programs to DOE and the Department of State. DOE became in charge of the MPC&A, IPP and related programs (including the soon to be created Nuclear Cities Initiative), which later were administered by the newly formed National Nuclear Security Agency (NNSA). The State Department was given responsibility for ISTC and related efforts in Ukraine and other nations.

From 1994 to 1998 major progress was made in improving the security of fissile material in Russia. Given the dangers presented by the Russian nuclear complex, the objectives of U.S.-Russian cooperation were to prevent proliferation by theft and diversion of materials, technologies, and scientists; to irreversibly eliminate excess fissile materials and warheads; and to downsize the complex in a rational manner.

Furthermore, in 1999, the Clinton administration unveiled the Expanded Threat Reduction Initiative, which requested expanded funding and extension of the life spans of many of the existing cooperative security programs. The U.S. and Russia joined to extend the CTR agreement, and in 2000, both countries signed a plutonium disposition agreement providing for the elimination of 34 t of excess weapons-grade plutonium by each country.

By the end of the decade, cooperative programs created equally important but less tangible benefits, including a better appreciation in Russia of the importance of nonproliferation; the development of deeper levels of trust between U.S. and Russian officials, military officers, and scientists; and the creation of important new political linkages and relationships not thought possible during the Cold War. These intangible benefits are hard to quantify in official reports, but they are a unique result of this work.

However, despite the progress made, members of Congress started to question the value and effectiveness of these cooperative programs as they wore on, and the Russian desire to continue collaborating with the U.S. was also eroding. For instance, over the years, tension continued over how much of the cooperative security budget was spent in Russia versus in the United States. Other political issues also emerged, such as the tendency of some U.S. officials treating collaboration with Russia as a client-donor relationship, with Russia acting as a subcontractor to the United States rather than as a partner. The cooperative agenda itself also continued to remain very sensitive, transcending its actual programmatic components and encompassing intangible issues such as recognizing the Russian need to maintain a feeling of national pride while participating in threat reduction efforts, and the need to sustain mutual respect and trust among both countries' participants. Without Russian participation, the cooperative nuclear security agenda would continue to wither in the subsequent years.

2.2 Nuclear Security Efforts After 9/11

U.S. Efforts After the Pentagon and World Trade Center attacks on September 11, 2001, numerous leaders and institutions underscored the need to prevent weapons of mass destruction from falling into the wrong hands. At the Slovenian Summit, both Presidents, Bush and Putin, affirmed this goal as their "highest priority."

In 2001, Congress increased the funds for critical threat reduction activities substantially above the requested amounts, including in the post-9/11 Supplemental Appropriations Act. Congress included in a $40 billion emergency appropriation an additional $120 million for nuclear material control, $15 million for alternative employment for weapons scientists, and $10 million for improving the safety and security of Soviet-era nuclear power reactors and facilities. The initiative came solely from the Congress. When the administration notified Congress of its priorities for this funding, it did not designate any of the funds for WMD security activities in Russia or the FSU, despite the anthrax attacks against Congress and the emerging danger of potentially nuclear-armed terrorists.

After a decade of U.S.-Russian threat reduction cooperation, much had been done to reduce the vulnerability of these stockpiles, but much needed to be accomplished. The National Intelligence Council Annual Report to Congress on the Safety and Security of Russian Nuclear Facilities and Military Forces of 2002 emphasized that even though Russia's nuclear security improved slowly due to U.S. efforts, risks continued to remain. These risks were easily quantifiable – roughly two-thirds of Russia's nuclear material continued to remain inadequately secure, and only a small percentage of its oversized weapons infrastructure was eliminated.

Political resources as well as financial ones were required to secure nuclear and radiological material more quickly and effectively. Suggestions were made to use the available funding to expand the MPC&A Program, increase security upgrades, improve Russian and FSU's borders and export controls, continue relevant programs, continue to prevent further proliferation of nuclear knowledge, facilitate fissile material disposition and elimination, and promote warhead and fissile material stockpile monitoring and transparency [10].

While the world hoped for a new spirit of cooperation to develop between the U.S. and Russia in the fight against terrorism, progress on the threat reduction agenda, at the time, lagged in key areas and work was at a virtual standstill. Cooperation under the CTR Program was suspended the Spring and Summer of 2002, over a dispute concerning Russia's chemical and biological weapons declarations.

In 2004, NNSA created the Global Threat Reduction Initiative and the International Nuclear Materials Protection and Control program that work with other countries outside the FSU to assist with material security. These programs were in essence the repackaging of already existing efforts. However, in a 2009 National Academy of Sciences report suggested that the overall effort needed to be updated from "CTR 1.0" to "CTR 2.0," and the programs needed to evolve to be more agile, flexible, and globally responsive while retaining their cooperative, results-focused core [11]. By concentrating on joint problem solving and cooperative approaches to mitigating dangers, CTR has achieved nuclear material security improvements that would not have been possible otherwise, validating the importance of this ad hoc approach.

Although the core of the nuclear material security initiatives is run by the NNSA, in recent years, programs within the Department of Homeland Security have also begun contributing to these efforts, including the Domestic Nuclear Detection Office which is charged with creating a global nuclear detection architecture.

G-8 Global Partnership Against the Spread of Weapons of Mass Destruction To supplement the bilateral U.S.-Russian nuclear security activities, the Group of Eight (G-8), composed of Canada, France, Germany, Italy, Japan, Russia, the United Kingdom, and the United States, at their 28th Summit in 2002, created the Global Partnership Against the Spread of Weapons of Mass Destruction (Global Partnership). This was a major step forward for multilateral efforts.

Under the new initiative, the G-8 nations committed to support specific cooperation projects, initially in Russia and the FSU to address non-proliferation, disarmament, counter-terrorism, and nuclear safety issues. This expansion was an unquestioned success. More funding and participation from countries other than the U.S. also provided a framework for thinking concretely about the future of threat reduction with Russia, FSU and other countries. The G-8 leaders called on all countries to join them in committing themselves to the six G-8 principles to prevent terrorists from acquiring or developing nuclear, chemical, radiological, and biological weapons; missiles; and related materials, equipment, and technology.

To generate real progress, the G-8 needed to prioritize and coordinate its activities to avoid duplicative spending. In the past, there was always a mismatch between U.S. and other nation's financial contributions to the non-proliferation agenda. Moreover, each country had their own agenda, for example, the U.S. planned to continue its non-proliferation activities and to possibly include new efforts to reduce excess nuclear materials. Canada's agenda was to prioritize the security and disposition of submarine fuel. For Germany, the top three priorities were to facilitate chemical weapons destruction at Kambarka, submarine dismantlement, and secure nuclear materials and waste. The U.K. committed another additional $750 million, to its original funding of $125 million, to spend on nuclear safety and security, plutonium disposition, and submarine dismantlement and disposition over 10 years.

Japan also played an important role in enhancing nuclear security in the FSU. Most of its funding for nuclear efforts went through nongovernmental organizations, and most of this cooperation consisted of information exchanges and delegations, and seminars on specific topics in both countries. Japan began government-to-government cooperation with Russia in the 1990s, it later pledged approximately $100 million to support the dismantlement of nuclear submarines in Russia, primarily focusing on the disposal of radioactive liquid waste. In June 1999, at the Cologne Summit, Japan pledged an additional $200 million for continued support of dismantlement of decommissioned submarines in the Russian Far East, conversion of Russian military resources to the private sector, and disposition of surplus weapons-grade plutonium removed from dismantled nuclear weapons.

In 2008, the Global Partnership's geographical focus was expanded beyond Russia and the former Soviet states to allow multilateral efforts wherever terrorism and proliferation risks existed. But the G-8 nations experienced some difficulty in shifting from their focus on Russian needs, and the majority of the funds were still spent in Russia.

Proliferation Security Initiative Launched in 2003, the Proliferation Security Initiative (PSI) aims to interdict WMD and related material in transit. As of 2015, 105 participating nations have endorsed the PSI Statement of Interdiction Principles and participated in meetings, workshops, and other exercises with other members to improve their capacities for breaking up black markets and detecting and intercepting material. PSI members relied on national and international legal authorities to impede WMD trafficking. In President Obama's April 2009 speech in Prague, he called for the transformation of PSI into a formal institution.

Global Initiative to Combat Nuclear Terrorism In October 2006, Russia and the United States created the Global Initiative to Combat Nuclear Terrorism (GICNT). The Global Initiative is a nonbinding forum for sharing nonproliferation expertise and information and for preventing nuclear terrorism. Unexpectedly, in 2009 the initiative grew from 13 to 76 member nations. There are also three official observers, the IAEA, European Union (EU), and International Criminal Police Organization (Interpol). In 2009, its members agreed to strengthen the group by promoting greater civil society and private sector involvement.

GICNT aims to *"strengthen global capacity to prevent, detect, and respond to nuclear terrorism by conducting multilateral activities that strengthen the plans, policies, procedures, and interoperability of partner nations"* [12]. GICNT is co-chaired by the United States and Russia and convenes biannual plenary meetings and various exercises and workshop activities. An Implementation and Assessment group is currently chaired by the Netherlands and is responsible for developing and executing GICNT priorities through its Nuclear Detection Working Group, chaired by Finland, Nuclear Forensics Working Group, chaired by Australia, and Response and Mitigation Working Group, chaired by Morocco. Global in scope, GICNT fosters information sharing, emphasizes multi-sectoral expertise, and integrates capabilities to bolster the global counterterrorism structure. The GICNT currently lacks firm institutional grounding, despite President Obama's 2009 call for making it a "durable international institution."

2.3   The International Atomic Energy Agency

The International Atomic Energy Agency is the primary international organization that deals with global nuclear issues and at present has 168 member states. The IAEA's central mandate is to promote the peaceful use of nuclear energy and ensure the technology is not used for military purposes. The IAEA is obviously one extremely important resource. Its assistance is not limited to countries that are signatories of the NPT. Any state that is an IAEA member can request assistance.

The IAEA plays a central role in supporting effective nuclear security. However, the IAEA is only allowed to produce recommendations and encourage states to take action on nuclear security matters. At present, it has no mandate to evaluate state performance in implementing or complying with its recommendations.

The most developed set of recommendations and guidance that the IAEA offers on the physical protection of nuclear materials and facilities can be found in Information Circular (INFCIRC) 225/Revision 5. The fifth revision of INFCIRC 225 was released in early 2011. It addresses the post-9/11 threat environment, as the previous revision was completed in 1999. The most recent version updates categorizations of nuclear material and clarifies site access and control areas. Other changes involve new licensing requirements, prevention of sabotage, interface with safety, interface with material accounting and control systems, and response to a malicious act.

The IAEA also has an Office of Nuclear Security with several responsibilities. It plays the leading role in planning, implementing, and evaluating the agency's nuclear security activities. It also produces Nuclear Security Series documents (15 of which have been published to date) and manages the Nuclear Security Fund which is used to prevent, detect, and respond to nuclear terrorism. This fund is largely reliant upon extra-budgetary contributions from member states, though it does receive some small funding from the regular IAEA budget.

In addition to the documents that the IAEA produces, member states can augment their domestic security protections by seeking in-country assistance. The IAEA's nuclear security advisory services include: International Nuclear Security Advisory Service (INNServ) missions which help identify a country's broad nuclear security requirements and measures for meeting them; International Physical Protection Advisory Service (IPPAS) missions which evaluate a country's existing physical protection arrangements; and IAEA State Systems for Accountancy and Control Advisory Services which provides recommendations for improving a country's nuclear material accountancy and control systems.

International Physical Protection Advisory Service The IAEA's International Physical Protection Advisory Service is one of the few peer review tools available. At the request of member states, a team of international experts will conduct investigatory missions to review and compare national nuclear security capabilities with international best practices and recommendations. Site visits to specified facilities are optional. The results of IPPAS reviews are shared with the member states and otherwise kept confidential. Follow-up missions can be scheduled to assess implementation of the IPPAS team's recommendations, but they are not required.

International Nuclear Security Advisory Services The IAEA also offers International Nuclear Security Advisory Services, which are complementary and compatible with IPPAS missions. Upon request by member states, INSServ missions assist states to review the general status of their implemented measures to protect against nuclear terrorism. The service covers areas such as relevant legislative and regulatory systems, physical protection measures, detection and response and illicit trafficking. INSServ missions generally result in country-specific Integrated Nuclear Security Support Plans, which can then be implemented with IAEA assistance.

IAEA Code of Conduct on the Safety and Security of Radiological Sources The IAEA Code of Conduct on the Safety and Security of Radiological Sources was approved by the Board of Governors in 2003. While the Code is not legally binding, many countries have written to the IAEA Director General to express their support for the Code, and its principles. The Code rests on the assumption that every state should take appropriate measures to ensure that radioactive sources within its territory are "safely managed and securely protected" and to promote safety and security culture. The Code addresses areas such as legislation and regulation, regulatory bodies and import and export guidance. The Code also categorizes radioactive material into Categories 1, 2, and 3.

Incident and Trafficking Database The IAEA maintains the Incident and Trafficking Database. The database offers states a digital forum for reporting instances where nuclear or radiological material is identified outside of regulatory control.

2.4   International Conventions and Agreements

Convention on the Physical Protection of Nuclear Material The 1980 Convention on the Physical Protection of Nuclear Material (CPPNM), a legally binding agreement to protect civilian nuclear materials, was amended in 2005, requiring states to protect their civilian nuclear facilities and materials, and expanding measures to prevent and respond to nuclear smuggling. The Amendment entered into force after 10 years.

The CPPNM requires that states establish and maintain a legislative and regulatory framework to govern physical protection, establish or designate an enforcing body to implement such a framework, and take other actions as necessary to protect material and facilities. It includes a review conference mechanism that was used to negotiate the 2005 Amendment, but has otherwise rarely been utilized. The IAEA acts as the Secretariat for the convention.

The International Convention for the Suppression of Acts of Nuclear Terrorism The International Convention for the Suppression of Acts of Nuclear Terrorism (Nuclear Terrorism Convention or ICSANT) was adopted by the United Nations (UN) General Assembly in April 2005 to ensure that states would criminalize the illicit possession or use of nuclear material or devices by non-state actors. Under the Nuclear Terrorism Convention, states must enact laws to investigate possible offenses and to arrest, prosecute, or extradite offenders. Countries are also called upon to cooperate and share information on nuclear terrorism investigations and prosecutions, protect radioactive material within their borders, and receive instruction on how to proceed if an illicit device or material is recovered from non-state actors. Unlike the CPPNM, the Nuclear Terrorism Convention applies to civilian and military material.

United Nations Security Council Resolutions Several UN Security Council resolutions (UNSCR), including Resolutions 1373, 1540 and 1887, passed in 2001, 2004, 2009, respectively, are aimed at preventing WMD terrorism.

In the weeks following the terrorist attacks of September 11, 2001, the UN Security Council unanimously passed UNSCR 1373. Though it focused on general counterterrorism mechanisms and enforcement measures, it specifically cites "*the threat posed by the possession of weapons of mass destruction by terrorist groups*" and "*illegal movement of nuclear, chemical, biological and other deadly materials*" [13]. Because the resolution was passed under the UNSC's Chapter VII authority, action is not voluntary. It requires members to take measures to combat terrorism. Despite its mandate for action, the resolution has loopholes, and its shortcomings were highlighted by the discovery of an international nuclear proliferation network run by the Pakistani scientist A.Q. Khan.

A more universal approach to WMD security, including fissile materials, was approved in 2004 in UNSCR 1540. For the first time, UN member states were bound to take and enforce measures against WMD proliferation and were required to report on their nuclear, chemical, and biological security status and nonproliferation activities. The resolution was primarily aimed at preventing WMD terrorism by non-state actors. It also requires nations to submit reports on their efforts, though compliance with this mandate has been inconsistent and irregular. It would be very useful for the Global Partnership's members to provide financial, technical, and manpower support to those countries that need to do a better job of reporting but do not have the resources.

By mid-2009, 148 states had submitted their reports and over 40 nations had not. These programs overall have achieved impressive results and have changed the methods by which nuclear security is approached. The traditional focus on treaties and international agreements has been supplemented with ad hoc and flexible bilateral and multilateral mechanisms. As Russia ceases to be the primary focus of securing nuclear materials, both for political reasons and because key objectives are being accomplished, the challenge of preserving and adapting this model to other global needs has arisen.

Of course, these programs are dealing with extremely sensitive materials, facilities, and personnel. Governments are naturally going to be cautious and security forces are going to have a prominent role in the process. But, the security challenges posed by vulnerable nuclear materials transcend domestic concerns and national borders. If fissile material were to leak from a nation or make its way into the hands of terrorists, that would be an international crisis, not a domestic concern. Therefore, the domestic political requirements need to be balanced against the need for international stability.

In September 2009, U.S. President Barack Obama chaired a session of the UNSC, during which UNSCR 1887 was unanimously adopted. UNSCR 1887 reaffirmed the threat of nuclear proliferation to global security and the need for multilateral actions to prevent it. The resolution highlighted the need for improving the security of nuclear materials to prevent nuclear terrorism and expressed support for the 2010 Nuclear Security Summit, the goal announced at that meeting of securing all vulnerable nuclear materials around the world within 4 years, minimizing the civil use of HEU, and multilateral initiatives such as the Global Partnership and the Global Initiative to Combat Nuclear Terrorism.

3 Nuclear Security Summits

3.1 The Prague Speech and Initial Recommendations

In his April 5, 2009, speech in Prague, President Obama outlined his arms control and nuclear nonproliferation objectives. At the top of the list was his assessment that terrorists are "*determined to buy, build, or steal*" a nuclear weapon, and to prevent this, the United States led an international effort to "*secure all vulnerable nuclear materials around the world within four years.*" As a step toward this goal, he pledged to convene a summit on nuclear security within a year to "*secure loose nuclear materials...and deter, detect, and disrupt attempts at nuclear terrorism*" [14].

With the growing global stockpile of nuclear and radiological materials and the increasing boldness of terrorists, international requirements for nuclear security were rapidly changing. These head-of-state level summits were unprecedented opportunities to drive the agenda and address the new challenges that were more geographically dispersed.

At the time, the world's stockpile of fissile material had been estimated at 1600 metric tons of HEU and 500 metric tons of plutonium. According to the IAEA, fissile material was located at 1131 facilities and locations; about half the world's fissile material was – and is – in military stockpiles and the other half, in civilian stockpiles. Even with these imprecise figures, there was – and still is – certainly enough material to manufacture 100,000 to 150,000 nuclear weapons [15, p. 8].

Before the convening of the first Summit, suggestions were made by nuclear security experts to create a new policy agenda that would build a new global framework to address twenty-first century nuclear security realities. These included creating a global nuclear material security road map based on measurable benchmarks of vulnerability and proven security upgrades; accelerating efforts to secure and eliminate global HEU and Pu stockpiles; minimizing and then eliminating the use of HEU around the globe; securing all radiological sources in hospitals; pursuing sufficient nuclear security funding for the IAEA and U.S. domestic nuclear security programs; establishing Regional Nuclear Training Centers in key regions of the world to cultivate local security culture and provide access to best nuclear security practices; and, establishing real-time monitoring of nuclear materials security.

3.2   Summit Accomplishments

Four head-of-state Nuclear Security Summits were organized within 6 years, Washington, D.C. (2010), Seoul (2012), The Hague (2014), and Washington, D.C. (2016). The summits drew international attention to the threat of nuclear and radiological terrorism, and the need to adequately protect weapon-usable nuclear material and radiological material around the globe. They allowed new initiatives to grow and sought to achieve goals within set timeframes. The summits also featured the top leaders from 50 nations and 4 international organizations – the EU, the IAEA, the Interpol, and the UN.

HEU and Plutonium Removals and HEU Reactor Conversions By the end of the summit process more than 1500 kg of HEU and separated plutonium was recovered or eliminated. Ten countries became HEU-free and 20 nuclear reactors were converted to low-enriched uranium (LEU).

The most ambitious objectives of the 2010 NSS Work Plan considered the consolidation of national sites where nuclear material is stored, the removal and disposal of nuclear material no longer needed for operational activities, and the conversion of HEU fueled reactors to LEU fuels. In 2012 at the Seoul Summit, several countries pledged to repatriate HEU in their territories to its country of origin. Belgium, France, the Netherlands, and the U.S. committed to support the conversion of European medical isotope production to non-HEU-based processes by 2015. In a second HEU-focused gift basket, Belgium, France, South Korea, and the U.S. committed to cooperating on a project to produce high-density LEU fuel to facilitate the conversion of more research reactors from HEU to LEU fuel.

The 2016 NSS gift basket had 22 signatories who committed themselves to combat the threat of nuclear terrorism by refraining from the use of HEU in new civilian facilities or applications; converting or shutting down HEU reactors; removing, downblending or disposing HEU stocks; using LEU for medical isotope production and reviewing their progress in 2018 at an international conference.

Italy, the Netherlands, Sweden, and Switzerland have also pledged to give up small stocks of separated plutonium. The IAEA offers Plutonium Management Guidelines (INFCIRC/549), but not all states with plutonium have agreed to them.

Only Belgium, China, France, Germany, Japan, Russia, Switzerland, the United Kingdom, and the U.S. have declared their plutonium holdings to the IAEA annually, and some of these reports are incomplete. Total civilian holdings of separated plutonium are approximately 250 MT and expected to rise.

Centers of Excellence A dozen or more countries have established Centers of Excellence (CoE) for nuclear security training, education, and research, including separate new institutions in China, Japan, India, Pakistan, and South Korea. Other CoEs, beyond Asia, include the IAEA's Nuclear Security Support Center network, the EU's Chemical Biological Radiological and Nuclear Center network and a growing number of institutes, organizations, and stand-alone centers that focus on or include curricula on nuclear security. The original center of excellence was the Russian Methodological and Training Center located in Obninsk, Russia which was formed as a joint project with the U.S.

The role of these centers is envisioned to be as repositories and disseminators of nuclear material security best practices. But they also can become advocates for improvements in the governance structure and work on some of the key questions that need to be addressed, including how to facilitate transparency to generate international confidence without revealing sensitive information, to better secure high intensity radiological sources, and to improve the independence of regulatory authorities.

In 2012, twenty-four NSS participants signed a gift basket, expressing their intention to collaborate on the development and coordination of a network of nuclear security CoEs. At the 2014 Hague summit, an updated CoE gift basket was presented by thirty-one countries to further the development of the CoE network. In 2016, twenty-eight participants issued another updated CoE gift basket, and only Jordan, Nigeria and Thailand were new signatories.

In addition, more than a dozen of workshops on issues including nuclear forensics, guard force improvement training, and detection of and response to nuclear smuggling have been held around the world. For example, France and Italy have incorporated nuclear security education into their academic curricula.

Nuclear Security Treaties and Conventions Ratifications Over the 6 years, more than 25 summit participants ratified the 2005 Amendment to the CPPNM and allowed to achieve its entry into force – a key summit goal – after the end of the summit process, on May 2016. Furthermore, more than 15 participants also ratified the ICSANT.

Before the summit processes, only 18 summit participants had ratified the 2005 Amendment to the CPPNM, and 24 ratified ICSANT. In 2010, the first communiqué recognized the importance of the CPPNM as amended and ICSANT, as essential elements of the global nuclear material security regime. Since then, NSS participants made national commitments to ratify the relevant nuclear security treaties.

Commitment Making and Progress Reporting Each participant country has made at least a few national commitments during the summit process, and around 90% of countries have also participated in multinational commitment making through the "gift basket diplomacy." Declared by the NSS participant countries' heads-of-state or senior officials, national commitments became politically binding.

At the 2010 Summit, individual national commitments became known as "house gifts." Throughout the summit process, more than 100 were made, but their implementation was not made mandatory.

The Seoul Summit introduced "gift baskets" or multilateral commitments. Thirteen gift baskets were issued in 2012, 14 at the 2014 Nuclear Security Summit and 20 at the last summit. Of these, more than half were updates to baskets issued in 2012 and 2014. The new ones in 2016 focused on cyber security and the strengthening of the global nuclear security architecture through the establishment of a Nuclear Security Contact Group (Contact Group).

Even though there was no formal requirement, countries also provided reports on the progress of their commitment implementation. The most active states in summit commitment making and reporting were: the three summit hosts – the United States, Korea, and the Netherlands – and Canada, Japan, Kazakhstan, Norway, Philippines, Spain, and the United Kingdom.

IAEA Recommendations Semi-Institutionalized The most prominent nuclear security governance gift basket that came out of the NSS process was the Strengthening Nuclear Security Implementation Joint Statement. The gift basket is a commitment to meet the intent of the IAEA's recommendations and guidelines contained in their Nuclear Security Series which outline best practices, suggested regulations, and a commitment to continuous improvement of the regime. This Trilateral Initiative goes further by underscoring that nuclear security is an international responsibility as well as a national one, and encouraging the signatories to "assess new ideas to improve the nuclear security regimes."

Thirty-five NSS participants signed onto this initiative. This gift basket has outlived the summit process because it was introduced into the IAEA as Information Circular 869, subsequently widening the scope of the gift basket to include non-summit states. Currently, three additional signatories – China, India, and Jordan – have been collected in support of the statement.

Requests for IAEA IPPAS review missions also increased since 2010. By 2016, 24 of the 29 countries that requested IPPAS review missions were summit participants.

Closer Cooperation Among Civil Society, Industry, Governments, and International Organizations In addition to the official NSS, the nuclear industry and civil society also held side summits. These were very important events bringing key stakeholder communities together and focusing them on the nuclear security agenda. They also brought the industry and civil society into closer cooperation. Such cooperation was, and is, necessary, if progress continues to be made.

In 2012, two parallel non-governmental events were convened at the sidelines of the summit: an expert symposium and a nuclear industry summit. The 2012 Nuclear Security Symposium for the expert community was cohosted by the Korea Institute of Nuclear Nonproliferation and Control and the Institute of Foreign Affairs and

National Security. The 2012 Nuclear Industry Summit was hosted by Korea Hydro and Nuclear Power. This meeting benefited from the advanced efforts of three working groups on HEU minimization, nuclear information security, and the interface of nuclear safety and security. The working group outcomes formed the basis for a joint statement which was released at the industry event [16].

In 2014, nuclear experts influenced the summit by assisting in the development and promotion of the nuclear governance gift basket, which was one of the most popular gift baskets produced. In 2016, for the first time, there was a joint session held between the industry and civil society summits. Nearly 200 experts from 50 countries participated to discuss the steps that governments, the nuclear industry, and nongovernmental community should take to strengthen the global system and prevent nuclear and radiological terrorism. The organizers of the 2016 Nuclear Knowledge Summit issued a statement where they identified "Five Priorities" to strengthen nuclear security. This was unprecedented event and could not have been anticipated when the NSS process began in 2010, as both communities were quite estranged from one another at that time.

High Level Political Support Although the objectives of these heads-of-state summits remained to be voluntary, they focused bureaucracies and raised public awareness. For a highly technical issue like nuclear security that was and is important.

For example, while the 2010 Summit focused almost exclusively on fissile materials, Seoul expanded the scope to include the interface between safety and security at nuclear facilities and the protection of high activity radioactive sources that can be used in "dirty bombs." This was in large part attributable to the interest of some key governments, and the result of the nuclear reactor accident at Fukushima in Japan.

These types of summits also drove progress, though not ever as much as is needed, in part because of the need to generate consensus. However, a multilateral pursuit of improved nuclear material security was promoted, giving some international legitimacy to non-universal action in support of nuclear material security.

4   Beyond the Summits

The developments achieved throughout the NSS process were positive and will further solidify the current foundation of the current nuclear material security regime. But, even if implemented completely and rapidly, they will not be sufficient to address the evolving nuclear terrorism threat. And, while the NSS processes had the political power to produce national commitments, they did not have the ability to drive the agenda and regime significantly beyond where it exists today.

Unfortunately, at the end of the NSS process the nations involved did not choose a particularly effective path to sustain and build upon the mission of the nuclear security summits. Instead they opted for the disaggregation of the NSS mission and offered pieces of the agenda to five different organizations and initiatives – the IAEA, United Nations, Interpol, GICNT and the G-8. In addition, a Contact Group of Sherpas and Sous-Sherpas was established for the post-summit process. However, participation at the 2016 Ministerial Meeting was downgraded, garnering less global attention and action to nuclear security. This approach may have created a very weak bridge for the achievement of future progress.

4.1   Closing the Gaps and Boosting the Nuclear Security System

Preserving Summit Innovations   The regime still lags behind the safety, nonproliferation, and arms control regimes. At the very least all of these other regimes require some element of transparency and/or verification of commitments. Without the summits, there is no mechanism in place for nations to make new commitments or to report on their implementation. There is also no forcing mechanism that places political and expectation pressure on countries to do more in this area.

To preserve the summit innovations, government-industry and civil society collaboration is essential and should be preserved. Government, civil society, and the private sector all play important roles in responding to twenty-first-century nuclear proliferation threats, and each sector offers a vital contribution the others lack. None of these stakeholders in isolation has the power to drive the agenda, but together they can. The problem is that the summits have provided the impetus for this collaboration and it could be difficult to restore it.

Continuing Political Engagement   With the end of the summits the nuclear security issue sank back from heads-of-state to heads of government departments or the Sherpas forming part of the Contact Group. That was a big step back, and may mean ending back where we were before the summit process began, with technocrats in control. In its initial meetings the Contact Group has been plagued with an inability to identify a forward-focused nuclear security agenda.

For this reason, it is essential to continue a political track, and not to revert back to the purely technocratic system of the pre-summit period. The whole point of the summits was to overcome this. The political track should allow for broader policy engagement among nations, the nuclear industry and civil society.

Five Priorities for Progress   What is needed is an international nuclear security regime that emphasizes transparency of action, shared standards, and confirmed performance and accountability by nations. While the Summit processes took the important step of establishing global fissile material security as a top-level international objective, a more robust, effective, and flexible twenty-first century nuclear material security architecture will require actions beyond the current mechanisms and international consensus. Building a more robust, effective, and flexible nuclear security architecture will require an evolution of global nuclear governance.

A legally binding global framework convention on nuclear security is one approach that could address this challenge. This new framework convention would strengthen and build on the current regime but fill the gaps and unify relevant international security agreements and recommendations. It would commit states to an effective standard of securing dangerous nuclear materials and give the IAEA the mandate to evaluate whether states are meeting their nuclear security obligations and provide assistance to states that need help in doing so.

In 2016 in advance of the final summit, a number of international experts issued "Five Priorities" that world leaders should act upon to strengthen the nuclear security infrastructure.

Comprehensive Countries must make the global nuclear security regime comprehensive by fully implementing all elements of the existing nuclear security regime. All nuclear weapons and weapon-usable materials must be effectively and sustainably protected against a full range of plausible threats. Those threats are not uniform and can vary based on country and region. Countries will need to demonstrate the effective security for all nuclear materials, including civilian and military.

The first line of defense for the security of nuclear materials resides with the country that manufactured or stores them. These materials are national possessions, and the laws and regulations of individual nations are the most relevant protections. Individual nations are very protective of this sovereign control. As a result, there is little information regarding the national laws and regulations governing nuclear security available to the international community. International confidence in nuclear security, therefore, must rely on international instruments and assertions of adequate national nuclear security. Unfortunately, compliance with these instruments is inconsistent.

Rigorous The current nuclear security regime is a patchwork of requirements, recommendations, and agreements. These contain no uniform requirements for implementation and no enforcement or penalty mechanisms for non-compliance. In addition, there is no consistency in the adoption of the elements of the regime by individual nations. There are roughly 55 separate components that nations could participate in or implement, and these approaches are not one-size-fits-all. There should be common standards and objectives developed, and the means to implement and verify them.

The goal of expanding and improving the nuclear security regime is to increase international confidence that security practices in all countries are as robust as possible, with adequate peer review, transparency, and certified staff. However, because of the sovereign nature of many nuclear activities, there also will need to be a balance between increasing global responsibility for nuclear operations and sovereign control.

Open There needs to be a process for building international confidence in nuclear security. At present, every nation develops its own nuclear security system and there is no requirement that they inform any other nation or international body of their system. This leaves major gaps and uncertainties. Countries should identify and exchange non-sensitive data about security practices, standards, and implementation. They should also accept regular peer reviews of security instruments and practices.

The value of transparency has been acknowledged as an important element in addressing complex transnational challenges in the twenty-first century. The importance of transparency has been an important element in addressing the global economic crisis, mitigating climate change, and targeting sanctions to limit financial transactions, among other issues. The goal of transparency is to increase international confidence and ensure that standards are being implemented. These are key issues in the nuclear security area.

Between the Hague and final Washington Summits, a simple checklist of the elements of the nuclear security regime has been produced by the international expert community. Countries were encouraged to fill out the checklist and provide general information of the regime elements they participate in or implement. The submission of the checklist relies on the integrity of the nation submitting it, and there may be some disincentive to make the checklist public (including highlighting weaknesses). If the checklists are adopted, then there would be some way to measure progress on paper. And, an increased comfort level with the checklist could lead to the submission of more detailed national reports. In addition to this type of reporting there are additional voluntary actions that could be undertaken and cultural changes that would need to be achieved in order to remove weak links in the nuclear security system.

Sustainable Nuclear security is a system that needs to respond to evolving circumstances. This could be achieved by establishing an integrated mechanism that ensures continued high-level attention and drives new nuclear security commitments and resources. But also, share best practices, build security culture, and constantly assess where improvements can be made.

Focused on Minimization Countries should consolidate materials and reduce the volume of materials, civilian and military. This will make the protection process easier and less expensive, and could be accomplished by creating a time-bound roadmap to eliminate, and minimize, civilian HEU and plutonium needs. There are more than 4 million pounds of fissile material remaining around the globe.

Empowering the IAEA With all of its useful and detailed products and services, and as a key member of the post-summit action plan process, the IAEA is indispensable and irreplaceable. But, its capacity and power are constrained by the voluntary nature of its recommendations, the consensus basis of its decision making, and the limits of its budget. Without doubt, the IAEA will remain at the center of the nuclear security agenda as a deep repository of expertise and continue to serve the very important function of achieving universality in the decisions and recommendations it produces. But, it does not and potentially cannot, have a monopoly on a dynamic nuclear security agenda, particularly if its member states do not provide it with greater power, latitude, and funding. There is an important requirement for a separate political track beyond the IAEA that is flexible, allows for greater policy innovation, is not bound by consensus and universality among the parties, and includes all stakeholders.

Securing Radiological Security Countries need a plan for inventorying, tracking, securing and substituting for these sources. Currently, there are high intensity radioactive sources used in every country for cancer treatment, food safety and energy exploration. The IAEA cannot give a definitive figure on how many sources there are around the globe, and sources regularly go missing. One step countries could take is to recognize the importance of the IAEA in the area of nuclear and radiological security, and request greater international political and financial support for its activities.

Cyber Security Cyber is a challenge for nuclear security as well as for many other systems that many rely upon. The cyber issue was addressed only in the last summit. However, this is an area that is in constant change and is an issue for reactors that could be breached and caused to melt down like Fukushima or to undermine physical security systems around sensitive materials. Flaws and gaps have been identified by recent analyses made by Chatham House and the Nuclear Threat Initiative.

Framing the Issue for Citizens There is a need to explain to people why nuclear security and nuclear issues in general are relevant to their lives. The Partnership for Global Security, with the collaboration of other experts from the civil society, has tried to do this through a video series under the "Five Priorities." Without public support, there is not the necessary multiple pressure points on governments that are required.

As Obama said, his greatest error as President was assuming that the correct policy would sell itself. Instead governments, industry and the civil society sectors need to sell the right policies, which is often overlooked.

5   Conclusion

One of the significant challenges in maintaining nuclear material security as a high global priority is that a number of nations do not see nuclear terrorism as a near-term threat. Nuclear terrorism scenarios are not far-fetched in light of the detailed and patient planning that went into the September 11 surprise attacks on the U.S., the political turmoil that continues in the Middle East and that may wash over other authoritarian-led nations, the continuing intelligence community assessment that terrorists are seeking these weapons materials, and the vulnerable locations of some of these materials around the globe. And nuclear crises can erupt without warning, with devastating results, and extremely high price tags as the emergency in Japan has underscored. In the case of nuclear terrorism, the cost of the damage and response would dwarf the price of prevention.

Outside of the connection between the Fukushima meltdowns and the possibility of terrorists creating the same conditions at another reactor, not much was said in the official summits about nuclear power. But there is an important connection between current and future nuclear power, the overall nuclear governance system, and its role in addressing climate change.

Nuclear alone is not the answer to climate change, but it is a significant contributor. However, as nuclear infrastructure grows and moves into new regions, so do the amount of nuclear material and facilities that need to be secured, potentially increasing opportunities for nuclear proliferation and terrorism. Major nuclear operators in the U.S. and Europe are not looking to build many new plants. But, China, Russia and India are.

Therefore, there is an urgency in taking action to ensure that the nuclear governance system minimizes security and proliferation concerns and global nuclear insecurity. This can be done but it will require leadership and the implementation of improvements on a continuum.

What the international regime needs are big goals that will help to drive the process forward at both the technical and political levels. As a first step, nations should act to universalize the implementation of the current regime elements. A second step should be the adoption of a principle of continuous improvement matched with voluntary actions that add improvements over time and are measured regularly through the IAEA and a parallel political process similar to the NSS processes. The final step is to adopt a comprehensive instrument for nuclear security that is comprehensive, flexible, and effective.

Acknowledgments   To Grecia Cosio, Research Associate at the Partnership for Global Security, for her significant research and contributions to this paper.

References

1. Bunn, M.: Securing the Bomb 2010: Securing All Nuclear Materials in Four Years. Belfer Center for Science and International Affairs, Cambridge (2010)
2. Weapon Material Basics. Union of Concerned Scientists (2009), http://www.ucsusa.org/nuclear-weapons/nuclear-terrorism/fissile-materials-basics#.WFwbPhsrK00
3. IAEA Incident and Trafficking Database. (2016), http://www-ns.iaea.org/downloads/security/itdb-fact-sheet.pdf
4. Lee, B., Schmerler D.: CNS Global Incidents and Trafficking Database: 2015 Annual Report. (2016), http://www.nti.org/media/documents/global_incidents_trafficking_report.pdf
5. Nuclear Security Series Glossary. International Atomic Energy Agency, Version 1.3, 18 (2015), http://www-ns.iaea.org/downloads/security/nuclear-security-series-glossary-v1-3.pdf
6. Hand, K., Lyudmila, Z.: Nuclear Smuggling Chains: Suppliers, Intermediaries, and End-Users. American Behavioral Scientist, vol.46, 822–844. Sage Publications (2003)
7. Schmid, A.P., Spencer-Smith, C.: Illicit Radiological and Nuclear Trafficking, Smuggling and Security Incidents in the Black Sea Region since the Fall of the Iron Curtain-an Open Source Inventory. Perspectives on Terrorism, vol.6 (2012)
8. Nunn, S., Lugar, R.G.: Dismantling the Soviet Arsenal. The Washington Post, A25 (1991)
9. Stern, J.E.: U.S. Assistance Programs for Improving MPC&A in the Former Soviet Union. The Nonproliferation Review, Winter, 17–32 (1996)

10. Luongo, K.N.: Improving U.S.-Russian Nuclear Cooperation. Issues in Science and Technology, vol. 58 (2001), http://www.issues.org/18.1/luongo.html
11. Global Security Engagement: A New Model for Cooperative Threat Reduction. Office of Senator Richard Lugar Press Release (2009)
12. Global Initiative to Combat Nuclear Terrorism. http://gicnt.org/
13. Security Council Unanimously Adopts Wide-Ranging Anti-Terrorism Resolution. United Nations Security Council press release SC/7158 (2001), http://www.un.org/News/Press/docs/2001/sc7158.doc.htm
14. Remarks by President Barack Obama. Office of the Press Secretary, The White House (2009), http://obamawhitehouse.archives.gov/the-press-office/remarks-president-barack-obama-prague-delivered
15. Global Fissile Material Report 2009: A Path to Nuclear Disarmament. International Panel on Fissile Materials (2009), http://www.fissilematerials.org/ipfm/site_down/gfmr09.pdf
16. Cann, M., Luongo, K.: Nuclear Security: Seoul, the Netherlands, and Beyond. U.S.-Korea Institute (2013), http://uskoreainstitute.org/wp-content/uploads/2013/10/USKI-NSS-Report_Full.pdf

Multidisciplinary DSS as Preventive Tools in Case of CBRNe Dispersion and Diffusion: Part 1: A Brief Overview of the State of the Art and an Example – Review

Jean-François Ciparisse, Roberto Melli, Riccardo Rossi, and Enrico Sciubba

Abstract The paper addresses some important issues related to the need for a timely, reliable and accurate tool for the early warning in case of CBRNe events. The state-of-the-art of the currently available tools is briefly presented in the first part of the two-papers set. While the accurate calculation of the dispersion of both lighter- and heavier-than-air contaminants in complex three-dimensional domains is definitely possible with commercially available CFD packages, the time needed to obtain a reliable numerical solution, under the pertinent atmospheric conditions prevailing at the time of the attack, exceeds the requirements of a first-aid intervention. Therefore, it would be advisable to combine these CFD packages with some sort of "intelligent" Decision Support System that makes use of multidisciplinary knowledge base and of some kind of detection-diagnostic-prognostic Expert System. The DSS could be interfaced with some standard early detection tools and ought to include an enhanced diagnostic/prognostic utility based on a specific series of local CFD simulations of dispersion events. Its use ought to be relatively easy for trained personnel. Since the database for the CFD dispersion calculation is by definition "local", detailed maps of the presumable target areas must be included in the database. The second part of this paper presents a detailed description and one example of application of such an Expert Assisted CFD dispersion calculation, named FAST-HELPS (Fast Hazard estimate of low-level particles spread).

Keywords DSS simulation software • CFD

J.-F. Ciparisse (✉) • R. Rossi
Department of Industrial Engineering, University of Rome "Tor Vergata", Rome, Italy
e-mail: jf.ciparisse@gmail.com

R. Melli • E. Sciubba
Department of Mechanical and Aerospace Engineering, University Roma Sapienza, Rome, Italy
e-mail: enrico.sciubba@uniroma1.it

© Springer International Publishing AG 2017
M. Martellini, A. Malizia (eds.), *Cyber and Chemical, Biological, Radiological, Nuclear, Explosives Challenges*, Terrorism, Security, and Computation,
DOI 10.1007/978-3-319-62108-1_12

257

Nomenclature

ρ	Density
$\vec{V}$	Velocity vector
p	Pressure
μ	Molecular viscosity
μ_T	Turbulent viscosity
k	Turbulent kinetic energy
ε	Turbulent kinetic energy dissipation rate
P_k	Turbulent kinetic energy production term
$C_{\varepsilon 1}, C_{\varepsilon 2}, C_\mu$	Turbulence model constants
υ_T	Turbulent kinematic viscosity
φ_d	Dispersed particles volume fraction
c_d	Dispersed phase mass fraction
ρ_c	Continuous phase density
ρ_d	Dispersed phase density
d_d	Dispersed phase particles diameter
$\vec{u}_c$	Continuous phase velocity vector
$\vec{u}_d$	Dispersed phase velocity vector
$\vec{U}_{slip}$	Slip velocity
$\vec{g}$	Gravity acceleration vector
φ_{max}	Maximum particles volume fraction
C_d	Particles drag coefficient
Re_p	Particle-based Reynolds number
Q	Breath volumetric flow rate
N_b	Number of spores in each endospores
n_b	Number of spores per volume unit
ψ	Number of inhaled spores
ξ	Infection probability
l	Lethality of the infection
CBRNe	Chemical, Biological, Radiological, Nuclear, explosive
BWA	Biological Warfare Agents
CFD	Computational Fluid Dynamics

1 Introduction

The present globalization trend is improving links between countries, leading to more uniform technology developments, easier information exchanges and life quality growth. Unfortunately, the great amount of sensible information obtainable through the open web links is often used to plan for intentional dangerous attacks to the society as a whole. Among these, Chemical, Biological, Radiological, Nuclear and explosive events (CBRNe) are potentially the most hazardous worldwide. These events could be classified in accidental and intentional (terrorism) events. While prevention of such acts is a task for Security Agencies, CBRNe defense is the science that aims to mitigate, protect and avoid CBRNe events once they have happened. The large variety of CBRNe events implies the necessity to work on multiple frameworks and to develop different types of emergency intervention tools.

In war situations, chemicals have been used for years and a large number of hazardous chemical agents is nowadays known; unfortunately, large quantities of some of these chemicals are known to be illegally stockpiled for possible damaging use by criminal and terroristic organizations. The first use of poison gas as a weapon dates to the 2nd battle of Ypres [1, 2], where Germans released chlorine gas. The Sarin, or GB, is one of the most dangerous gases: it is an organophosphorus compound and is available in the form of colourless and odourless very volatile liquid [3, 4]. It is so lethal to be considered a weapon of mass destruction. Discovered in 1938 by German scientists, it has been used several times, such as in the Tokyo subway Sarin attack (1995) [5], in Iraq (2005) [6] and in the recent Syrian civil war in the Ghouta chemical attack (2013) [7]. The brutality and the extreme non-selectivity (they target everything in the area of release and not only military objectives) of chemical agents led to the Chemical Weapons Convection (CWC). The CWC was born with the aim of prohibiting "the development, production, acquisition, stockpiling, retention, transfer or use of chemical weapons by State Parties" [8].

Uses of biological agents are significantly increasing in the last decades, especially for terrorist acts [9, 10]. Furthermore, non-intentional pathogen spreads are also today a threat, to the latest example being the Ebola virus infection in Africa [11, 12]. Several are the possible diffusion techniques of pathogens, such as dispersion in the environment, water and food contamination [13]. The importance of biological emergency management has been ever more frequently discussed [14–16] and tools to help targeted groups in making fast protection decisions are necessary [17].

Several human fields, such as industry, medicine and agriculture [18], use a lot of radiological materials that can be used to develop Radiological Disperse Devices (RDDs), also known as "dirty" bombs [19]. Since these materials are often not handled under sufficiently stringent safety and security measures, radiological events must also be taken into account in the terrorism scenario [20]. Furthermore, radiological dispersion events may also accidentally occur, being caused, for example, by explosions or protection-isolation leakages [21, 22].

In order to face these events, it is necessary to develop tools able to aid first responders in emergencies and to improve the knowledge about CBRNe physics.

2 Software Tools to Improved CBRNe Defence

Recent progress in Computer science allows for the development of tools able to aid CBRNe operators in making fast and correct decisions. These Software tools can be classified into two main different categories:

- *Numerical simulation software;*
- *Decision support software.*

Numerical simulation software tools are numerical packages able to more or less accurately reproduce the physics of the event. They directly solve the equations that describe the phenomenology of pollutants dispersion. These packages cannot be used in case of emergencies since they usually require a large amount of computing time. Quite on the opposite, decision software are often based on semi-empirical solutions that lead to a fast but not completely accurate response. Thanks to their computational speed, though, these programs are used in emergencies to assist emergency managers to make informed decisions.

There are several software packages that can be used to analyse CBRNe events. In this first part of the paper, the authors examine only the best-established tools in the CBRNe framework and discuss some of the most recent work done in this area.

2.1 Decision Software Tools

Decision software tools are computer codes able to give information about a CBRNe event in an acceptable time for response (<15 min in case of chemical and radiological releases). Among them tools, the most used are:

- HotSpot;
- ALOHA;
- WISER;
- STEM.

HotSpot HotSpot is a free-license code developed by the National Atmospheric Release Advisory Center [23]. HotSpot is able to reproduce the radiological dispersion of radionuclides in the atmosphere. It estimates the radiation dose in short range (maximum 10 km) and short time (24 h). The estimate is conservative, i.e. it usually overestimates the dose. The code can handle a large number of radionuclides and the user can even create a specific mixture of the radionuclides available in the code library.

HotSpot uses the Gaussian Plume Model (GPM). The advantage of the GPM model is the acceptable accuracy in an extremely short computational time. Anyway, the model gives sufficiently accurate results only under simple meteorological and terrain conditions [24]. HotSpot code models the GPM parameters via several external user-specified conditions, such as the Atmospheric Stability Classification and terrain conditions. Details of the model can be found in [25].

The software also performs an evaluation of the scenarios by computing the radiological risk. It analyses the dose ratio through the methods recommended by the US Environmental Protection Agency (EPA) [26–28] and the International Commissions on Radiological Protection (ICRP) [29].

Thanks to its computation velocity, acceptable accuracy and a wide range of applicability, HotSpot allows CBRNe operators to make fast decisions. Furthermore, the software is also adopted by researchers to validate and understand the applicability limits of the tool: Di Giovanni et al. [30] simulated two scenarios of radionuclide release (Cs-137, Sr-90), analysing the influence of different stability classes in the GPM. Cacciotti et al. reproduced the Caesium local diffusion due to Chernobyl accident, showing an acceptable agreement between experimental and numerical data [31].

ALOHA ALOHA is a tool for the computation of chemical dispersion in the atmosphere. It can reproduce various dispersion models such as toxic and flammable gas clouds, pool and jet fires. On the basis of user-specified information about the chemical release, ALOHA performs several analyses and generates a threat estimation [32]. Through MARPLOT [33], the software can draw threat zones on real maps.

Many examples of ALOHA simulations can be found in literature: Cameo's document [34] shows several events caused by two different sources: tank and direct source. Furthermore, it contains some guidelines to simulate different sources: toxic vapor cloud, pool fire, BLEVE, vapor cloud explosion and jet fire. Kulynych and Maruta [35] simulated a pipeline failure with a consequent hydrogen sulphide release, to estimate the concentration of chemicals as a function of time and distance. Bhattacharya and Ganesh Kumar [36] analysed the hazardous chemical release of various substances in different cases such as butanol vapor dispersed by a storage tank leak and chlorine leakage from a tonner.

WISER Wiser is a tool developed to support first responders in case of hazardous mass release incidents. It consists of a library of materials with information on their physical characteristics, effects on human health, containment, identification and suppression guidance. The software is developed by the U.S. National Library of Medicine and its library contains over 460 substances. Newer versions implement also a radiological (21 radioisotopes) and biological support. It runs on a wide variety of devices, such as PCs, tablets and smartphones and works with several operating systems (Windows, Apple, Android and BlackBerry) [37]. Wiser is a very useful tool, and by integrating it with a detection system or with a dispersion simulation, first responders are able to characterise the degree of hazardousness of the target area, the best protection tools and the best way to react.

STEM in the framework of biological infection diffusion, the Spatiotemporal Epidemiological Modeler (STEM) open-source tool is one of the most widely used software. STEM is a tool capable of helping in the development of models to predict the evolution of epidemics. The software can work both statistically and deterministically. It implements several numerical solvers to solve different sets of modeled disease diffusion equations [38].

The tool needs some information about the disease such as infection rates, mortality rates, incubation period, recovery-and transmission rate. Then, on the basis of specific (user-supplied) information about countries, transportation and other linked environmental factors, it develops numerical models to determine the disease spread rate. Through the insertion of proper information, STEM can predict a large range of events. It is able to take into account different population models: standard populations, insect vectors, migratory birds and demographic models. It can also replicate foodborne diseases [38, 39].

2.2 Numerical Simulation Software Tools

A different kind of tools to prevent CBRNe events are those that solve the mass-, momentum- and energy equations applicable to the simulated event. The most popular techniques to solve these equations are the Finite Elements- (FEM), the Finite Differences- (FDM), the Discrete Elements (DEM) and the Finite Volume Method (FVM). All of them are based on a discretization of the target area into small domains and on the subsequent solution of a very large set of non-linear algebraic equations that can be derived by the corresponding discretization of the non-linear differential mass-, momentum- and energy equations. As for the time dimension, it is also discretized in time intervals of possibly different size (time steps). Since dispersion problems are by definition unsteady, different time marching schemes are used (e.g. explicit, implicit and Crank-Nicolson methods), each one of them having different accuracy and stability characteristics and different resource requirements. Schemes are usually classified in:

The Finite Element Method solves the discretized in each cell of the grid taking into account the neighboring cell values, which constitute the boundary condition for the i-th cell. FEM is currently the most used method to simulate structural mechanics problems [40], but has many applications in fluid flow computations.

The Finite Difference Method uses a Taylor series expansion to discretize the continuum problem. There are several numerical schemes to solve the resulting set of non-linear algebraic equations. Until recently, like FEM, FDM was avoided in solving the advection-diffusion equation because of stability issues [41]. All commercial codes appear though to have successfully solved this problem.

The Discrete (or Distinct) Element Method is useful to compute the ensemble characteristics of a large number of particles. It is a model with a better accuracy compared with other codes but it is strongly limited by the maximum number of particles it can handle. In fact, the needed computing resources arise with the cube of the number of particles [42].

The Finite Volume Method is currently the preferred technique to simulate compressible fluid dynamic problems. It discretises the entire domain in sub-volumes. The main advantage of the method consists in the fact that the divergence term is converted into surface integrals through the divergence theorem so that the corresponding terms are calculated as fluxes and the method is conservative. As for the FEM and FDM, recent developments in Numerical Analysis have effectively cured most of the instability issues [43].

The main steps in a numerical problem solving procedure are, in sequence:

- Identification of the computational domain
- Construction of the relevant geometry;
- Construction and verification of the mesh;
- Iterative solution of the discretized equations;
- Post-processing of the results.

The first step is the creation of the geometry. Often the software used to solve the equations does not contain a tool to draw the geometry and a CAD software, as SolidWorks or AutoCAD, must be used.

Then, the domain must be divided into sub-domains where the numerical scheme solves the discretised equations. This operation is called "meshing". There are several types of mesh, macroscopically classified in structured, unstructured and hybrid grids. The shape of the cells can be triangular or quadrilateral for two-dimensional geometries and tetrahedral, pyramidal, prismatic-triangular and hexahedral for three-dimensional geometries [44].

The equations solving is performed via the numerical scheme specific for each tool and is usually proprietary, since it represents the core of the tool and the commercial success of the package depends on its efficiency and accuracy.

Post-processing consists in data handling and plotting to visualise the results in the desired way.

We limit our discussion to three of the most popular CFD codes:

- OpenFOAM;
- ANSYS;
- COMSOL Multiphysics.

OpenFOAM OpenFOAM is a free open source software that can be applied in a broad variety of fields, such as solid mechanics, electromagnetism, fluid dynamics and chemical reactions. It has several sub-modules that can be linked to each other to solve different physical problems. The most relevant advantage of using OpenFOAM is the fact that it is open source. In fact, it allows user intervention on the core code and on each sub-module model, thus permitting ad hoc or experimental simulations. On the other hand, OpenFOAM is the less user-friendly software in our list. It has a raw user interface and it is intended for experienced users the k [45].

ANSYS The suite ANSYS is released by ANSYS Inc. [46]. The company develops many software packages devoted to different type of problems, such as CFX and Fluent for fluid dynamics, Mechanical Enterprise for mechanics and HFSS for electromagnetic fields. ANSYS has a set of tools that allows the user to construct the geometry and mesh it, solve the equations and post-process the data. For greater generality, though, the software contains proper utilities that allow for a relatively easy interface with other pre- and post-processors. Compared to OpenFOAM, the software has a much better user interface and it is easier to use. In the other hand, ANSYS source code is inaccessible to the user and the package is quite expensive.

COMSOL Multiphysics COMSOL Multiphysics is a platform integrated with geometry and mesh generator, physic equations solver and post-processing tools. It includes over 40 modules that can interact through the multi-physics option. The fields of application are electrical, structural & acoustic, fluid & heat and chemical problems. Furthermore, there are other modules to interface the platform with other software (AutoCAD, SolidWORKS, etc.). As the other two packages, COMSOL contains its own material library completely integrated with the modules. It has a well-structured and simple user interface. COMSOL allows modifying some equations inside of the modules. The package is available on a paid license basis.

The above described tools allow sufficiently expert users to simulate the sort of events we are interested here. In the case of radiological, chemical or biological dispersion, through the proper model, the user can obtain results accurate enough to plan for an almost immediate intervention. Anyway, as mentioned before, this approach requires a long computational time to simulate even an event of brief duration. Therefore, numerical simulations can be used to study and reproduce possible scenarios and can help in some design decision. Lazaroaie et al. [47] showed the capability of CFD to study the aerodynamics in the vicinity and inside a system of collective protection (COLPRO). Through numerical simulation, it is possible to understand not only the fluid dynamics of this structure but also dosimetric characteristics (through post-processing) and mechanical resistance, providing useful insight for the design, positioning and protection of the structure. Similarly, these tools can be applied to study the operating characteristics of individual pieces of equipment, and extract operational details. Y-C. Su and C-C Li [48] studied fluid dynamics in homogenous media to improve industrial-grade gas mask canisters. J-F Ciparisse et al. [49] presented numerical simulations of hazardous resuspensions: a loss of vacuum accident inside a cylindrical camera, validated by experimental data; tungsten powder in a cavity removed by high-speed (100 m/s) airflow; toxic gas release (chlorine) in air from a vertical pipe.

As mentioned above, numerical simulations are not restricted only to dispersion and structural issues but can solve other kinds of natural or human-induced disasters. For example, T. Baba and his group [50] modelled the far-field tsunami caused by the 2011 Tohoku earthquake, calculating the water heights (experimentally validated) in neighbouring regions. A tool with these features could help in building design, emergency plans and protection improvement [51–53].

3   An Example: CFD Simulation of an Anthrax Attack in a Subway Station

"Anthrax endospores were used in the past to attempt to the life of civilians, like in the failed attack in Tokyo in 1993 or the 22 cases, including five deaths, occurred in USA in 2001. Because of the high case-fatality rate (CFR around 50% for the respiratory form of anthrax), of their strong resistance to adverse environmental conditions and of their high mobility in the near-ground atmosphere, anthrax endospores represent the ideal agent to perpetrate an offensive attack. Subway stations are densely populated places and are therefore a likely objective for terrorists. The aim of this work is to determine, by means of Computational Fluid Dynamics simulations of the airflow dispersion of anthrax endospores released on the platform of a subway station, the time evolution of their concentration, and to estimate the related health risk. COMSOL Multiphysics(TM) software was used to simulate the multiphase flow.

Fig. 1 The computational domain (This figure corresponds to Figure 1 of [55])

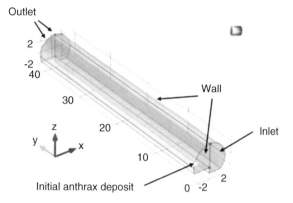

3.1 Aim of the Study

The CFD code COMSOL Multiphysics(TM), was used to simulate the dispersion of anthrax endospores in a subway station. The event was modeled as follows: less than a cubic centimetre of anthrax is dropped on the platform near the tunnel exit (yellow star), as shown in Fig. 1 (all the reported dimensions are in metres).

When the train arrives in the proximity of the station, it generates a flow that blows the powder, making it disperse in the air. People the platform are therefore contaminated by both inhalation and skin contact. In this paper, only the inhalation route was considered for the health effects, due to its much greater infectious effectiveness with respect to the skin contact. As it has been explained, this attempt does not require to access the ventilation system to disperse the anthrax endospores. The results will show how much this stealth and very easy technique is effective and therefore how dangerous it is.

3.2 Simulation Settings

The anthrax endospores features are reported in Table 1. This table corresponds to Table 1 of [55].

In order to simulate the flow, a multiphase turbulent model (Euler-Euler) has been used to take into account the interaction between the solid dispersed particles and the surrounding air (continuous phase). The full set of equations governing the flow is:

$$\rho = \rho_c \cdot (1 - \varphi_d) + \rho_d \cdot \varphi_d{}^\circ \tag{1}$$

$$\partial \rho / \partial t + \nabla_- \cdot (\rho V_-) = 0 \tag{2}$$

Table 1 Anthrax endospores features

Diameter (μm)	Density (kg/m3)	Shape	Number of spores per cluster[a]
5	1100	Spherical	10

[a]Due to the presence of humidity, single spores tend to form little clusters

$$P \cdot \left(\frac{\partial \vec{V}}{\partial t} + \left(\vec{V}\vec{\nabla} \right)\vec{V} \right) = \vec{\nabla} \cdot \left(\begin{array}{l} \left(-p - \frac{2}{3}\left(\mu + \mu_T \right)\left(\vec{\nabla} \cdot \vec{V} \right) - \frac{2}{3}\rho k \right) \cdot I \\ + \left(\mu + \mu_T \right)\left(\vec{\nabla}\vec{V} + \left(\vec{\nabla}\vec{V} \right)^T \right) \end{array} \right)$$

$$- \vec{\nabla} \cdot \left[\begin{array}{l} \rho \cdot c_d \cdot \left(1 - c_d \right) \cdot \left(\vec{u}_{slip} - \frac{D_{md}}{1 - c_d} \cdot \frac{\vec{\nabla}\varphi_d}{\varphi_d} \right) \\ \cdot \left(\vec{u}_{slip} - \frac{D_{md}}{1 - c_d} \cdot \frac{\vec{\nabla}\varphi_d}{\varphi_d} \right)^T \end{array} \right] + \rho \cdot \vec{g} \tag{3}$$

$$\left(\rho_c - \rho_d \right) \cdot \left\{ \vec{\nabla} \cdot \left[\varphi_d \cdot \left(1 - c_d \right) \cdot \vec{u}_{slip} - D_{md} \cdot \vec{\nabla}\varphi_d \right] - \vec{\nabla}\left(\varphi_d \cdot \vec{u}_d \right) \right\} + \rho_c \cdot \vec{\nabla} \cdot \vec{V} = 0 \tag{4}$$

$$\vec{u}_d = \vec{V} + \left(1 - c_d \right) \cdot \vec{u}_{slip} - D_{md} \cdot \frac{\vec{\nabla}\varphi_d}{\varphi_d} \tag{5}$$

$$c_d = \frac{\rho_d \cdot \varphi_d}{\rho} \tag{6}$$

$$D_{md} = \frac{\mu_T}{\rho \cdot \sigma_T} \tag{7}$$

$$3 \cdot \cdot \frac{C_d \cdot \rho_c}{4 d_d} \cdot \left| \vec{u}_{slip} \right| \cdot \vec{u}_{slip} = -\left(\rho - \rho_d \right) \cdot \left(-\frac{\partial \vec{V}}{\partial t} - \vec{V} \cdot \left(\vec{\nabla} \cdot \vec{V} \right) + \vec{g} \right) \tag{8}$$

$$\mu = \frac{\mu_c}{\left(1 - \frac{\varphi_d}{\varphi_{max}} \right)^{2,5 \cdot \varphi_{max}}} \tag{9}$$

$$C_d = \begin{cases} \dfrac{24}{R_{ep}} \cdot \left(1 + 0.15 \cdot R_{ep}^{0.687} \right) \text{for } R_{ep} < 1000 \\ 0,44 \text{ for } R_{ep} > 1000 \end{cases} \tag{10}$$

$$R_{ep} = \frac{d_d \cdot \rho_c \cdot \left| \vec{u}_{slip} \right|}{\mu} \tag{11}$$

$$\rho \cdot \left(\frac{\partial k}{\partial t} + \left(\vec{V} \vec{\nabla} \right) k \right) = \vec{\nabla} \cdot \left[\left(\mu + \frac{\mu_T}{\sigma_k} \right) \vec{\nabla} k \right] + P_k - \rho \epsilon \qquad (12)$$

$$\rho \cdot \left(\frac{\partial \varepsilon}{\partial t} + \left(\vec{V} \vec{\nabla} \right) \varepsilon \right) = \vec{\nabla} \cdot \left[\left(\mu + \frac{\mu_T}{\sigma_\epsilon} \right) \vec{\nabla} \epsilon \right] + C_{\epsilon 1} \frac{\epsilon}{k} P_k - C_{\epsilon 2} \rho \frac{\epsilon^2}{k} \qquad (13)$$

$$P_k = \mu_T \cdot \left[\vec{\nabla} \vec{V} : \left(\vec{\nabla} \vec{V} + \left(\vec{\nabla} \vec{V} \right)^T - \frac{2}{3} \cdot \left(\vec{\nabla} \cdot \vec{V} \right)^2 \right) - \frac{2}{3} \rho k \cdot \left(\vec{\nabla} \cdot \vec{V} \right) \right] \qquad (14)$$

$$\mu_T = \rho C_\mu \frac{k^2}{\epsilon} \qquad (15)$$

(1) average density equation; (2) continuity equation; (3) Navier-Stokes equation; (4) mass balance between the two phases equation; (5) dispersed phase velocity vector; (6) dispersed phase mass fraction; (7) dispersed phase turbulent diffusion coefficient; (8) momentum balance between the two phases equation; (9) effective viscosity equation; (10) drag coefficient calculation (Schiller-Naumann model); (11) particle-based Reynolds number; (12) turbulent kinetic energy conservation equation; (13) turbulent kinetic energy dissipation rate conservation equation; (14) turbulent kinetic energy production rate; (15) turbulent viscosity calculation.

It has been assumed that the air has a density of 1.29 kg/m³ and that its viscosity is 1.78e-5 Pa*s.

Initially, the air is at rest and a t a pressure of 1 atm; the anthrax endospores are put on the platform in the marked point. The powder deposit has a cylindrical shape, with a radius equal to 0.5 cm and a thickness equal to 0.25 cm. The endospores have a maximum volume packing factor of 0.62. The wind enters at 20 km/h, and the pressure at the exit of the domain is set to 1 atm.

The actual volume of the anthrax clusters is therefore:

$$v_a = \varphi_{in} \cdot \pi R^2 h = 0.196 \, cm^3 \qquad (16)$$

The number of endospores contained in the deposit is:

$$N_a = \frac{v_a}{\frac{4}{3} \pi \left(\frac{d_d}{2} \right)^3 \cdot N_b} = \frac{6 \varphi_{in} \cdot R^2 \cdot h}{d_d^3 \cdot N_b} = 1.86 \cdot 10^8 \qquad (17)$$

3.3   Simulation Results and Discussion

The number of anthrax endospores per volume unit, n, is derived from their volume fraction as follows:

$$v_a = \varphi_{in} \cdot \pi R^2 h = 0.196\, cm^3 \quad n = \frac{N}{V_{tot}} = \frac{V_d \cdot \frac{\rho_d}{m_d}}{V_{tot}} = \phi_d \cdot \frac{\rho_d}{m_d} \tag{18}$$

Where, N is the number of particles, V_{tot} is the volume of the air-particles mixture, V_d is the volume of the dispersed phase and m_d is the mass of each particle. As m_d is calculated as follows:

$$m_d = \rho_d \cdot \frac{4}{3} \pi \cdot \left(\frac{d_d}{2} \right)^3 \tag{19}$$

We finally get:

$$n = \frac{6\varphi_d}{\pi d_d^3} \tag{20}$$

As each endospore contains N_b spores, the number of bacteria per volume unit is:

$$n_b = N_b \cdot n = N_b \cdot \frac{6\varphi_d}{\pi d_d^3} \tag{21}$$

It is assumed that $N_b = 10$.

In Fig. 2, the spatial distribution of the spores' concentration (in logarithmic scale) and the streamlines are shown after 0.6, 1.2, 1.8, 2.4 and 3 s.

As it can be seen, the formation of a trapped vortex behind the wall occurs just after the wind has started blowing. The powder is entrained by this vortex, which inseminates the main flow.

In order to evaluate the local effect of the attempt, the number of inhaled endospores during the simulation time was calculated in each point of the computational domain. It was assumed that a person breaths at a flow rate $Q = 10$ L/min and that all the endospores which entered in the lungs remain inside them. The number of inhaled endospores is therefore:

$$\psi(x,,y,,z) = \int_0^{\Delta t} n_b(x,,,y,,,z,,,\tau) \cdot Q \cdot d\tau = \frac{6Q \cdot N_b}{\pi d_d^3} \int_0^{\Delta t} \varphi_d(x,,,y,,,z,,,\tau) d\tau \tag{22}$$

In Fig. 3, the number of inhaled anthrax endospores during 3 s is shown in logarithmic scale:

In order, the determine the local effectiveness of the attempt, the infection probability, p, has to be correlated to ψ. An empirical law (the equation was obtained by linear regression from the graph reported by Toth et al. [54]) has been used to calculate the local probability of an infection:

$$\xi\left(x,,y,,z\right)=p\left(\psi\right)=e^{\frac{-a}{\psi}}=\exp\left[\frac{-a}{\dfrac{6Q\cdot N_b}{\pi d_d^3}\cdot \int_0^{\Delta t}\varphi_d\left(x,,,y,,,z,,,\tau\right)d\tau}\right] \qquad (23)$$

Fig. 2 n_b at 0.6 s (**a**), 1.2 s (**b**), 1.8 sec (**c**), 2.4 s (**d**) and 3.0 s (**e**) with streamlines ($\log_{10}(n_b*1m3)$) (This figure corresponds to Figure 2 of [55])

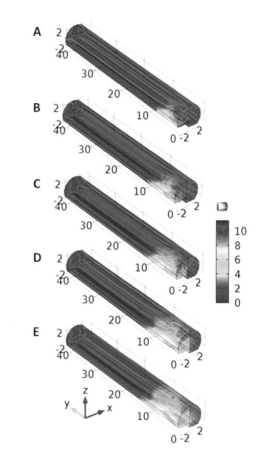

Fig. 3 ψ ($\log_{10}\psi$) – (This figure corresponds to Figure 3 of [55])

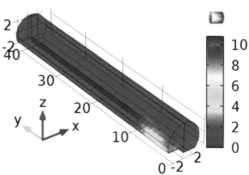

where a = 6694. Now, assuming the lethality l of an infection is meanly about 50% (FDA 2015), the local mortality will be:

$$\omega\left(x,,y,,z\right)=1\cdot\xi\left(x,,y,,z\right)=1\cdot\exp\left[\cfrac{-a}{\cfrac{6Q\cdot N_b}{\pi d_d^3}\cdot\int_0^{\Delta t}\varphi_d\left(x,,,y,,,z,,,\tau\right)d\tau}\right] \quad (24)$$

It is so possible to have an idea of how much an area is dangerous for the people staying there. To determine the risk distribution over space, ω has been evaluated on three horizontal planes, 0.8, 1.3 and 1.8 m above the platform (nose level for children, adolescents and adults), as shown in Fig. 4:

As it can be seen, the mortality is very high in the first 10 m, i.e. in the part of the subway station which is blown by the wind in the first 3 s, and especially near the platform, that means that the zone below 1 meter above the platform is the most dangerous. The simulation demonstrates that the simple air flow generated by the incoming train is fully sufficient to deadly harm many people staying on the plat-

Fig. 4 ω at z = 0.8 m (**a**), z = 1.3 m (**b**) and at z = 1.8 m (**c**) (This figure corresponds to Figure 4 of [55])

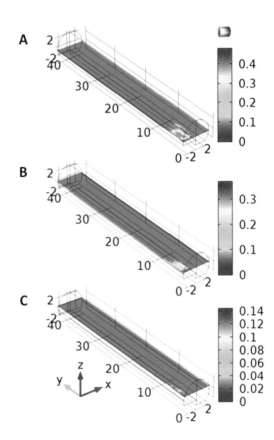

form in the surroundings of the deposited anthrax endospores powder. The simulation was carried out only for a 3-s duration, but, as it can be imagined, the endospores continue moving (and infecting people) even after the train has stopped.

3.4   Conclusive Remarks About the Simulation

The CFD approach has led to the determination of the endospores mobilisation speed. It has been found that, due to the extremely small diameter of the anthrax endospores, the slow wind generated by a train approaching the station is fast enough to disperse almost all endospores initially put on the platform. That means that such a diffusion can easily be made with discretion and is likely to be fully successful in spite of its simple realisation. Further refinements of the model can take into account the moisture influence on the solid particles mobilisation, in order to predict in a more realistic way the actual consequences of this kind of attempt" [55].

4   Conclusion

In the last decades, there was an increase of CBRNe events and their dangerousness. Chemical, biological and radiological releases and dispersions have been improved and their use took place not only in wars but also in terroristic attempts and accidental events. These events have involved the development of new tools, networks and approaches to face the various risks. Between these new techniques, several software tools have been developed to aid first responders and decision makers.

These software can be divided into two main classes: decision and numerical simulation tools. The decision tools are software able to give results in very short time with a limited accuracy. At the contrary, numerical simulations directly solve the governing equations of the phenomenon. Therefore, these have a large accuracy but usually require a very large time of computation.

In this works, the authors showed some of the most spread tools in these fields. The decision tools described are HotSpot, ALOHA, WISER and STEM. Hotspot is a code developed to reproduce the radiological dispersion of radionuclides through a Gaussian plume model. It is used to make fast results in critical situations. ALOHA is used to reproduce chemical dispersion in the atmosphere. It allows the use of several models such as toxic vapor clouds, pool fires and explosions. WISER is a huge library of materials with several information: physical characteristics, containment, identification, suppression and effects on human health. It can be downloaded onto several devices, such as smart phones, personal computers and tablets. The last decision software is STEM, a tool used in biology to simulate the diffusion of biological infections. It is able to provide many information about the infection and its critical aspects.

The numerical simulation tools described are three: OpenFOAM, ANSYS and COMSOL Multiphysics. OpenFOAM is an open source software where a large number of physics can be simulated: fluid dynamics, solid mechanics, electromagnetics, etc. It allows to enter the code and modify it, in order to achieve the desired solution. Anyway, it is very hard to use and it requires experience to simulate also simply events. ANSYS is a suite of several models that allows simulating a many different events and physics. It is not open source but the are several options concerning the models that allow to personalize the simulation and achieve good results. COMSOL is an integrated platform where it is possible to build the simulation from its geometry to the post processing. Furthermore, it has a friendly user interface that makes the software easier to use. It gives also the possibility to create applications.

The case which was considered in the numerical simulation carried out with COMSOL is an anthrax dispersion in a subway station. After solving the unsteady flow which is produced by the incoming train and which blows a little *Bacillus anthracis* endospores dropped on the platform, a clinical model was used to elaborate the information outcoming from the simulation and to get the mortality field in the station. It was so shown that the combination of CFD and biology/medicine knowledge is not only possible, but also highly recommendable to predict the effects of an intentional or accidental release of harmful agents, as bacteria. This approach has to be preferred to the one relying on decision tools when the need of accuracy and activities planning is much more important than the rapidness of the results obtaining.

References

1. Love, D., The Second Battle of Ypres, Apr-1915. Sabretasche, 26(4), (1996)
2. Connelly, M., The ypres league and the commemoration of the ypres salient, 1914–1940, War in History, 16(1), 51–76 (2009)
3. Singer, B. C., Hodgson, A. T., Destaillats, H., Hotchi, T., Revzan, K. L. and Sextro, R. G., Indoor Sorption of Surrogates for Sarin and Related Nerve Agents, Environ. Sci. Technol. 39, 3203–3214 (2005)
4. National Institute of Standards and Technology., http://webbook.nist.gov/cgi/cbook.cgi?Name =sarin&Units=SI&cMS=on
5. Okumura, T., Suzuki, K., Fukuda, A., Kohama, A., Takasu, N., Ishimatsu, S. and Hinohara, S., The Tokyo subway sarin attack: disaster management, Part 1: Community emergency response, Acad. Emerg. Med. 5(6) , 613–617 (1998)
6. MSNBC, Bomb said to holddeadly sarin gas explodes in Iraq., http://www.nbcnews.com/id/4997808/#.WI9E0_nhBPZ
7. Medecins Sans Frontieres, Syria: Thousands suffering neurotoxic symptoms treated in hospitals supported by MSF., http://www.msf.org/article/syria-thousands-suffering-neurotoxic-symptoms-treated-hospitals-supported-msf
8. OPCW, Convention on the prohibition of the development, production, stockpiling and use of chemical weapons and on their destruction, (2005)
9. Meyer, M. T., Spinella, P. C. and Cieslak, T., Agents of Biological and Chemical Terrorism, Pediatric Critical Care Medicine, 645–656 (2014)

10. Bologna, M., Biological Agents and Bioterrorism, in Detection of Chemical, Biological, Radiological and Nuclear Agents for the Prevention of Terrorism, 2914, 1–10
11. Beeching, N. J., Fenech, M. and Houlihan, C. F., Ebola virus disease, British Medical Journal, 349 (2014)
12. Laupland, K. B., Ebola virus disease, J. Infect. Dis. Med. Microbiol. 25(3), 128–129 (2014)
13. Shweta, S. and Jagtar, S., Classification, Causes, Control Measures and Acts of Bioterrorism, International Journal of Applied Biology and Pharmaceutical Technology 7(2), 342–354, (2016)
14. Cenciarelli, O., Pietropaoli, S., Malizia, A., Carestia, M., D'Amico, F., Sassolini, D., Di Giovanni, D., Rea, S., Gabbarini, V., Tamburrini, A., Palombi, L., Bellecci, C., and Gaudio, P., Ebola Virus Disease 2013–2014 Outbreak in West Africa: An Analysis of the Epidemic Spread and Response, International Journal of Microbiology 2015, 12 (2015)
15. Cenciarelli, O., Pietropaoli, S., Frusteri, L., Malizia, A., Carestia, M., D'Amico, F., Sassolini, D., Di Giovanni, D., Tamburrini, A., Palombi, L., Bellecci, C., and Gaudio, P., Biological Emergency Management: The Case of Ebola 2014 and the Air Transportation Involvement, J. Microb. Biochem. Technol. 6(5), 247–253 (2014)
16. Cenciarelli, O., Gabbarini, V., Pietropaoli, S., Malizia, A., Tamburrini, A., Ludovici, G., Carestia, M., Di Giovanni, D., Sassolini, A., Palombi, L., Bellecci, C., and Gaudio, P., Viral bioterrorism: Learning the lesson of Ebola virus in West Africa 2013–2015, Virus Res. 210, 318–326 (2015)
17. Carrol, L. N., Au, A. P., Detwiler, L. T., Fu, T. C., Painter, I. S., and Abernethy, N. F., Visualization and analytics tools for infectious disease epidemiology: A systematic review, Journal of Biomedical Informatics 51, 287–298 (2014)
18. Canadian Nuclear Safety Commission, Types and sources of radiation, http://nuclearsafety.gc.ca/eng/resources/radiation/introduction-to-radiation/types-and-sources-of-radiation.cfm
19. Green, A. R., Erhardt, L., Lebel, L., Duke, M. J. M., Jones, T., White, D., and Quayle, D., Overview of the Full-scale radiological dispersal device field trials, Health Physics, 110(5), 403–417 (2016)
20. Cornish, P., The CBRN System – Assessing the threat of terrorist use of chemical, biological, radiological and nuclear weapons in the United Kingdom, Chatham House, 2007
21. Malizia A., Poggi, L. A., Ciparisse, J.-F., Rossi, R., Bellecci, C., and Gaudio, P., A Review of Dangerous Dust in Fusion Reactors: from Its Creation to Its Resuspension in Case of LOCA and LOVA, Energies 9(8) (2016)
22. Steinhauser, G., Brandl, A., and Johnson, T. E., Comparison of the Chernobyl and Fukushima nuclear accidents: A review of the environmental impacts, Science of The Total Environment, 470–471, 800–817 (2014)
23. National Atmospheric Release Advisory Center, HotSpot ~ Health Physics Codes for the PC, https://narac.llnl.gov/hotspot
24. Richter, C., State of the art atmospheric dispersion modelling: should the Gaussian plume model still be used?, Kerntechnik 81(5), 559–564 (2016)
25. Homann, G. and Aluzzi, F., HotSpot – Health Physics Codes – User's Guide, Livermore (2015)
26. Eckerman, K. F., Wolbarst, A. B. and Richardson, A. C. B., Limiting values of radionuclide intake and air concentration and dose conversion factors for inhalation, submersion, and ingestion: Federal guidance report No.11, Environmental Protection Agency, Washington DC (1988).
27. Eckerman, K. F. and Ryman, J. C., External exposure to radionuclides in air, water, and soil, EPA (Environmental Protection Agency), Washington DC, 1993
28. Pawel, D., Leggett, R. W., Eckerman K. F. and Nelson, C., Uncertainties in cancer risk coefficient for environmental exposure to radionuclides. An uncertainty analysis for risk coefficients reported in federal guidance report No.13, EPA (Environmental Protection Agency), Washington DC (1999)
29. International Commission on Radiological Protection (ICRP), Basic Anatomical & Physiological Data for use in Radiological Protection: The Skeleton, Elsevier, (1996)

30. Di Giovanni, D., Luttazzi, E., Marchi, F., Latini, G., Carestia, M., Malizia, A., Gelfusa, M., Fiorito, R., D'amico, F., Cenciarelli, O., Gucciardino, A., Bellecci, C. and Gaudio, P., Two realistic scenarios of international release of radionuclides (Cs-137, Sr-90) – the use of the HotSpot code to forecast contamination extent, in WSEAS Transactions on Environment and Development (2014)
31. Caciotti, I., Aspetti, P. C., Cenciarelli, O., Carestia, M., Di Giovanni, D., Malizia, A., D'Amico, F., Sassolini, A., Bellecci, C. and Gaudio, P., Simulation of Caesium-137 (137Cs) local diffusion as a consequence of the Chernobyl accident using HotSpot. Defence S. and T. Technical Bulletin 7(1), 18–26 (2014)
32. EPA – Environmental Protection Agency, ALOHA Software, https://www.epa.gov/cameo/aloha-software
33. EPA – Environmental Protection Agency, MARPLOT Software, https://www.epa.gov/cameo/marplot-software
34. National Oceanic and Atmospheric Administration and U.S. Environmental Protection Agency, ALOHA – Example Scenarios, (2016)
35. Kulynych, V. and Maruta, M., ALOHA – modern tool for modeling the risks associated with the spread of volatile pollutants in extraction of hydrocarbons, AGH Drilling, Oil, Gas, 33(2), 315–322 (2016)
36. Bhattacharya, R. and Ganesh Kumar, V., Consequence analysis for simulation of hazardous chemicals release using ALOHA software, International Journal of Chem Tech Research 8(4), 2038–2046 (2015)
37. U.S. National Library of Medicine, "WISER – Wireless Information System for Emergency Responders, https://wiser.nlm.nih.gov/about.html
38. Eclipse, Spatio-Temporal Epidemiological Modeler – STEM., http://wiki.eclipse.org/STEM
39. Sevilla, N., Global Biodefense https://globalbiodefense.com/2016/01/28/open-source-disease-modeling-a-tool-to-combat-the-next-pandemic/
40. Dhatt, G., Touzot, G., and Lefrançois, E., Finite Element Method, Wiley (2012)
41. Smith, G.D., Numerical Solution of Partial Differential Equations, Clarendon Press, (1985)
42. Luding, S., Introduction to Discrete Element Methods. European Journal of Environmental and Civil Engineering 12, 785–826 (2008)
43. Schaler, M., Finite-Volume Methods, in Computational Engineering – Introduction to Numerical Methods, Springer, 77–105 (2006)
44. Bern, M. and Plassmann, P., Mesh generation, in Handbook of Computational Geometry, pp. 291–333, ELSEVIER SCIENCE Amsterdam (2000)
45. OpenFOAM, The Open Source CFD Toolbox – User Guide (2017)
46. ANSYS., http://www.ansys.com/
47. Lăzăroaie, C., Zecheru, T., Său, C. and Cherecheş, T., Airflow Modeling and Simulation Required in CBRN Collective Protection Design, International conference KNOWLEDGE-BASED ORGANIZATION 22(3), 649–653 (2016)
48. Yin-Chia, S. and Chun-Chi, L., Computational fluid dynamics simulations and tests for improving industrial-grade gas mask canisters, Advances in Mechanical Engineering, 7(8), 1–14 (2015)
49. Ciparisse, J.-F., Malizia, A., Poggi, L. A., Cenciarelli, O., Gelfusa, M., Carestia, M., Di Giovanni, D., Mancinelli, S., Palombi, L., Bellecci, C. and Gaudio, P., Numerical Simulations as Tool to Predict Chemical and Radiological Hazardous Diffusion in Case of Nonconventional Events, Model. Simul. Eng., (2016)
50. Baba, T., Sebastien, A., Hossn, J., Cummins, P. R., Tsushima, H., Imai, K., Yamashita, K. and Kato, T. Accurate numerical simulation of the far-field tsunami caused by the 2011 Tohoku earthquake, including the effects of Boussinesq dispersion, seawater density stratification, elastic loading, and gravitational potential change, Ocean Modelling, (2017)
51. Cacciotti, I., Aspetti, P. C., Cenciarelli, O., Carestia, M., Di Giovanni, D., Malizia, A., D'Amico, F., Sassolini, A., Bellecci, C. and Gaudio, P., Simulation of Caesium-137 (137CS) local diffusion as a consequence of the Chernobyl accident using HOTSPOT. Defence S&T Technical Bulletin 7, 18–26 (2014)

52. Di Giovanni, D., Luttazzi, E., Marchi, F., Latini, G., Carestia, M., Malizia, A., Gelfusa, M., Fiorito, R., D'Amico, F., Cenciarelli, O., Guicciardino, A., Bellecci, C., Gucciardino, A., and Gaudio, P., Two realistic scenarios of intentional release of radionuclides (Cs-137, Sr-90)–the use of the HotSpot code to forecast contamination extent. WSEAS Trans. Environ. Dev. 10, 106–122 (2014)
53. Baldassi, F., D'Amico, F., Carestia, M., Cenciarelli, O., Mancinelli, S., Gilardi, F., Malizia, A., Di Giovanni, D., Soave, P.M., Bellecci, C., Palombi, L. and Gaudio, P., Testing the accuracy ratio of the Spatio-Temporal Epidemiological Modeler (STEM) through Ebola haemorrhagic fever outbreaks. Epidemiology and Infection, 1–10
54. Toth, D. J. A., Gundlapalli, A. V., Schell, W. A., Bulmahn, K., Walton, T. E., Woods, C. W., Coghill, C., Gallegos, F., Samore, M. H. and Adler, F. R., Quantitative models of the dose-response and time course of inhalational anthrax in humans. PLoS Pathog, 9(8), (2013).
55. Jean-François Ciparisse, Orlando Cenciarelli, Sandro Mancinelli, Gian Marco Ludovici, Andrea Malizia1, Mariachiara Carestia, Daniele Di Giovanni, Carlo Bellecci, Leonardo Palombi & Pasquale Gaudio, A CFD simulation of anthrax diffusion in a subway station, Journal of Mathematical Models and Methods in Applied Sciences (2016)

Migration and Terrorism: A New Approach to Consider the Threat

Orlando Cenciarelli, Sandro Mancinelli, Gian Marco Ludovici,
and Leonardo Palombi

Abstract Migration is part of human history since ancient times; this phenomenon can result from several reasons, through which are delivered, as well as individuals, the main characteristics of the societies of origin. Currently, several theories relate the spread of international terrorism with migration, particularly by supporting the close relationship between migration, religious extremism and terrorist events. Other literature considers terrorism as a phenomenon with characteristics similar to a disease, for many common aspects with specific illness, such as cancer or psychiatric pathologies. In this review, the correlations between migration and terrorism are analysed and the different theories that consider terrorism similar to a disease are considered.

Keywords Migration • Terrorism • CBRNe • Disease • Threat

1 Introduction

Since the beginning of the twenty-first century, a globalisation of infectious diseases, which occur with unprecedented speed and frequency, was observed. In the globalised world, many factors – such as travels, migrations and international economy – play an important role in the emergence and re-emergence of infectious diseases, requiring a coordinated global response [1]. Emerging infectious diseases (EIDs) are defined as pathologies recently appeared or that have increased their

O. Cenciarelli (✉)
Department of Industrial Engineering, University of Rome Tor Vergata, Rome, Italy
e-mail: orlando.cenciarelli@uniroma2.it

S. Mancinelli • L. Palombi
Department of Biomedicine and Prevention, School of Medicine and Surgery, University of Rome Tor Vergata, Rome, Italy
e-mail: sandro.mancinelli@uniroma2.it; palombi@uniroma2.it

G.M. Ludovici
International Master Courses in Protection Against CBRNe Events, University of Rome Tor Vergata, Rome, Italy
e-mail: gianmarco.ludovici@gmail.com

© Springer International Publishing AG 2017   277
M. Martellini, A. Malizia (eds.), *Cyber and Chemical, Biological, Radiological, Nuclear, Explosives Challenges*, Terrorism, Security, and Computation,
DOI 10.1007/978-3-319-62108-1_13

geographical distribution and frequency [2]. EIDs and reemerging infectious diseases (ReIDs) arise from several factors, which need to be faced with a coordinated approach among the various sectors of society including public health, environmental, economics, and public policy [1]. However, human and environmental factors make these actions difficult to manage; just to mention some of that issues, the human susceptibility to infection, demographics, trade, environmental change, economic development, wars, efficiency of public health infrastructure, an evolution of the pathogen (e.g., antimicrobial resistance) or its adaptation, and migrations [3] contribute to increase the challenge against EIDs and ReIDs.

Movement of humans plays a fundamental role in the occurrence of infectious diseases. Migrations, throughout history, have been and they still constitute an emergency, increasing the frequency and the spreading of infection in geographical areas and among populations [4]. Currently, significant migrations, in terms of volume and speed, are occurring; the impact of these events on the population and on the ecosystem cannot be neglected. Migrations associated with wars or violent conflicts, as well as environmental disasters, can lead to the burst of outbreaks of infectious diseases (e.g. cholera and typhoid fever), which are the consequence of the overcrowding population, poor sanitation and lack of basic medical care [5]. One of the diseases most frequently associated with these situations is cholera (e.g. during 2010 in Haiti) [5]. During twentieth and twenty-first centuries, large migrations caused by wars, genocide, ethnic cleansing, civil unrest, economic crises and natural disasters have happened [6]. At the end of the twentieth century, around 150 million people lived elsewhere by their country of origin; of these, about 15 million were refugees [6]. Considering these numbers, it is easy to understand how the large migrations create opportunities for the dissemination of infectious diseases. The diseases imported after human movements can be classified as (I) imported cosmopolitan diseases, when they are outsourced globally and (II) imported tropical diseases, which they exist in tropical and subtropical areas, but they are almost inexistent in developed countries because the lack of a carrier or thanks to eradication [2, 7].

2   The Human Migrations

Humans, in their long history, have shown a strong propensity to move from their place of origin in order to seek new territories [8]. Just as an example, when a gain-of-function genetic mutation occurred, or after an innovation resulting in the increase the population in a territory, a part of the population was stimulated to move away from the source group to go looking for new spaces with more resources [9]. During the history, migrations resulted in dramatic effects, such as the smoothing out of the genetic differences, up to lead to the formation of a single human species on the entire planet [9, 10]: the great migrations of the past have homogenized our species and, to some extent, they accelerated the current social organization, that without the constraints of migrations, would not have developed as we know it today [11, 12]. In more recent times, from the late nineteenth century,

Europe began to experience a mass migration, especially toward the America [13, 14]. In particular, it was estimated that $\approx$ 60 million Europeans migrated in the period 1820–1940, 38 million moved to the United States. The migration from Europe took place on a voluntary basis, however, migration flows to the Americas due to slavery (mainly from Africa) need to be considered. The most large-scale migration flows that occurred during the nineteenth century covered a wide spectrum of cultural backgrounds [14].

Europe, in recent time, experienced a radical change in terms of migrations [15]. The old continent, once the centre of emigration, is become the primary destination of large-scale immigration. After World War II the reconstruction process and the subsequent industrial development attracted millions of people from the countries of southern Europe to the North and Center nations (especially Germany, Belgium, Switzerland and France) [15]. Some decades later, since the seventies of the last century, these movements ceased and nations such as Portugal, Spain, Italy and Greece, which for decades were exporters of labour, experienced not only the return of many of their countrymen but also the exhaustion of the phenomenon [15]. With the end of the "internal" migration, more and more massive migrations from Africa, Asia and the Eastern European countries was experienced by the Western Europe [16]. Currently, the geopolitical situation in the North Africa and the Middle East has led to what has been regarded as the first great migration of the third millennium, which has as its epicentre the Europe [15].

2.1 Management of Migration Flows: Causes, Migrants and Refugees

Today international migrations reached dimensions unknown in previous centuries, in part due to the increasingly rapid development of means of transport [17]. At the institutional level, the need for labour on the part of some countries has been a strong factor of the attraction of migrants. In many countries, several business sectors depend almost entirely on the presence of immigrant workers and, in some cases, many of these immigrants have been encouraged or recruited to fill positions available in periods of economic expansion [18]. Frequently immigrant workers perform hard works, low paid, with less social protection and in times of economic difficulties, and are also the first to be expelled from the production process [18]. In addition to the more traditional factors of attraction, the globalisation of the economy in its current form has also led to a globalisation of the labour market, despite the restrictive measures adopted by many governments of industrialised countries to limit migration to their countries [19]. Focusing on some trends relating to international migration, it is important to highlight as the increasing in migration all over the world is mainly due to the increasing economic difficulties and to the collapse of economic balance, political, social and environmental factors that allow people to live and stay in their own countries [19]. Moreover, the number of people who are forced to emigrate due to armed conflicts or political persecution (which occur in most of individual countries and less between a country and another), or because disastrous effects of some natural phenomena (floods, hurricanes, droughts, desertification, etc.) [19] is increasing. A very eloquent example is the migration flow coming from the Balkans as a result of various conflicts that have occurred in the area in the past but in recent times [20]. An alarming trend in the countries hosting the migrants and refugees is the increase in racist and xenophobic hostility against them, seen as scapegoats in a series of social problems, unemployment, and crime in the sense of personal insecurity and loss of social protection (lower welfare), in particular in urban areas [21]. This tendency is most tangible in industrialised countries and, in the case of these countries in Western Europe, the negative portrayal of migrants and hostilities arising therefrom have become part of the political lines and the public discourse of many parties [19].

2.1.1 Internally Displaced Persons (IDPs) and Refugees

Refugees are defined as people forced to leave their country. Among them, the UNHCR (United Nations High Commissioner for Refugees) considers refugees (internally displace persons – IDPs – and refugees) anyone who, as a result of a very well-founded fear of being persecuted for racial, religious and national, or because belongs to a particular social group or professing certain political opinions, lives outside the territory of their country and is unable and even, because of fears, refuses to make use of the homeland security [22].

The boundaries between emigrated (voluntary choice), IDPs (forced for various reasons) and refugees (forced because persecuted) are not always defined [23]. Several countries increasingly tend to consider the refugees and the IDPs as economic migrants, in order to subject them to the immigration rules, which allows them to reject or expel the newcomers, rather than to asylum seekers [24]. According to estimates of the United Nations, at the beginning of 1998, there were more than 120 million people living in countries other than those of origin and 13 millions of them were recognised as refugees by UNHCR [25]. The International Labour Organization (ILO) estimates that international migration affecting about 18 million people in Africa, 7.5 million in Asia, 20 million in Europe, 16 million in North America, 10 million in Central and South America and 7 million in West Asia (Arab states) [26].

3   Could the Migration Be a Vehicle for Terrorism Diffusion?

The international migrations play a key role in the safety program of several countries. The effect of immigration on security, economy, along with the question of whether the economic benefits of immigration exceed its costs represent major concerns [27]. The link between immigration and terrorist attacks has been (and it is still now) widely debated between policy makers and public institutions, resulting in spots for or against the theory [28]. Reports of the International Organization for Migration (IOM) highlighted several areas where migration policies and national security designators intertwine, suggesting that a massive migration in a specific country could result in increased risks for the national security [29]. The same reports warn against links too close between migration and national security [29]; the perception of a link between terrorism and immigration, however, has led in the past, and even now, to legitimise the implementation and application of stricter migration laws, regulations and controls. The United Nations Security Council Resolution 1373, issued after the terrorist attacks on the World Trade Center on Sept. 2001, which encourages states to "prevent the movement of terrorists or terrorist groups by effective border controls and controls on issuance of identity documents, and through measures for preventing counterfeiting, forgery or fraudulent use of identity papers and travel documents" generated an opportunity for the countries to "securitize" the migration, in particular in the United States [30]. The processes of globalisation and the increased mobility of people across borders, however, have made the relationship between migration and security difficult to overlook [31]. In 2006, Adamson examined qualitatively different ways by which the international migration reshapes the security environment, and how migration can affect particularly the security interests of states [27]. International migration flows provide opportunities for new forms of action between different nations, which are used both by political movements, that by terrorist organisations [31]. Nixon Center has recently provided a rating, stressing that: "immigration and terrorism are linked; not because all immigrants are terrorists, but because all, or nearly all, terrorists in the West have been immigrants" [32].

How migration and terrorism could be related? The migration flows affect the patterns of social interaction, making more likely that ties are developed between individuals and international terrorist groups [33]. Starting with these premises, it is conceivable that the migrants could function as a vehicle for the spread of terrorism. In 2004, Sageman describes the process of joining the Jihad and in general the terrorist activities as a three-step process: social belonging, progressive intensification of beliefs and faith, and the formal acceptance [28]. Through these steps, the social ties play a key role as they provide both emotional and social support, making possible the development of a common identity, and encouragement to adopt a new faith. The putative group of terrorists is constituted by clusters of friends or members of the same family, united together by a strong bond: this improves social cohesion, same point of views and loyalties, and a strong sense of community [34]. However, the presence of a pre-existing social framework is a necessary requirement to join in terrorist groups; in many cases, these networks exist long before all members engage in terrorist activities [33]. Migrants can provide such social bonds, and terrorist organisations can exploit them for their own ends. In terrorist groups, each insider is linked to the other through a complex network of direct exchanges and mediated [34]. They are self-organized as they lack a comprehensive recruitment strategy, which implies that the terrorist organisations need to build pre-existing links, nodes, and the networks to pursue their goals: the migrations can provide those linkages [33, 34]. In 2011, Sageman examined the circumstances in which people joined the global Islamic terrorism and found that being an expatriate was a common feature [35]. Join a terrorist organisation depends mainly on overcoming various obstacles to the mobilisation, which often can be achieved thanks to the links between individuals [28].

Leiken and Brooke, in their study of immigration model, reported that the decision to migrate is usually influenced by the presence of relatives and friends which can provide assistance in finding a place to live, work, etc. [36]. Under this umbrella, the potential use of these links by terrorist organisations for the recruitment of new potential members among migrants, and the involvement of already active terrorists as brokers, need to be considered when migration inflows are related to the terrorism diffusion [33]. One relevant example of this process is the Hamburg cell, a group of radical Islamists that became operational after the attacks of 9/11 [33, 35]. The flows of migrants from countries where terrorism is radicalised, may be also related to the emergence of new terrorist movements, as they help the creation and formation of social identity through a process of interaction and socialisation [27].

Moreover, the joining of a new member to a terroristic group is considered as a bottom-up process, where most of the recruits want to participate actively, without knowing how. In this context, originating from the same country facilitates this process [33].

4   Face the Terrorism: The Disease Approach

The result of many analyses suggested that terrorism acts like a disease, as an example the cancer. It has been observed that the challenges to prevent and combating cancer are similar to those associated with the fight against terrorism [37]. Terrorism analysts shown similarities with oncologists: the studies to face terrorism and its actions may be compared with those against cancer; every day something more on the mechanisms is discovered, and new approaches are developed to face the threat [37]. The similarities between terrorism and cancer are startling, from the endogenous and exogenous causes of both events to their behaviour. Cancer is a disease that occurs as a result of mutations in healthy cells, as well as terrorists, are can be considered socially mutated having a completely distorted version of reality, compared to the vision that belongs to common people [37]. Currently, even after the great advances in science, medical doctors still do not have a clear way to distinguish between benign tumours and those highly aggressive, which makes it very difficult to find an early and accurate diagnosis [38]. Likewise, psychiatrists, psychologists and analysts who study terrorism have thus far been unable to identify pathologies in terrorists [39]. According to Horgan, every terrorist has some peculiarities that would allow us to predict the probability, the risk of involvement and engagement in any particular person or social group that is either valid or reliable over a meaningful period of time [40].In many cases, the distinction between cancerous and healthy cells can be as difficult as differentiating between terrorists and non-violent extremists [41]. Therefore, both analysts and oncologists have the problem of identifying which cells and individuals are more susceptible to "mutate" [37]. Any person can develop cancer, however, some individuals are more predisposed to the disease than others, due to mechanisms as hereditary mutations, deficiency of the immune system, and many other factors [42]. Mutations, however, can occur also by environmental factors such as exposure to carcinogens, or overexposure to ultraviolet rays and radiation [42]. The same trends can be observed in terrorism; in this phenomenon, "mutation" should be replaced by "radicalization", and like in cancer, all individuals may be susceptible [43]. There are individuals, more than others, who are social, emotionally, and mentally unstable that may be more susceptible to acting on extremist ideology and propaganda. Other examples relate to individuals who do not have these initial vulnerability but become violent extremists when they are persuaded by extremist groups [44]. Like for cancer, the exact causes which radicalise terrorists are several and unpredictable.

One of the reasons why cancer is such a difficult disease to fight is its ability to metastasize unpredictably. Metastases occur when the primary tumour cells spread in other areas of the body through, for example, the lymphatic system or bloodstream [45]. The application of these concepts to terrorism is intuitive. We might imagine that the cancerous cells are the terrorists, and the hosting community of non-violent citizens represents the non-cancerous cells in which the tumour could grow. The characteristics of the non-cancerous cells can often dictate how easy or difficult it is for the tumour to grow and spread.

In terms of prevention, terrorism and cancer showed many similarities: the development of both of them cannot be currently prevented and both are chronic problems in society. However, over the years, strategies to reduce the development of cancer (e.g., reducing excessive intake of alcohol, smoking, etc...) were developed [46]. Likewise, countries can reduce the development of terrorist groups by enacting similar macro-level behaviours and policies, like providing to the minority groups the access to the political system [47]. The main tactic in the fight the terrorism can be divided into two basic points: (I) preventing radicalization before its occur and (II) eradicate the terrorists without causing collateral damage to civilian populations or creating new terrorists [48].

Terrorism, as well as being considered one of the "body disease", can also be defined as a "brain" disease. Generally, are defined insane people who are guilty of particularly violent and extreme crimes [49]. Literature reports that there is no notable scientific evidence about the correlation between the mental disorder and the terrorist violence [50]. Analysing the phenomenon, from the point of view of psychopathology, we would be immediately aware that we are faced to a spectrum of different terrorist mentalities united by the use of violence, including against civilians, in deliberate intent to achieve ideological, political or psychological goals [51].

Several theories identify a difference between separatist/independence terrorism and terrorism linked to religious extremism [52]. In terrorism linked to religious extremism, there is a differentiating element: the suicidal behaviour of the assailant. Such suicides must be inserted in a cultural (not psychopathological) context in which the martyrdom is set in the foreground: death will give a reward in an afterlife

[53]. People who choose for martyrdom have already had the opportunity to conceptualise about it, participation in the holy war is a consequence of a reflection upstream [52]: suicide is part of martyrdom, the renunciation of earthly life for a greater good. All these factors can be considered as a cultural illness, not a mental illness [54]. Using a psychological approach, an inelastic mental structure, unable to analyse different points of view, can be identified in terrorists; this approach operates selecting what confirms their theories and discarding what causes a crisis, with a tendency to jump to the conclusions without analysing alternatives. However, all these elements highlight a cultural structure rather than psychopathological aspects [53]. The European Counter Terrorism Centre in July 2016, contrariwise, assessed the potential mental illness in several terrorists in the gestures performed by them [55]. In December 2014 two attacks in a similar way have occurred in France. In both cases, the authors seemed to be suffering from mental illness [56]. Olivieri, in 2016, reported that about 35% of the authors of solitaire attacks occurred between 2000 and 2015 have suffered from some kind of mental health disorder. In the Europol report on changes in the modus operandi of IS was also reported that a significant amount of foreign fighters were affected by mental health problems [55]. Recently, the neurologist Kathleen Taylor of Oxford University said that religious fundamentalism and other ideological forms harmful to society, in the near future will be treated as a mental illness [57]. Taylor explains that a general practice carried out before entering in a terrorist organisation is a brainwashing; the perpetrators of this practice acts on the unconscious to induce mental paths that make the brainwashed unable to think back to situations and to recognise these routes in subsequent periods. Brainwash works best on young people, which explains why most of the suicide terrorists are between 20 and 30 years [57]. The latest techniques of brainwashing ranging from challenging the individual belief system to insinuate doubt, from isolating the individual to the dissemination of propaganda by controlling access to information. A similar approach has proved extremely effective, considering the growing number of adherents to the various terrorist groups. The terrorist is no longer being recruited and thought as a weak subject, amorphous, amassed, indeed is chosen for its being a unique individual and the manipulation techniques play precisely on the elements that exalt individualism: the challenge, the self-celebration, the leadership [57].

5   Conclusions

Migratory flows can facilitate the spread of terrorism by providing a network of relationships and trust among migrants, that share a common background, mainly determined by the country of origin or from the same ethnic group. Terrorist organisations can deliberately take advantage of these links for the recruitment of new potential members [35]. This argument leads to the hypothesis that terrorism is more likely to spread to countries with the largest migrant inflows. Cancer and terrorism are both insidious and devious events, characterised by exogenous and endogenous causes, still poorly understood; the medical doctors that try to fight cancer have made many of the mistakes that policymakers have made in countering terrorism [37, 47, 48]. Neurologists considered terrorism a kind of psychological deviance rather than a real mental disorder; they support that the anti-terrorism strategy of the future is the neuroscience, as the developments in research show that one day we will be able to control the mind preventing extreme forms of convictions [57].

Acknowledgements Special acknowledgement for the realisation of this work goes to the International Master Courses in Protection Against CBRNe Events (http://www.mastercbrn.com).

References

1. Mackey, T.K., Liang, B.A., Cuomo, R., Hafen, R., Brouwer, K.C., Lee, D.E: Emerging and Reemerging Neglected Tropical Diseases: a Review of Key Characteristics, Risk Factors, and the Policy and Innovation Environment. Clin. Microbiol. Rev. 27, 949–979 (2014)
2. Soto, S.M.: Human Migration and Infectious Diseases. Clin. Microbiol. Infect. 15, 26–28 (2009)
3. Mackey, T.: Emerging and Reemerging Neglected Tropical Diseases: Review of Key Characteristics and Risk Factors. In 2015 APHA Annual Meeting & Expo. APHA (2015)
4. Tatem, A.J., Rogers, D.J., Hay, S.I.: Global Transport Networks and Infectious Disease Spread. Adv. Parasitol. 62, 293–343 (2006)
5. Mack, A., Choffnes, E.R., Relman, D.A.: Infectious Disease Movement in a Borderless World: Workshop Summary. National Academies Press (2010)
6. Cottle, S.: Global Crises in the News: Staging New Wars, Disasters and Climate Change. Int. J. Comm. 3, 24 (2009)
7. Lee, K.S., Lo, S., Tan, S.S.Y., Chua, R., Tan, L.K., Xu, H., Ng, L.C.: Dengue Virus Surveillance in Singapore Reveals High Viral Diversity Through Multiple Introductions and in Situ Evolution. Infect. Genet. Evol. 12, 77–85 (2012)
8. Johanson, D.: Origins of Modern Humans: Multiregional or Out of Africa? AIBS (2001)
9. Vegvari, C., Foley, R.A.: High Selection Pressure Promotes Increase in Cumulative Adaptive Culture. PLoS ONE 9, e86406 (2014)
10. Stock, J.T.: Are Humans Still Evolving? EMBO Rep. 9, S51–S54 (2008)
11. Bigo, D., Tsoukala, A.: Terror, Insecurity and Liberty: Illiberal Practices of Liberal Regimes After 9/11. Routledge (2008)
12. Zeleza, P. T.: The challenges of studying the African diasporas. African Sociological Review/ Revue Africaine de Sociologie, 12 (2008)
13. Coatsworth, J. H.: Globalization, Growth, and Welfare in History. Globalization: Culture and education in the new millennium, 38–55 (2004)
14. Skeldon, R.: Migration and Development: A global Perspective. Routledge (2004)
15. Geddes, A.: The Politics of Migration and Immigration in Europe. Sage (2003)
16. Castles, S., Miller, M.J.: Migration in the Asia-Pacific Region. Migration Information Source (2009)
17. Czaika, M., Haas, H.: The Globalization of Migration: Has the World Become More Migratory? Int. Migr. Rev. 48, 283–323 (2014)
18. Hirschman, C., Mogford, E.: Immigration and the American Industrial Revolution From 1880 to 1920. Soc. Sci. Res. 38, 897–920 (2009)
19. Olwig, K.F., Sorensen, N.N.: Work and Migration: Life and Livelihoods in a Globalizing World. Routledge (2003)

20. Drent, M.E., Hendriks, R.J., Zandee, D.H.: New Threats, New EU and NATO Responses. Clingendael, Netherlands Institute of International Relations (2015)
21. Knox, K., Kushner, T.: Refugees in an Age of Genocide: Global, National and Local Perspectives During the Twentieth Century. Routledge (2012)
22. Loescher, G., Betts, A., Milner, J.: The United Nations High Commissioner for Refugees (UNHCR): The politics and Practice of Refugee Protection Into the 21st Century. Routledge (2008)
23. Turton, D.: Who is a Forced Migrant. Development Induced Displacement: Problems, Policies and People 14, 20 (2008)
24. Martin, S.F.: Women and Migration. In Consultative Meeting on "Migration and Mobility and how this movement affects Women" (2004)
25. Dal Lago, A.: Confini, Guerre, Migrazioni. InTrasformazione: Rivista di Storia delle Idee 4, 32–36 (2015)
26. Bonfiglio, A., Hugo, G., Hunter, L., Khoo, S.E., Massey, D., Rees, P.: The Future of International Migration. World Population and Human Capital in the Twenty-First Century, 333 (2014)
27. Adamson, F.B.: Crossing Borders: International Migration and National Security. Int. Security 31, 165–199 (2006)
28. Bove, V., Böhmelt, T.: Does Immigration Induce Terrorism? J. Polit. 78, 572–588 (2016)
29. International Organization for Migration (IOM): International Terrorism and Migration, 2nd ed. Geneva (2010)
30. Bandyopadhyay, S., Sandler, T.: Immigration Policy and Counterterrorism. J. Public Econ. 110, 112–123 (2014)
31. Neumayer, E.: Unequal Access to Foreign Spaces: How States Use Visa Restrictions to Regulate Mobility in a Globalized World. Trans. Inst. Br. Geogr. 31, 72–84 (2006)
32. Leiken, R.S.: Bearers of Global Jihad? Immigration and National Security After 9/11. Nixon Center (2004)
33. Hoffman, A.M.: Sageman, Marc. Understanding Terror Networks. Philadelphia: University of Pennsylvania Press, 2004. J. Conflict Stud. 27 (2007)
34. Perliger, A., Pedahzur, A.: Social Network Analysis in the Study of Terrorism and Political Violence. Political Sci. Politics 44, 45–50 (2011)
35. Sageman, M.: Leaderless Jihad: Terror Networks in the Twenty-First Century. University of Pennsylvania Press (2011)
36. Leiken, R. S., Brooke, S.: The Quantitative Analysis of Terrorism and Immigration: an Initial Exploration. Terror. Political Violence 18, 503–521 (2006)
37. Schneiderman, N., Ironson, G., Siegel, S.D.: Stress and Health: Psychological, Behavioral, and Biological Determinants. Annu. Rev. Clin. Psychol. 1, 607–628 (2005)
38. Cheng, T., Zhan, X.: Pattern Recognition for Predictive, Preventive, and Personalized Medicine in Cancer. EPMA J. 1–10 (2017)
39. Weenink, A.W.: Behavioral Problems and Disorders Among Radicals in Police Files. Perspectives on terrorism, 9 (2015)
40. Horgan, J.: The Psychology of Terrorism (Cass Series: Political Violence) (2005)
41. Richards, A.: Conceptualizing Terrorism. OUP Oxford (2015)
42. Anand, P., Kunnumakara, A.B., Sundaram, C., Harikumar, K.B., Tharakan, S.T., Lai, O.S., … Aggarwal, B.B.: Cancer is a Preventable Disease That Requires Major Lifestyle Changes. Pharm. Res. 25, 2097–2116 (2008)
43. Kotia, E., Abdallah, M.: An Evolving Security Dilemma: Adopting a Comprehensive Approach to the Changing Dynamics of Terrorism in Africa. Int. Relat. 4, 85–94 (2016)
44. Bakker, E., van Zuijdewijn, J.D.R.: Jihadist Foreign Fighter Phenomenon in Western Europe: A Low-Probability, High-Impact Threat. Terrorism and Counter-Terrorism Studies (2015)
45. Cabral, H., Makino, J., Matsumoto, Y., Mi, P., Wu, H., Nomoto, T., … Kano, M.R.: Systemic Targeting of Lymph Node Metastasis Through the Blood Vascular System by Using Size-Controlled Nanocarriers. ACS Nano 9, 4957–4967 (2015)

46. Singh, A.R., Singh, S.A.: Diseases of Poverty and Lifestyle, Well-Being and Human Development. Mens Sana Monogr. 6, 187 (2008)
47. Fenstermacher, L., Leventhal, T., Canna, S.: Countering Violent Extremism: Scientific Methods & Strategies. AIR FORCE RESEARCH LAB WRIGHT-PATTERSON AFB OH (2011)
48. Kydd, A.H., Walter, B.F.: The Strategies of Terrorism. Int. Security. 31, 49–80 (2006)
49. Hodgins, S., Alderton, J., Cree, A., Aboud, A., Mak, T.: Aggressive Behaviour, Victimisation and Crime Among Severely Mentally Ill Patients Requiring Hospitalisation. Br. J. Psychiatry 191, 343–350 (2007)
50. Borum, R.: Radicalization Into Violent Extremism I: A Review of Social Science Theories. J. Strateg. Security 4, 7 (2011)
51. Ganor, B.: Trends in Modern International Terrorism. In To Protect and To Serve pp. 11–42. Springer New York (2009)
52. Gregg, H.S.: Defining and Distinguishing Secular and Religious Terrorism. Perspectives on Terrorism, 8 (2014)
53. Feldmann, K.: Suicidology Prevents the Cultivation of Suicide (2014)
54. Corte, L.D.L., Giménez-Salinas, A.: Suicide Terrorism as a Tool of Insurgency Campaigns: Functions, Risk Factors, and Countermeasures. Perspectives on Terrorism, 3 (2010)
55. Oliveri, M.: Internal Security in the European Union: Research Programmes and Cooperation Policies (Bachelor's thesis, Università Ca' Foscari Venezia) (2016)
56. Paulussen, C.: Repressing the Foreign Fighters Phenomenon and Terrorism in Western Europe: Towards an Effective Response (2016)
57. Taylor, K.: The Brain Supremacy: Notes From the Frontiers of Neuroscience. OUP Oxford (2012)

Meeting Growing Threats of Misuse of Toxic Chemicals: Building a Global Chemical Safety and Security Architecture and Promoting International Cooperation

Krzysztof Paturej and Pang Guanglian

Abstract We witness today a global spread of threat of use of toxic chemicals as a means of warfare or terror. The recent use of chemical weapons and chemicals as weapons in Syria, and terrorist attacks against chemical infrastructure are visible confirmations of a growing threat of misuse of chemicals. The threat is today coming not only from a few rough states but mainly from the groups operating in territories not controlled by the governments and non-state actors. While the globalization and spread of Chemical, Biological, Radiological and Nuclear (CBRN) industries and materials are rapid and very dynamic, international responses and regulatory mechanisms in the area of prevention, preparedness and response against misuse of these agents, especially chemical agents, remain weak. The answer is to move towards global chemical safety and security architecture. A global chemical safety and security architecture, free from political limitations should be built within two core processes. A first process is a continued development of a global chemical safety and security culture. The second one is the development of the global market for chemical safety and security.

K. Paturej (✉)
President of Board of the International Centre for Chemical Safety and Security,
Warsaw, Poland
e-mail: k.paturej@iccss.eu

P. Guanglian
Vice Secretary-General & Director of International Affairs China Petroleum and Chemical
Industry Federation (CPCIF), Beijing, China
e-mail: cpcif.pang@gmail.com

M. Martellini, A. Malizia (eds.), *Cyber and Chemical, Biological, Radiological,
Nuclear, Explosives Challenges*, Terrorism, Security, and Computation,
DOI 10.1007/978-3-319-62108-1_14

1   Global Spread of Chemistry Requires Global Chemical Safety and Security

In the realm of globalization and growing interdependence in all areas of international cooperation, chemical safety and security and environmental protection have become leading factors in shaping international security, as well as economic cooperation and development policies. They have also become an important element in bilateral and multilateral relations in the spheres of security, economy and development.

We are witnessing a global spread of the threat of use of toxic chemicals as a means of warfare or terror. The recent use of chemical weapons and chemicals as weapons in Syria, and terrorist attacks against chemical infrastructure, are visible confirmations of a growing threat of the misuse of chemicals. Today, the threat comes not only from a small number of rough states but mainly from groups operating in territories not controlled by governments and non-state actors. While the globalization and spread of Chemical, Biological, Radiological and Nuclear (CBRN) industries and materials are rapid and very dynamic, international responses and regulatory mechanisms in the area of prevention, preparedness and response against misuse of these agents, especially chemical agents, remain weak.

In the process of renovating and accommodating international agendas in terms of today's and tomorrow's threats, which are increasingly chemical or ecological in nature, the international community should embrace chemical and ecological safety and security. Chemical and ecological safety and security should become an important vehicle to assist governments, societies and industries to reduce chemical threats and increase international cooperation.

The answer is to move towards global chemical safety and security architecture and promote international cooperation.

A global chemical safety and security architecture, free from political limitations, should be built within three core processes. The first such process is the building of a global movement to reduce chemical threats, free from political biases, with all relevant international and national stakeholders as leaders of these efforts. The second process is the continued development of a global chemical safety and security culture, and the third is the development of a global market for chemical safety and security and of bottom-to-top chemical safety and security, through CHEM-FRIENDLY acknowledgment.

To promote international cooperation in order to meet the challenges of growing chemical threats, we have to move beyond traditional arms control and disarmament divided by political views and groupings and build a global consensus that chemistry should not be an area of conflict and political competition. An integrated system of chemical safety and security should be promoted as part of international security, which includes environmental and health security. We need to develop a global action to reduce chemical threats and enhance chemical security, which requires a whole-society or whole-government approach.

Efforts to develop a global approach to chemical safety and security were given a momentum with the global Chemical Safety and Security Summit, the first global multi-stakeholder process to address chemical safety and security solutions throughout the chemical supply chain. The first global summit, CHEMSS2016 (www. chemss20126.org), organized from 18 to 20 April 2016, adopted a Summit Declaration on the development of international cooperation to enhance chemical safety and security and promote a global chemical security culture, which laid down goals, guidelines and principles of global cooperation against chemical threats.

The challenge to move towards a global regime for chemical security and safety, dealing also with chemical hazards and related technological matters of concern in compliance with the Chemical Weapons Convention (CWC), is a hard, complex task. Indeed, States-Parties to the CWC have different national implementation mechanisms (National Authorities), different facilities, different methodologies to enhance and strengthen the chemical security and safety culture, and different arrangements to enhance private and public partnership. Thus, it is vital to start from the concrete experiences developed by some CWC/OPCW Member States in order to build upon the above vision. The authors, playing relevant roles in their respective organizations, will deal with this thematic in the framework of the experiences developed by Poland and China. This bottom-to-top approach, namely going from national experience to the global regime, is fundamental to produce concrete, operative action plans, such as those developed during the CHEMSS summit in 2016, which will be followed by a second summit in 2017.

2 Building a Global Movement to Reduce Chemical Threats

2.1 Chemical Security and Safety: Go Global

Facing worldwide development of the chemical industry and trade, as well as global access to chemicals, the safety and security of manufacturing, infrastructure and supply chain of chemicals and energy sources and carriers have become a priority for governments, chemical industry and communities in which chemical activities are conducted, and for the world of science and non-governmental organisations. Globalization of chemical production means that global solutions are required to implement chemical security and safety and prohibit the illegal use of chemicals.

Many countries have in recent years implemented legal and administrative frameworks, regulatory measures and operational programs to reduce chemical threats and chemical hazards. Chemical companies have also developed corporate policies and sound practices to strengthen government efforts. Furthermore, civil society provides competent advice and actively supports the goals of safe and secure access to and use of chemicals. These efforts should be implemented and operationalized internationally, particularly in developing countries and countries with emerging economies. Chemical safety and security should go international.

While the globalization and spread of CBRN industries and materials are rapid and very dynamic, international responses and regulatory mechanisms in the area of prevention, preparedness and response against misuse of these agents remain weak. We deal with chemical safety and security nationally or within the relevant chemical industries and major companies, while chemical production, transportation, use, and chemical threats are global in reach.

3   Polish-Chinese Leadership in Moving Global Chemical Safety and Security Forward

Poland and China have taken the lead in the development of comprehensive approaches to introducing and integrating chemical safety and security in national security and economic policies and to sharing their experiences worldwide. They cooperate closely through industry organizations and civil society. These fruitful contacts have resulted in the development of closer Polish–Chinese co-operation and in the preparation and conduct of the second global summit on chemical safety and security, CHEMSS2017.

Poland's existing capacities and comparative advantages in the strengthening of chemical safety and security.

Poland's active engagement in chemical safety and security is based on a combination of several characteristics. Poland has introduced effective national legal measures and operational instruments to provide safety and security and environmental protection. Thanks to its growing position in the European Union, Poland is creating a more than attractive environment for economic development in Central Europe that includes modern solutions in chemical security and environmental protection. Poland is an active partner in the development of international security and a leading participant in the efforts to ban chemical weapons, to enhance chemical safety and security worldwide and to strengthen environmental protection.

Poland has in place an effective system of regulation of installations that create risks of major accidents to people and the environment. This regulation is conducted by the organs of the Inspectorate for Environmental Protection and the State Fire Service. The national system for the prevention of major accidents meets the requirements of European Council Directive 96/82/EC of 9 December 1996 on the control of major-accident hazards involving dangerous substances ("Seveso II Directive") and Directive 2003/105/EC of the European Parliament and of the Council of 16 December 2003 amending Council Directive 96/82/EC on the control of major-accident hazards involving dangerous substances. Furthermore, national legislation was recently adopted to implement the Directive of the European Parliament and of the Council 2012/18/UE of 4 July 2012 on the control of major-accident hazards involving dangerous substances ("Seveso III").

The inspection of environmental protective measures in Poland is a modern and professionally managed public institution with relevant instruments to conduct effective environmental policy and to protect the environment for present and future generations. Poland has developed and implemented modern solutions for the disposal of toxic substances and waste.

Additionally, Poland has introduced effective crisis response mechanisms and critical infrastructure protection. Polish government institutions and private companies educate and employ competent staff who are well prepared and equipped to implement chemical safety and security requirements and environmental protection. There are also a number of leading Polish research and development entities that provide modern solutions and conduct active international cooperation in the areas of chemical safety and security and environmental protection.

Risks of major accidents, as well as threats of terrorism in chemical industry and in the transportation of chemicals in the region, call for cooperation among neighbouring countries and the strengthening of cross-border security. In cooperation with Ukraine, Poland initiated regional cooperation to enhance chemical security. Within the task of reducing tensions and fostering peace, stability and security, the International Centre for Chemical Safety and Security (ICCSS) participates in the development of the Integrated Chemical Safety and Security Program and in addressing other CBRN risks in Ukraine. The integrated program accords priority to strengthening safety and security in chemical activities at chemical plants, transportation of chemicals and energy carriers, security of airports and sea-ports and the protection of the critical infrastructure; these are regarded as important tasks in preventing the hostile use of chemicals and supporting economic development in Ukraine.

Polish solutions and approaches to strengthening chemical safety and security may serve as an example for enhancing chemical security both in other countries and at the regional level. The ICCSS in Warsaw (www.iccss.eu) offers chemical security solutions based on sustainability, continuity and modern management, throughout the range of chemical activities, including storage, transportation and transmission of chemicals and energy carriers. Leading international partners, governments, organisations and chemical industry have actively supported the Centre, jointly developing and introducing international programs to enhance chemical security at the national and regional levels. The ICCSS actively promotes the global chemical security culture.

4   China: Global Leadership in Chemical Development and Production

China's chemical industry, having undergone more than 30 years of rapid development, has become a strategic and fundamental raw material industry with a consolidated foundation, sophisticated portfolio and advanced technical level. Today, China's chemical industry ranks number 1 in the world. There were over 29,000 manufacturing enterprises above designated size in China's petroleum and chemical industry in 2015, and over 300,000 enterprises are related to hazardous chemicals. There are 380 chemical industrial parks of the provincial level and above and most of the economically developed areas along rivers and the coastline are key petrochemical areas.

By 2015, China's chemical industry was realizing sales revenue of 8.84 trillion yuan and a national total output of 550 million tons of major chemical products annually. The country has established a full-range petrol and chemical industry system, from oil and gas, to petroleum refining, to downstream processing. The Chinese chemical industry has made a great contribution to meeting the needs of the 1.3 billion Chinese people in terms of food and clothing, housing, transportation and other survival and development needs, thus improving the living standards of the Chinese people.

In the course of the rapid development of China's chemical industry, chemicals are entering into everyday life at an unprecedented rate. It has not only become a topic of public concern, but also the future of sustainable chemical industry development has produced higher requirements. Therefore, the Chinese Government has made the green, sustainable development of the chemical industry a national strategy. The Chinese Government has set up special safety and environmental regulatory authorities, to enhance chemical safety and environmental laws and standards.

There is still a considerable loophole in Chinese chemical safety management, however. Although many large companies have set good examples by applying strict safety standards, we still see incidents such as the explosion on August 12, 2015, at a hazardous goods warehouse in Tianjin Port, where over 170 people died or were missing. Since the Tianjin explosion incident, relevant government departments have further reinforced safety inspection and safety supervision for hazardous chemicals enterprises, and carry out special inspections and controls for hazardous chemicals transportation, packaging and storage enterprises specifically to address serious issues resulting from the fast development of hazardous chemicals logistics, lagging supervision, weak basis and frequent accidents in recent years.

The Chinese Government continuously improves its governmental supervision system with clear responsibilities and full coverage, including building a high-level safety standards system consisting of special laws and uniform management regarding chemical safety, improving the establishment of basic data and setting up an information-sharing mechanism.

As China's chemical industry is developing expeditiously, while relevant safety assurance conditions have not been developed accordingly, China's chemical safety is also faced with huge challenges:

1. Low intrinsic safety level of enterprises and frequent occurrence of safety accidents:

    There is a "polarization" among hazardous chemicals enterprises in China. On the one hand, large chemical enterprises with large devices, complex processes, intensive industrial development, and strong technology and capital support are doing a better job in safety management, but due to the higher intrinsic danger, severe and disastrous accidents occur from time to time. On the other hand, medium and small chemical enterprises with less advanced equipment, lower employee competence level and a lower level of safety management tend to have more accidents. Thus, the basis for the safe production of hazardous chemicals is weak, and as a result, severe and disastrous accidents occur from time to time.

2. Regulations and supervision mechanism are not yet adequate:

There is weakness in the supervision system, resulting in many management and supervision blind spots and a lack of resources and capabilities in supervision and enforcement. There is no hazardous chemicals safety information-sharing mechanism, and no complete and uniform national hazardous chemicals database. The legislation level of the regulations is low, and the standards are sometimes redundant and sometimes non-existent.

3. Inadequate emergency assurance system and insufficient information sharing:

Laws or regulations on emergency management are inadequate and the responsibilities of relevant parties are not clearly defined. The hazardous chemicals emergency rescue mechanism is not adequate, and there is no collaboration mechanism for handling accidents among various departments. Emergency rescue capability planning and layout are not rational and emergency rescue personnel lack in systematic, professional and actual field training. The existing emergency equipment is unsophisticated, and there is a serious lack of emergency material reserve.

4. Transport of dangerous chemicals is still the focus:

Many accidents and damage to the environment are caused by illegal and unsafe transportation methods and behaviors.

5. Public information and response to public sentiment still requires improvement:

With the development of the economic society and the instant information transmission enabled by the internet, public health and safety standards are becoming increasingly high; attention and sensitivity to hazardous chemicals accidents are increasing, while tolerance is decreasing; and public awareness for protecting personal legal rights and interests is increasing. In recent years, mass disturbances caused by petrochemical construction projects such as the "PX Incident" have demonstrated the inadequacy in publicizing information and in guaranteeing public participation and monitoring, and the systematic, objective and there is no rational information dissemination or guidance for the public.

In recent years, the Chinese Government, industrial organizations and enterprises have worked together to take a series of measures to increase the intrinsic safety level of the chemical industry.

As a bridge between the government and enterprises, the China Petroleum & Chemical Industry Federation (CPCIF) has been focusing on promoting the green, safe and sustainable development of the industry, and increasing the chemical safety and security level. Firstly, it participates in the formulation of policies of relevant government departments, provides technical support, and provides feedback to enterprises' demands. Secondly, it implements Responsible Care®, and increases the safety and security awareness of the industry. Thirdly, it supports governmental departments in their endeavors to establish and improve the emergency response system, and strengthens emergency-rescue capacity building. Fourthly, it carries out publicity, exchange and training relating to chemical safety and security, increases the safety management capabilities of the industrial workforce, proactively raises the public's awareness of chemical safety, and creates favorable public opinion for the scientific and health-and-safety development of the industry. Fifthly, it promotes international cooperation, and enhances China's chemical safety and security level by learning from international best practices.

To meet the challenges of the increasing use of chemistry and to fulfill the requirements of environmental protection and chemical safety and security at a local level, the CPCIF and ICCSS, in cooperation with local Chinese authorities, initiated the development and implementation of a program entitled: Local Awareness and Responsibility in Chemical Safety and Security. The main goal of the program is to create a local policy of sustainable development in broadly defined chemical activity, managing chemical compounds and toxic waste, planning and responding to threats connected with this activity and with the transport of dangerous materials. The programme will also support the partnership between the administration, self-government, citizens and users of chemical units at a local level, develop the awareness of safe usage of chemical compounds, identify stored or transported dangerous materials, and help in preparing guidelines for local authorities in creating and introducing emergency response plans. The programme will offer training for the administration and local self-government representatives, chemistry users and civil society.

Chemical safety and security and environmental protection are becoming increasingly interlinked with internal and external security, public health protection, economic and trade cooperation, border protection, as well as development and humanitarian policies. Communities worldwide, especially in developing countries, are more and more dependent on chemical products – from fertilizers and petrochemicals to electronics and plastics – for economic development and improving livelihoods. Sound chemicals management is as valid an area as education, transport, infrastructure, direct healthcare services and other essential public services. The growing access to chemicals increases threats of misuse of toxic chemicals or their terrorist use, which could undermine national and international security, and economic and social stability. There is a need to assist developing countries in strengthening chemical safety and security.

The ICCSS, with international stakeholders, including Dow Chemicals, has developed an international program for chemical safety and security in Kenya. The implementation of the program started in August 2014, thanks to a donation from Norway, which provided finance to implement the first part of the Kenyan program. During the first part of the Kenyan project an education module was introduced, which created solid Kenyan internal expertise and trained personnel to implement the project in different areas of chemical activities. A priority area in the first phase included enhancing expertise in chemical security within the border controls, customs controls and transportation of chemicals. One of the core deliverables of the education module was delivery of the best practices and capacity building to the Kenyan government institutions and stakeholders to prevent, prepare for and respond to misuse of toxic chemicals, including for law enforcement. African Member States, at their 2015 annual meeting within the OPCW, offered their support for the Kenyan project to be implemented in the other African countries.

5 Going Global to Reduce Chemical Threats: Leading Outcome of the Global Chemical Security Summit – Chemss2016

Efforts to globalize chemical safety and security and enhance the safety and security culture began in earnest with an international meeting in November 2012 in Tarnów, Poland, co-organized by the ICCSS and OPCW. This was followed by the OPCW's Third Review Conference in April 2013, which provided consensus support for efforts to promote a global chemical safety and security culture. The Global Partnership Against WMD Proliferation provided a broad international policy framework for the Chemical Security Sub-Working Group, which offers strategic advice and coordinates the provision of expertise and resources. The Warsaw Conference on chemical and ecological safety and security, held in May 2014, proposed that the ICCSS and its partners organize a global chemical safety and security event.

CHEMSS2016 (www.chemss2016.org) organized by the ICCSS from 18 to 20 April 2016 in Kielce, Poland, was the first multi-stakeholder event conducted within the CHEMSS global chemical safety and security process. The CHEMSS process is dedicated to addressing chemical and environmental safety and security solutions in the supply chain of raw materials, production, infrastructure, transportation and use of chemicals in all areas of chemical activities, and to promoting the development of a global chemical safety and security culture.

CHEMSS2016 was attended by over 400 participants from 47 (mainly developing) countries. The largest delegations at CHEMSS2016 were from China, the USA, Ukraine and Poland. The CHEMSS2016 Summit Declaration on the development of international cooperation to increase chemical safety and security adopted at the CHEMSS2016 Final Session is the first document of this kind in the world, and linked process safety with measures against misuse of chemicals and environmental protection, stressing the need for the development of a global chemical safety and security culture. The Declaration endorsed the role of the ICCSS as an international leader in chemical safety and security.

CHEMSS2016 stressed the importance of UN Security Council Resolution 1540 (2004) for the development of chemical safety and security as an important element of security at the national and international levels. The role of the OPCW as a platform for cooperation in reducing the chemical threat was highlighted. Support for the Paris Agreement on Climate Change was expressed.

During the African Forum, participants declared that chemical safety and security is an important element in the economic growth of security in each country and in the region as a whole and expressed readiness to implement legal and organizational measures to increase the security in economic activities. The African Forum highlighted the role of the Kenyan Program for Chemical Safety and Security, implemented by the ICCSS and financed by the Government of Norway, as a model solution for enhancing chemical and ecological safety and security in Africa.

The China Dialogue on Developing Polish-Chinese Cooperation in Reducing Chemical Threats and Enhancing Chemical and Ecological Safety confirmed that the "Local Awareness and Responsibility" program coordinated by the ICCSS will be implemented in China. Chinese partners confirmed their strategic partnership in the preparation and conduct of future CHEMSS summits.

During the Forum on Jordan's Preparedness against Chemical Threats, a proposal was presented for a joint initiative to prepare a working document about measures to build trust in chemical safety and security in the Middle East.

The Forum on Building Local Awareness, Responsibility, Partnerships and Response to Mitigate Chemical Safety and Security Threats Voluntary Programs confirmed that traditional responses to crises are not sufficient to meet new challenges brought about by the rise of chemical accidents and catastrophes, regardless of whether they have natural or human-led causes. The Forum expressed its support and stressed the importance of the "Local Awareness and Responsibility in Chemical Safety and Security" program, initiated by the ICCSS, as a model program for introducing chemical and ecological safety and security at a local level.

The session on Global Partnership Against the Spread of Weapons and Materials of Mass Destruction and the meeting of the Chemical Security Sub-Working Group confirmed support of the members of the G7 Global Partnership for the development of chemical safety and security on a global scale.

The Seminar on Integrated Impact Assessment for new and existing industrial enterprises indicated the role of the integrated impact assessment in the evaluation of the extent to which investment will influence the environment and whether there is a possibility to decrease this impact.

During the summit, the Chemical Security Milestone Awards were officially presented. They were funded by the ICCSS and awarded for personal accomplishments in the development of chemical safety and security in the world. The Chemical Security Milestone Awards were given in the following categories: Personality (Amb. B. Jenkins, US Department of State), Public Administration (M. Borowski, Head of the Technical Inspection Office, Poland), International Organization (Amb. V. Verva, OBWE in Kiev), Industry (T. Scott, DOW Chemicals), Science (Prof. M. Martellini, Italy) and Civil Society (O. Sadovsky, Ukrainian Chemists Union). The Chemical Security Milestone Awards Special Distinctions were granted to: Amb. K. Mworia from Kenya; D. Wulf, US Department of State; Amb. H. Farajvand from the Islamic Republic of Iran, and A. Mochón, CEO of Targi Kielce.

6 The Chemical Threat Should Be Regarded as a Leading Issue by International Structures

Today, many countries are at the nexus of transnational terrorist threats, science and technology capacity building, and WMD-usable materials and expertise, including in the chemical area. As we move increasingly into the area of chemical safety and security in this changing landscape, international structures, including the United Nations, should shift their agenda from chemical weapons destruction or non-proliferation to a post-chemical weapons phase, with a focus on chemical safety and security. In the process of modernizing the international agenda and attuning it to today's and tomorrow's needs and challenges, we have to embrace chemical safety and security. Chemical security should not be an objective in itself. Chemical security should be a vehicle which will allow international structures to assist countries in reducing the chemical threat and increasing international cooperation.

7 Roles of the United Nations

A global response is not possible without an effective multilateral system and cooperation. A core task is to combine efforts to globalize chemical security within the United Nations as the backbone of the global multilateral system. An effective multilateral system with global chemical security can respond to security threats in many ways. Chemical security and efforts against terrorism are interconnected.

What does this mean in practice? Firstly, greater chemical security means fewer opportunities for non-state actors, including terrorists, to misuse chemicals. Secondly, a multilateral system will provide legitimacy to global chemical security efforts and constitutes a recognized international norm of behavior. Chemical security, when implemented, requires states and industries to strengthen controls on toxic chemicals, including measures to prohibit chemical weapons and implement national controls. Thirdly, the existence of effective multilateral cooperation through incorporating global chemical safety and security best practices and capacity building, and incorporating and implementing export controls, reduces chemical threats globally.

A crucial element in building an effective and up-to-date international system of chemical safety and security is recognition and understanding of the global character of chemical threats. There is no global understanding regarding the character of the chemical threats and risks. Moreover, we are far from consensus on how we should deal with them. The unanimous United Nations Security Council Resolution 1540 (2004) is of enormous importance for the efforts to stem chemical threats.

8   Role of the OPCW

The full and effective implementation of the CWC represents in itself a significant contribution to the peaceful and safe use of toxic chemicals, and thus a contribution to chemical safety and security. According to the CWC, chemical safety and security means the measures to prevent non-deliberate releases of toxic chemicals. It comprises disciplines such as environmental safety, occupational safety and transportation safety, etc. Generally speaking, the legal and regulatory framework for chemical safety is well-established, with many international agencies working on it. Therefore, the OPCW aims to be the key complementary partner in this area internationally. On the other hand, chemical security refers to measures to prevent deliberate release of toxic chemicals and to mitigate the impact if such events occur. The relevant regulatory framework is under development both at the national and international levels. Therefore, the OPCW attempts to be the leading organization in such efforts.

Given the broad focus of OPCW activities and their impact, the OPCW's contribution to enhancing chemical safety and security can be emphasized and implemented in three key areas:

1. Promoting the OPCW as a platform for consultation and cooperation (facilitating the identification of good practices and making these available to States Parties);
2. Catalyzing international partnerships; and
3. Providing advice to States Parties, ideally together with other relevant international and regional entities.

The increasing needs of the States Parties are best indicated in the decisions taken by the OPCW Third Review Conference, which recalled that:" chemical safety and security, while being two distinct processes, are the prime responsibilities of States Parties. It encouraged the promotion of a safety and security culture regarding chemical facilities and of transportation of toxic chemicals. It noted that capacity building activities in these fields are one of the elements of the decision on components of an agreed framework for the full implementation of Article XI adopted by the Conference at its Sixteenth Session (C 16/DEC.10). The Third Review Conference noted the initiatives taken by States Parties and the Secretariat to promote activities in the areas of chemical safety and security, and welcomed the role of the OPCW as a platform for voluntary consultations and cooperation among the States Parties and the relevant stakeholders, including the private sector and academia, to promote a global chemical safety and security culture."

The main outcome of these decisions is to set the OPCW as a platform of multi-stakeholder and voluntary cooperation in enhancing chemical safety and security. The platform should promote international cooperation in peaceful uses of chemistry, enhancement of security at chemical plants, and support national capacity for prevention, preparedness and response against misuse of toxic chemicals. The platform should engage all relevant stakeholders, including government agencies, chemical industry associations, academia and the scientific community, and the relevant international organizations. The platform should facilitate and promote the comprehensive and synergetic implementation of the provisions of the CWC and voluntary cooperation in the areas of chemical safety and security.

Working together with the chemical industry associations, relevant government agencies, and international organizations, the OPCW could support the reduction of the chemical threat by cooperating with governments and chemical industry associations in raising awareness and improving chemical security and safety best practices. States Parties should be fully abreast of developments in the sphere of chemical security and safety best practices, by seeking and disseminating information and views from the National Authorities and relevant national agencies, the chemical industry (for example, via chemical industry associations and the Responsible Care® program) and the academic community (for example, via the International Union of Pure and Applied Chemistry (IUPAC)). Enhancing the chemical security culture will provide greater assurance that the national chemical security systems will accomplish their functions of preventing, detecting and responding to the theft, sabotage, unauthorized access and illegal transfer of chemical material and the associated facilities and transport.

The OPCW should continue to build engagement in chemical safety and security as a steady process of developing relations, partnerships and program activities, as appropriate, with relevant regional and international organizations, including international organizations related to chemical safety, chemical industry associations, the private sector and civil society, in order to enhance global chemical safety and security. It will serve as an important element in the implementation of the objectives and purposes of the CWC and support the transformation of the OPCW.

Articles VI, VII, X and XI of the CWC together provide the rationale for considering the current and future contribution of the OPCW to chemical safety and security. At the same time, according to the decisions of the OPCW policy-making organs, there are increasing expectations from the States Parties that the OPCW should devote attention to the subject of chemical safety and security. Strengthening the role of the OPCW in matters of chemical safety and security is indeed part of its broader mandate to prevent the re-emergence of chemical weapons.

One of the goals of the OPCW chemical security program could be the provision of guidance and assistance to help Member States establish a strong chemical security culture. The chemical industry and relevant international and national partners will be crucial partners in promoting chemical safety and security at the OPCW.

It is crucial to identify the boundaries beyond which the OPCW activities should not expand – so called red lights. Therefore, the OPCW should avoid developing an independent role for itself, since there are no such provisions either in the CWC or in the decisions by the OPCW's policy making organs. Chemical safety and security should not be part of OPCW inspection and verification activities. The OPCW should not develop OPCW regulatory measures, legal guidance or advice on chemical safety and security issues.

9 Continued Development of a Global Chemical Safety and Security Culture

Acknowledging that safety measures and security measures have in common the aim of protecting human life and health and the environment, we affirm support for a coherent and synergistic development, implementation and management of chemical safety and security culture, including through bilateral and multilateral cooperation. All stakeholders, including the government, regulatory bodies, industry, academia, nongovernmental organizations and the media, should fully commit to enhance a chemical safety and security culture and to maintain robust communication and coordination of activities.

The basis for the continued development of a global chemical safety and security culture is the CHEMSS2016 Summit Declaration on the development of international cooperation to increase chemical safety and security.

The CHEMSS2016 Summit Declaration is the first international document to offer a comprehensive and integrated approach to chemical safety and security culture. Chemical safety and security culture is no longer limited to the facility or even industry level and thus needs to be practiced creatively throughout continuously expanding and increasingly vulnerable supply chains with a multitude of diverse and overlapping players. The manifestation of chemical safety and security culture goals is achieved only if there are adequate inputs from the national and international levels. Chemical safety and security culture could be considered as an assembly of beliefs, attitudes, patterns of behavior, and solutions which assist in upgrading chemical safety and which strengthen and/or complement hard (equipment) and soft (rules and regulations) tools in their mission to ensure a high level of security. Safety culture is an integral part of technology emergence and involvement, while security culture offers a response to new inside and/or outside risks. Chemical security requires active measures to prevent, detect and respond to theft, sabotage, unauthorized access, illegal transfer or other malicious acts involving chemicals, or chemical sites. While safety and security cultures have a shared foundation of accountability and compliance, the ways in which these goals are achieved may differ significantly.

The effectiveness and continuous evaluation of a chemical safety and security system can be conducted only by personnel imbued with chemical safety security culture. In this sense, security culture connotes not only the technical proficiency of the people directly and indirectly involved in security, but also their willingness and motivation to follow established procedures and comply with regulations. The existence of culture should encourage personnel to take the initiative when unforeseen circumstances arise. Safety and security culture enables a person to respond to know and indefinite process safety and security risks out of carefully tuned and proactive habit rather than improvised effort.

The human factor plays a key role in ensuring the security and safety of chemical industries from outside as well as inside threats. Most security lapses at facilities result from human failings such as inadequate skills, negligence, miscalculation, or malice. Security systems, procedures, and practices are designed to stop an adversary whose goal is to defeat the system itself.

A comprehensive approach toward promoting chemical safety and security goes beyond chemical plants and should include the introduction of safety and security culture and its promotion to all relevant stakeholders. The stakeholders include governments and relevant national agencies; industries, including chemical associations and private industries; laboratories; regional and international organizations involved in the international safety and security and peaceful application of chemistry; and civil society, including academia, non governmental organizations, independent experts, and media.

Strict chemical safety and security measures must be applied throughout the full chain of production, infrastructure, transportation, storage, use, and disposal of chemicals in order to support effective barriers against misuse and diversion of chemical agents and materials.

The basic principles of chemical security culture include shared values, beliefs, and behavior patterns leading to the promotion, use, and development of safety management systems for humans; and environmental protection and security measures, including prevention, protection, detection of, and response to theft, sabotage, unauthorized access, illegal transfer, or other malicious acts involving the materials that can be used for unconventional terrorist purposes as well as their associated facilities.

A number of countries and organizations have promoted new program activities and announced plans to enhance chemical safety and security at the national and regional levels. They intend, inter alia, to develop national and international programs and centers on chemical safety and security, and to make use of the regional Centers of Excellence.

For more than 25 years, the Responsible Care® program of the chemical industry has promoted safety and, since 2001, also security practices that safeguard our workplaces, communities, and the broader environment. Corporate social responsibility and best environmental and public health practices are key parts of these efforts. These national, international and industry initiatives should receive support from all relevant stakeholders to advance chemical safety and security at the national level and in all relevant spheres of chemical activities.

The establishment of the ICCSS serves the purpose of the practical promotion and development of a chemical safety and security culture. The Centre offers both national and international partners a venue for cooperation in capacity-building, training, exchange of best practices, socially responsible management, and cooperation between professionals in the area of chemical safety and security.

The CHEMSS2016 Declaration encouraged the promotion of chemical safety and security by:

- putting the issue on the agenda of national policy-makers;
- improving efficiency in capacity building and exchanging information on best practices with regard to chemical safety and security;
- building and improving national and international coordination of chemical safety and security actors;

- promoting the application of research outcomes in assessing impacts on the environment and public health; and
- building awareness-raising programs for the public and for policy makers on issues relevant to chemical safety and security.

The goals of the CHEMSS2016 Declaration include:

- promotion of mechanisms for national, regional, and international networking to share experiences in these areas among all relevant stakeholders;
- promotion of solutions which will be continued, sustainable, affordable, and accountable globally, with due regard for the needs of developing countries and countries with economies in transition;
- establishment and/or promotion of mechanisms to identify and analyze countries' needs for assistance, upon their request, in chemical safety and security, and build partnership and cooperation with relevant national and international capacities to accommodate these needs and programs; and
- engagement of private sectors to build public-private partnerships to enhance chemical safety and security worldwide.

In order to systemize global efforts, the CHEMSS2016 Declaration established a set of general guidelines:

- to seek a comprehensive and synergistic approach among all relevant international treaties and agreements (including the CWC, Basel, Minamata, Rotterdam, Stockholm, and other relevant conventions) for capacity building that develops chemical safety and security in all areas pertaining to chemicals and wastes;
- to facilitate universal adherence to, and full national implementation of all relevant conventions;
- to provide demand-driven assistance that is respectful of recipient countries' needs, identification, and ownership of decision-making;
- to create networking to facilitate information exchange and to ensure that all partners are aware of each other's work in the area of chemical safety and security;
- to coordinate efforts to maximize interoperability, and to utilize the role of governments, as well as the complementary roles of multilateral partners; and
- to structure finances in a way that is more predictable, reliable, and transparent and could enhance mutual accountability.

The CHEMSS2016 Declaration also defined general principles of enhancing chemical safety and security:

- to focus on enhancing awareness and best practices through training, seminars, workshops, and other education processes in broad areas of chemical security and safety issues;
- to strengthen international cooperation to promote a chemical safety and security culture globally. These efforts should also engage researchers from academia and private industry;
- to promote implementation of the World Health Organization (WHO) International Health Regulations (IHR 2000) so as to improve global abilities to detect, assess, report, and respond to health events of international concern;

- to promote and work in line with the United Nations, the UN Environmental Programme (UNEP), the Strategic Approach to International Chemical Management (SAICM), the Inter-Organization Programme for Sound Management of Chemicals (IOMC), and national and international activities towards sound management of chemicals;
- to promote the implementation of the provisions of the CWC and the development of the OPCW as a platform for cooperation among all stakeholders to reduce chemical threats;
- to promote the comprehensive implementation of the provisions of United Nations Security Council Resolution 1540 (2004); and
- to promote the implementation of the Paris Agreement on Climate Change.

10 Development of the Global Market for Chemical Safety and Security

According to the International Council of Chemical Associations, some 20 million people around the globe have a job connected to the chemical industry, either directly and indirectly. The world chemical industry has an estimated annual turnover value of over 3 trillion dollars and is expanding. The chemical industry is the foundation upon which all other industries are built. It is present in every nation in the world, whether in operating plants, distribution routes or consumer use of end-products.

There is a growing global demand for modern technologies and equipment, as well as adequate administrative and legal solutions for chemical safety and security and environmental protection, safe disposal of a growing volume of toxic substances and wastes, protection of chemical facilities, and storage and transportation of energy sources. Providers of technology and chemical compounds increasingly frequently ensure that their customers are provided with state-of-the-art chemical safety and security solutions, leading to minimisation of the risk of breakdowns and cost reductions.

Therefore, activities related to chemical safety and security and environmental protection are becoming a profitable export, for example, services and products in such areas as science, development, manufacturing and services.

11 Chem-Friendly on Chemical Safety and Security: Building Confidence in Chemical Safety and Security at the Grass-Roots Level

The development, production, use and disposal of toxic chemicals are under the growing scrutiny of all relevant stakeholders, with a growing interest from consumers. Today, the internet, social media and communication platforms are full of information on toxic values of chemicals. Societies and individuals are increasingly threatened by founded or unfounded information about the dangers of chemicals or chemical products. For this reason, in the age of universal access to information, consumers openly ask for transparency from companies, knowledge about their operations and confirmation that their products are created as a result of safe processes. They expect continued confirmation that supplied chemical products are safe and secure and that chemicals will not be misused for illegal purposes.

At the same time, despite the fact that the chemical industry is one of the foundations of human, social and economic development, the confidence index towards the chemical industry remains at a very low level. While the large, global producers follow the growing number of regulations and are able to demonstrate that their products meet the relevant chemical, environmental and health regulations, the small and medium-sized companies, which provide the vast majority of chemical producers, often face a visibility and credibility problem, especially in developing countries. There should be a change in approach by those who develop, produce, transport, trade and use chemicals. They have to invest in developing confidence and trust that chemicals and chemical products are not harmful and are protected from being misused. Our answer is the acknowledgment of CHEM-FRIENDLY in chemical safety and security. CHEM-FRIENDLY is a declaration that a company conducts production processes and manages chemicals in a safe and responsible way. The acknowledgment of CHEM-FRIENDLY in chemical safety and security was developed at the First Global Summit in Chemical Safety and Security (CHEMSS2016). The aim of CHEM FRIENDLY in chemical safety and security is to build a positive image of the chemical producer, trader, user, by declaring that their chemical products are produced, delivered and used in accordance with relevant safety and security measures.

CHEM-FRIENDLY in chemical safety and security creates individual, societal and business responsibility, not through the increase in standards, norms or certificates, but through knowledge sharing, exchange of best practices, cooperation and training, with a view to create a positive image of chemistry and the enterprises engaged in this sector. Building a friendly image provides important development, production, storage, and trade leverage, not only for users but also for suppliers. Receipt of the CHEM-FRIENDLY acknowledgement will be confirmation that the company is aware of the requirements and acts responsibly in the area of chemical safety and security. At the same time, it supports the chemical safety and security culture and promotes its strengthening globally, along with the Declaration of CHEMSS2016. The CHEM-FRIENDLY acknowledgment is available as a recognizable logo (visible on product labels, brochures and websites) for consumers and business partners. The CHEM-FRIENDLY acknowledgment supports the development of global chemical safety and security culture, since it promotes the implementation of relevant national and international chemical safety and security measures, as well as the CHEMSS2016 Declaration. Through CHEM-FRIENDLY acknowledgment, consumers will be able to identify the company and confirm the company's credibility in the area of safety and security. This builds a positive image of the company and takes into account the consumer's needs, thus increasing the customer's confidence. Through CHEM-FRIENDLY acknowledgment the company's partners have increased confidence that they are operating on a similar level of safety and security. This facilitates access to customers globally.

12 Conclusions: Chemss Process – Overcoming Existing Barriers and Building a Global Chemical Safety and Security Architecture

Changing the environment of chemical threats makes it imperative to explore and support the worldwide efforts to globalize chemical safety and security. The CHEMSS process actively promotes codes of conduct, ethics, the exchange of best practices, and international, national or industry capacity building and other similar sources of voluntary commitments as increasingly effective tools of chemical safety and culture promotion throughout the chemical production and supply chain.

Efforts to strengthen chemical safety and security are weakened by several obstacles within the chemical industry. Firstly, chemical security is separated from chemical safety based on the approach that safety is the business of industry and security is the business of government.

Secondly, chemical security is built from the top down. Governments wish to regulate chemical security with legal, penal and administrative regulations and introduce growing numbers of directives. The regulations are known and implemented by major companies. Small and medium-sized companies, especially in developing countries, have neither the resources nor the capacities to implement the growing number of safety and security norms and regulations.

Thirdly, chemical production is leading for safety and security efforts, while the growing chemical problems relate to the development, transportation, storage and trade of toxic chemicals. Among these areas of chemical activities, the transportation of chemicals is the most vulnerable area.

To overcome these obstacles and go international and move towards improved global chemical safety and security management, the following approaches should be introduced:

Firstly, chemical security should no longer be separated from chemical safety. While this concept could be understood at the level of government, it has a deep impact at the local level – chemical security is forgotten.

Secondly, chemical safety and security management should be integrated into national security, economic and environmental plans, in a comprehensive manner. Currently, the responsibilities are usually distributed across several agencies, leading to fragmented and ineffective responses.

Thirdly, a multi-stakeholder approach should be applied in the development and implementation – there is a need to combine the knowledge and expertise of the governments and corporations, producers and users that have information on chemicals and technical capacity. Civil society should participate actively in enhancing chemical safety and security.

Fourthly, governments should develop and promote policies that concentrate on risk prevention where the responsibilities of the public and private sector are shared (prevention).

Finally, the lack of relevant expertise and the cost of effective management of chemical safety and security in developing countries should be countered with capacity building and the exchange of best practices. Developed countries and their chemical industries should share their extensive knowledge and expertise, in order to enhance chemical safety and security. Developing countries should offer investment opportunities for attracting innovative approaches and effective solutions.

The challenges posed by the production of chemical agents, including chemical weapons, through biological processes demand more attention in the area of convergence of chemistry and biology, to make them compliant with the Biological Weapons Convention (BTWC) and the CWC. The adoption of a harmonized approach lies far in the future, due to the different nature of the regimes. Some moves in that direction have been taken in the past by the OPCW, and recently, for the first time, by the G7GP Biological Security Working Group-Chemical Security Working Group Joint Session, under the G7 Italian Presidency. This issue, together with the identification of ways to engage the biological-chemical industries in this area, needs to be further explored. However, CHEMSS2017 could launch this challenge by identifying both chemical and biological corporations willing to support this future narrative and to promote chem-bio safety and security.

The underlying belief of the CHEMSS process is the view that effective chemical and environmental safety and security are essential conditions in the research, manufacturing and trade of chemicals. They are increasingly linked with internal and external security, public health protection, economic and trade cooperation, development and humanitarian policies. Strengthening chemical safety and security is also seen as an important element of critical infrastructure protection. The CHEMSS process focuses on enhancing chemical safety and security in developing countries.

The substantive basis for the global CHEMSS process is the CHEMSS2016 Summit Declaration on the development of international cooperation to enhance chemical safety and security and the promotion of a global chemical security culture.

The second global summit on chemical safety and security, CHEMSS2017 (www.chemss2017.org), will be held in Shanghai, China, on 19 and 20 September 2017. The China Petroleum and Chemical Industry Federation (CPCIF) and the International Centre for Chemical Safety and Security (ICCSS) will be the co-organizers of CHEMSS2017. They agreed that the CHEMSS2017 should remain a global event, free from political exclusion, which offers coherent approaches that involve leaders and practitioners in all disciplines of chemical safety and security and from all stakeholder communities, such as government, international organizations, industry, academia and civil society

CHEMSS2017 will be followed by 2017 China Petroleum & Chemical International Conference (CPCIC) from 20 to 22 September 2017. The organisation of CHEMSS2017 consecutively with the CPCIC is confirmation of a synergy of efforts to globalize chemical safety and security with Chinese approaches to meet the highest standards in chemical safety and security, and is also evidence of the global reach of the Chinese petroleum and chemical industries.

CHEMSS2017 aims to promote, operationalize and internationalize existing national and international best practices, training, and technical solutions in chemical and environmental safety and security, and to promote a positive image and approach on the part of chemical stakeholders. The leading CHEMSS2017 thematic streams include a combination of chemical and environmental topics which together constitute the global agenda: Facilities Safety and Security – Safety and security of chemical facilities and reduction of the risks of chemicals to local communities, including regulatory programs and industry approaches; Logistics – Chemical safety and security in the transportation and storage of chemicals and petrochemicals; CHEMSS Cyber Security – Increasing resilience and preventing chemical and petrochemical sites from cyber-attacks; Responsible Agriculture – Safe and secure handling of fertilizers; Youth Forum – Students for chemical and environmental safety and security; Human Resources – Developing skills and talent acquisition in developing countries; and CHEMSS Environmental Impact Assessment.

There are three main innovative concepts behind the CHEMSS process. Firstly, it is not a government-led initiative. It is an initiative coming from the grass-roots level, from those who are directly engaged in meeting chemical threats. Governments support the initiative but they are partners of industry, academia, and civil society in the preparation and running of the global summit. The multi-stakeholder ownership of the summit is evidence of an emerging whole-government or whole-society approach in meeting chemical threats. Secondly, the CHEMSS process is a truly global event, free from political limitations. The annual CHEMSS summits are another step towards building global responses to the global threat of misuse of toxic chemicals. Thirdly, the CHEMSS process does not seek new regulations or standards. It promotes, operationalizes and internationalizes existing national and international best practices, training, practices and technical solutions.

These three innovative concepts behind the CHEMSS process constitute the basis for the ongoing development of the global chemical safety and security architecture and promoting international co operation.

'We're Doomed!' a Critical Assessment of Risk Framing Around Chemical and Biological Weapons in the Twenty-First Century

Giulio Maria Mancini and James Revill

Abstract 'Risk' and 'risk assessment' rhetoric has become pervasive in twenty-first century politics and policy discourses. Although a number of different meanings of 'risk' are evident, the concept frequently purports to be an objectively, quantifiable and rational process based on the likelihood and consequences of adverse events. However, using the example of chemical and biological weapons (CBW), this chapter argues that security-related risks are not always objectively analysable, let alone quantifiable. Moreover, the process of risk assessment is not always 'rational'. This is, first, because efforts to quantify CBW-related risks normally require a body of data from which to inform assessments of probability when in fact there are limitations in data pertaining to the human dimension of CBW terrorism; with considerable gaps in knowledge of CBW incidents and a need for caution because of the emotive power of allegations of association with CBW. Second because the consequences of a CBW event are often informed by a wide range of variables, which make such weapons highly unpredictable. Third because conclusions that are drawn from any dataset often depend on the questions asked and the assumptions and values that 'subjectify' risk calculations, not least depending on if and how 'expertise' on risk is defined. This is not to say that risk assessment is not important, but that CBW risks might require a combination of a more rational phase of risk characterization with a more 'subjective' process of risk evaluation that acknowledges uncertainty of probabilistic modelling, deals with ambiguity, and opens-up the questions and assumptions that inform the risk assessment process to wider scrutiny and to the consideration of social and other factors.

The content of this chapter does not reflect the official opinion of the European Union. Responsibility for the information and views expressed in the chapter lies entirely with the author(s).

G.M. Mancini (✉)
Directorate-General of Migration and Home Affairs of the European Commission,
Brussels, Belgium
e-mail: giulio.mancini@ec.europa.eu

J. Revill
Harvard Sussex Program, SPRU, University of Sussex, Brighton, UK
e-mail: j.revill@sussex.ac.uk

© Springer International Publishing AG 2017
M. Martellini, A. Malizia (eds.), *Cyber and Chemical, Biological, Radiological, Nuclear, Explosives Challenges*, Terrorism, Security, and Computation,
DOI 10.1007/978-3-319-62108-1_15

Keywords Biological weapons • Chemical weapons • Risk • Uncertainty • Risk assessment • Bioterrorism • Chemical terrorism • Security • Threats • Scientific advice

1 Introduction

'Risk' and 'risk assessment' rhetoric has become pervasive in twenty-first century politics and policy discourses whether it is being applied to appraisals of science and technology, security, or their interactions.[1] For scholars such as Beck and Giddens the growing salience of risk is rooted in a tendency towards 'rational' decision-making processes in postmodern Western cultures. Indeed, although a number of different meanings of 'risk' are evident [2], the concept frequently purports to be an objectively, quantifiable and rational process based on the likelihood and consequences of adverse events [3, 4, 5]. This process is normally portrayed as undertaken by specialists and thereby presented as reflecting an expert analysis of the evidence resulting in reliable results. As such, risk is founded on positivist assumptions, with models derived from quantitative risk assessment applications in the nuclear and other engineering safety sectors [6], and looks at risks as largely an objectively observable, natural phenomenon. This approach to risk is used as a powerful tool in shaping policy options and validating policy decisions in relation to both science and security, as Williams notes "Risk has come to capture the minds of policy makers and public alike" [7].

However, using the example of chemical and biological weapons (CBW), this chapter argues that risks are not always objectively analysable, let alone quantifiable. Moreover, the process of risk assessment is not always 'rational', but frequently a combination of rationally comparable analysis *and* socially-mediated activity in which risks are socially constructed, and their "importance" subjectively evaluated or constructed. That is not to suggest that "anything goes" and risks are plucked out of the ether; but that the process of assessing risks is a human activity and informed by socially mediated assumptions, interests and (the limits of) knowledge. In this context, whilst some hazards will involve known risks that can be characterized in terms of probabilities and impacts, there will also be cases where there is uncertainty as to the 'likelihood' of a risk, ambiguity as to its potential consequence, and/or "ignorance, where we don't know what we don't know, and the possibility of surprise is ever-present" [8].

The chapter seeks to illustrate the relevance and limitations of risk assessment in relation to CBW through the application of critical thinking around risk assessment of emerging technologies – as developed by Andy Stirling and others- to the processes of looking at risks surrounding CBW in the twenty-first century. The first

---

[1] As Edmunds points out, the UK 2012 National Security Strategy (NSS) employs the term 'risk' no fewer than 545 times [1].

section of this chapter outlines some of the limitations in efforts to quantitatively model the risks posed by CBW, drawing attention to the limitations in available aggregate data, the challenges of determining consequences in a meaningful manner, the difficulty in effectively quantifying likelihood, and the limits of 'experts' in risk assessors. The chapter then elaborates on the social construction of security-related risks generally, and CBW risks specifically. The chapter then discusses possibility of amalgamating these two approaches to risk assessment, suggesting that an integrative, rather than exclusive, approach could be explored. In this model, the two approaches are not seen as mutually exclusive but complementary, with rational risk characterization and constructed risk evaluation forming the process of risk assessment. Essentially this would apply expert judgement through Bayesian techniques, and could be valuable in generating meaningful assessments of CBW risks which could be used to inform decisions around risk mitigation measures, even in the absence of a precise estimation of the baseline risk level. It should also be taken into account that the relative weight of the two framings within each assessment could vary. It is noted however that to maintain a minimal rational value, not all risks could be described with a sound characterization, largely depending on the level of uncertainty on likelihood estimation based on unknown factors related to the context or an intelligent threat,[2] even when impact can be relatively more clearly characterized basing on 'objective' characteristics of the hazards.

2 Risk in the Security Discourse

Over the course of the Cold War the process of conducting a threat assessment was relatively easy, as Dasse and Kessler state: "The enemy was known, its military capability was identified and its intentions understood – or so it was believed" [9]. Since the collapse of the Soviet Union and the easing of bipolar tensions, it is risks, rather than 'threats', which have grown in salience in both academic and policy discourses.

Indeed, in terms of the academic literature, risk has emerged as a nascent field of study within IR, influenced by the work of Beck on the concept of the risk society and the preoccupation in late modernity with the question of "how the risks and hazards systematically produced as part of modernisation can be prevented, minimised, dramatized, or channelled".[3] In terms of the policy discourse, Williams

---

[2] In the chapter "threat" is used to mean an intelligent (potential) perpetrator with intention to cause harm, i.e. a person or group of people, including a State or non-State actor.

[3] Beck's "risk society" is the post-industrial one that self-creates, through modernization, new risks that despite being created are less predictable than "classical", external risks. The "constellation in which new knowledge serves to transform unpredictable risks into calculable risks, but in the process it gives rise to new unpredictabilities, forcing us to reflect upon risks" is what Beck called "reflexivity of uncertainty" [10]. At the same time it is a society that no longer relies on the guidance of traditional or natural laws. The risk society uses decision-making tools such as risk assessment or risk mitigation. Furthermore, with the evolution from the "risk society" to the

suggested that "as a result [of 9-11] America became paranoid about possible security risks ... transatlantic relations truly entered the age of risks" [7]. As much is evident in the praxeology of a number of Western institutions in the twenty-first century, NATO for example has shifted from looking at nation-State threats (in the form of the Soviet Union) to "security challenges and risks", with the Alliance's 1991 strategic concept explicitly stating "[i]n contrast with the predominant threat of the past, the risks to Allied security that remain are multi-faceted in nature and multi-directional" [7]. In Europe, discourses largely dominated by specific conceptualizations of "threats" have evolved to integrate images of security risks. The European Security Strategy suggested that threats faced by Europe were "more diverse, less visible and less predictable"; and that threats, such as terrorism and the proliferation of WMD, put Europe and Europeans "at risk" [11]. Later, the EU Internal Security Strategy considered "threats" (terrorism, organised crime, cybercrime, as well as adverse events of a largely safety nature but with security implications) and "challenges" with the potential to generate risks for the Union and its citizens. The 2015 European Agenda on Security [12] employs risk framing in relation to border security, radicalisation, and disasters.[4]

2.1 Quantitative Security Risk Assessment?

Indeed, risk language and 'risk assessment' have become a preoccupation amongst policy makers seeking to respond to phenomena that could lead to adverse events, and which must be identified and measured for probability and consequence. In this approach to risk assessment, a risk can be considered as a function of the likelihood and consequences of a specific adverse event associated with specific (natural) hazards and/or (human) threats [3, 13, 14, 15]. The model of risk identification and characterization by Kaplan and Garrick is based on a trio of "fundamental questions", including: what could go wrong? How likely is it that that will happen? If it does happen, what are the consequences? In this sense, the first question relates to risk identification, and is a creative activity of exploring possible undesirable

---

"world risk society", Beck introduced a series of innovations specific for the international nature of risk society in the twenty-first century, including risk as (globalized) anticipated catastrophe and, especially relevant for the security discourse, transnational terrorism as an entire new category of global risk subverting calculations with "intention" in the place of "chance". A type of global risk that is even more peculiar when coupled with cutting-edge technologies that are continuing, as predicted 20 years earlier, to contribute to uncertainty. "Those responsible for well-intentioned research and technological development will in future have to do more than offer public assurances of the social utility and the minimal 'residual risk' of their activity. Instead, in the future the risk assessments of such technological and scientific developments will have to take into account, literally, intention as well as chance, the terrorist threats and the conceivable malicious uses as well as dangerous side effects" [10].

[4]As a mere indication, the word "risk" was used 4 times in the 2003 European Security Strategy, 15 times in the 2010 Internal Security Strategy, and 31 times in the 2015 European Agenda on Security.

scenarios, [13] that would be consistent with other work that has sought to apply the same framework to chemical, biological and other risks [16, 17, 18]. The second and third questions in the model would correspond to risk characterization, looking respectively to likelihood and impact.

Purportedly "scientific approaches" to risk all seem to share this vision. Although scholars such as Stirling associate the event itself with its impact, in other explanations the "adverse event" is distinguished from the "impact/consequences" it can have. As Kates and Kasperson note, "risks are measures of the likelihood of specific hazardous events leading to certain adverse consequences" [4]. A risk assessment process would start by identifying all the reasonably foreseeable possible adverse events, in order to answer the question "what could go wrong?" [3] This is the process of risk identification and would be followed by a process of risk characterization in which an analysis would be undertaken of all the factors that may influence the likelihood and/or consequences of the identified adverse events. Popular ways to perform this analysis include assigning values to the various factors and Multi-Criteria Decision Analysis (MCDA), which characterizes, relatively, various risks factors using qualitative definitions.[5]

Such purportedly "rational" or "objective" risk framings have been applied to assess a number of natural hazards as well as security-related risks associated with the deliberate misuse of science and technology by actors (person, group or nation-State) intending to cause harm, including using chemical and biological weapons. For example risk characterization related to chemical and biological weapons would include an analysis of factors pertaining to the nature, mode and context of dissemination [5]; the nature of the target [19]; and the motives [20], intensity [21], known values and beliefs [22], skills and objectives of possible perpetrators.[6]

The appeal of such approach is, in part, that it purports to be founded upon rational, 'sound science', 'expert analysis' or, in the case of the UK, "Subject-matter experts, analysts and intelligence specialists" – described by the then Prime Minister David Cameron as "all the relevant people" – who frequently serve as the definitive authority for decisions taken; and in part because the process serves to reduce complex political problems into 'single "definitive" technical or expert interpretations' upon which policy makers can act – and be seen to be acting – in a 'rational' manner [23].

Such a practice is evident in the development of the UK National Security Strategy which formed the "first ever National Security Risk Assessment (NSRA)" in which the UK's National Security Council identified: "the full range of existing and potential risks to our national security which might materialise over a five and 20 year horizon. All potential risks of sufficient scale or impact … were assessed, based on their relative likelihood and relative impact." Upon the unveiling of the National Security

---

[5] Methods are used to inform decisions in situations of limited and evolving knowledge from multiple sources. MDCA methods are based on weighted sum algorithms of multiple factors evaluated against each other [18].

[6] Such objectives include can include killing but also economic sabotage, media attention and prestige, incapacitation, crime, destabilization, disruption, deterrence and denial.

Strategy in the UK, Cameron suggested, "We have had a proper process-a national security process" and that: "the review has been very different from those that went before it. It has considered all elements of national security, home and abroad… It has been led from the top with all the relevant people around the table" [23].

In some respects the NSS was different in the sense that it recognizes the hazards and dangers posed by environmental change and new wars as well as the limitations of military means alone in responding to such challenges. Yet in other respects the National Security Strategy, continued to pursue an approach based on national security which used the language and practice of risk to mask the socially mediated assumptions, interests and (the limits of) organisational knowledge that were at play in the determination of the UK's National Security Risks [24].

3   Challenges with CBW-Related Risks Assessment

Looking at risks and trying to make sense of hazards is no bad thing. One of the issues with the approach to risk framing outlined above however is that there are significant limits as to how "true" and "reliable" the results of the seemingly rational analysis are; and by implication, the appropriateness of ensuing risk mitigation measures. Risk framing is often associated with 'sound science' and terms such as rational, objective, quantitative, probabilistic. Yet while these approach share a vision of "representing reality" they may actually mean different things and reflect different levels of confidence and certainty in part because of the limits of data sets of relevance to CBW and in part because of the limits of 'experts'.

3.1   Limits of Datasets

Indeed, efforts to quantify CBW-related risks normally require a body of data from which to inform assessments. This is relatively straight forward in areas such as engineering failures or car accidents where there is an aggregate body of data on events from which to inform probabilities; but even then data is often simplified, masking complexity and a certain amount of uncertainty (or ambiguity) as different and complex parameters are reduced and aggregated [25]. With new complex systems or little-known chemical or biological agents, this becomes even more difficult as there is frequently going to be a lack of aggregate data from which to meaningfully determine probabilities. In the absence of sufficient data risk assessment can become vulnerable to whimsy.

This is compounded by limitations in data pertaining to the human dimension of CBW terrorism, specifically information about motives, means and objectives of different groups. There have been several datasets of such information created, including public datasets such as the Global Terrorism Database (GTD) developed by the START consortium and various chronologies of the use of

CBW. Such datasets are useful sources of information on past cases of chemical and biological weapons adoption or use. However, a number of issues remain with these and indeed any dataset pertaining to CBW.

First, there remain considerable variations in definitions & assumptions surrounding CBW and CBW-related incidents. For instance, what is a chemical or biological weapon? Do such weapons include only pathogens and toxins, and only when explicitly optimized for a hostile purpose, or do they include any chemical compound or biological organisms that is used to cause some sort of harm? Are "munching insects" such as Thrips Palmi, or invasive species that can cause economical damage, biological weapons? Does an attack on chemical facilities, or the throwing of acid at people, constitute chemical weapons? An overly broad a definition of chemical and biological weapons can render the term meaningless. An overly narrow definition can also be unhelpful as it skews the focus around only those more significant incidents which in themselves may be anomalies in how chemical or biological weapons have been adopted or used.

The second factor is that there remain considerable gaps in knowledge of CBW incidents; more comprehensive datasets, such as the GTD or the POICN database, seeking to capture a broad range of incidents have acknowledged as much, with significant percentages of certain variables, including on inter alia the perpetrators, the agents used, the motivations and indeed the validity of some reported cases, omitted. Datasets also frequently omit or overlook seemingly validated events that are perhaps useful, but fall outside of key criterion or time frames. For example the use of CBW against animals or plants as was the case with the Mau Mau in 1950 [26] and the reported threat to use biological agents against crops by the Tamil Tigers circa 1982 [27], are not always included in datasets, despite providing useful illustrative examples of how CBW could be used.

Third, there is a need for caution in some of the cases included. Non-events or naturally occurring phenomenon have been mistaken for – or deliberately misrepresented as – chemical and biological weapons use. Fourth there is a need for caution in the reliability of datasets that are largely based on past data as past events are not necessarily useful in predict future ones, particularly as biotechnology evolves and chemical and biological sciences converge. Whilst scenario building exercises can be useful in this regard, there are challenges with departing from known events and moving from facts to fictions in risk appraisal.

Fifth, and yet perhaps most significant, is that there is a need for caution with data on CBW-related adoption because of the emotive power of allegations of association with CBW. Some reported incidents of CBW use in criminal or terrorist contexts are just that – reports which are unvalidated and in some cases unlikely, but nonetheless serve the interest of powerful actors as they can be used to demonise individuals, groups or countries. In this regard it is worth noting Robinson's remark that "Accusations of association with [CBW] have for centuries, even millennia, been used by well-intentioned as well as unscrupulous people to vilify enemies and to calumniate rivals" [28]. This presents major issues for those seeking to undertake in objective risk assessment drawing on past data, as it requires careful separation and a degree of judgment in separating reality from powerful 'alternative facts'.

3.2   Limits of Quantifying Likelihood

Such factors necessitate that the aggregate data required to quantify the 'likelihood' of chemical and biological weapons can prove difficult to acquire and potentially misleading. Even basic calculations of the frequency of CBW events become highly contested (and contestable) depending on definitions and data selected: for example, the inclusion of failed cases of agro-bioterrorism and the use of acid as a weapon, will generate very different frequency calculations to a data-set focusing on successful lethal biological attacks against humans. Even in circumstances where there is agreement on definitions and criterion, there will remain uncertainty over certain cases in open source datasets, and most likely closed, classified information on cases too. To some extent, uncertainty could be acknowledged and mitigated by the integration of uncertainty factor into calculations of frequencies, however this potentially creates another potentially subjective factor in the calculation.

3.3   Limits of Quantifying Consequences

Yet it is not only the likelihood of risks which are difficult to assess, so too are the consequences of a CBW event. CBW are frequently capricious weapons and vulnerable to factors such as *inter alia*, atmospheric stability, convective forces, ground cover (mist or fog), mechanical forces (terrain roughness) and rainout [29]; not to mention the public health, immunity and detection and response capacity of the target population factors. Such factors mean that the impact of the use of an agent can vary by orders of magnitude depending on the environment. Such issues of predictability are more than academic musing, but had a bearing on the selection of agents in Cold War CBW programmes.

   For example, Anthrax is arguably the archetypal biological weapon and has been considered in many state biological weapons programs in part because of the relative hardiness of the spores and the considerable knowledge of the agent. Yet for all the data on the characterisation of anthrax, the extrapolation of lethality data from animal test subjects to humans proved difficult in the case of the US program. This was compounded by the apparent variance in estimates of LD50 of Anthrax with LD50 calculations for humans ranging from 1000 to 6000 spores. As such, Anthrax was standardised not for use in strategic weapons, but as a weapon for use by special forces (the M2 munition). The US had greater success with the Tularemia based M210 Biological Warhead for the MGM-29 Sergeant Missile, however as Kirby has illustrated, the limitations on the weapon were considerable, with logistical factors, such as the half-life decay of the agent, and environmental conditions spelling the difference between mass effect, and negligible effect [29].

   One could argue that highly contagious biological agents could mitigate such logistical and environmental difficulties. Yet such weapons too are limited in their predictability. As has been noted "epidemics involve two dynamics; the first is the course of the disease in the individual, and is biomedical. The second is the spatial contact process among individuals, and is social". The latter in particular makes predictability difficult "random effects can be dramatic, spelling the difference between large-scale epidemics and abortive ones that never take off" [30].

Chemical weapons provide relatively more predictable effects, yet chemical weapons too are influenced by environmental factors. As Carus has remarked:

> Chemical agents are highly unpredictable. They are very sensitive to weather conditions, including temperature, wind, and atmospheric pressure. Even with high quality weather forecasting it is difficult to ascertain accurately the specific conditions that will exist at a particular place. [31]

For example, Botulin ranks amongst the most lethal agents know to humans and has been considered in several state chemical (and biological) weapons programs. However, devoid of complex stabilisation processes, botulin was highly unpredictable with "[e]xtremes of temperature and humidity will degrade the toxin … Depending on the weather, aerosolized toxin has been estimated to decay at between less than 1 % to 4 % per minute".

The point is not that consequences can or should somehow be ignored in risk assessment. Nor is this to suggest that the consequences cannot be estimated under certain conditions. However, any attempt to neatly quantify the consequences of biological and to a lesser extent chemical weapons needs to be heavily caveated; and for all the advances in science and technology, precisely predicting the outcome of CBW attacks is "the prerogative only of the ignorant" [32].

3.4   Limits of "Expertise"

The realist approach to risk typically places much greater importance on "experts", a category delineated from "lay people"; with the former regarded as neutral actors employing an objective and replicable measurement of risk, and the latter often viewed as unable to correctly assess risk and led by whimsy. As much is implicit in what Erik Millstone has termed the 'technocratic model' of science advice in policy making that has served as the dominant narrative for much of the last 60 years [33]. In this model, "policy is based (only) on sound science", with technocracy implying "that public administration by impartial experts should replace governance by those with particular interests because only the experts possess the relevant understanding and knowledge" [33]. The underpinning assumption of this model is that "the relevant scientific knowledge is objective, politically neutral, readily available and sufficient" [33]. However, as Millstone and others have indicated, knowledge is often "incomplete, uncertain or equivocal"; and experts are not always impartial and immune to bias [33] (Fig. 1).

Indeed, the technocratic model of scientific advice to policy makers has begun to weaken in several issue-areas over the last couple of decades, in part because of the recognition of uncertainty and in part because of the rise of freedom of infor-

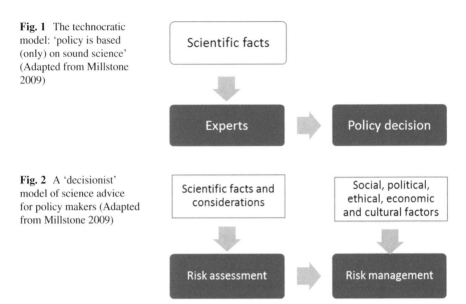

Fig. 1 The technocratic model: 'policy is based (only) on sound science' (Adapted from Millstone 2009)

Fig. 2 A 'decisionist' model of science advice for policy makers (Adapted from Millstone 2009)

mation act requests. Certainly, in relation to the former, uncertainty is increasingly recognised in scientific assessments. For example US legislation related to food and drugs has "acknowledged scientific uncertainties and provided federal agencies such as the US Food and Drug Administration (FDA) with guidance on how they should interpret and respond to such uncertainties" [33]; similarly European bodies, such as the European Food Safety Authority (EFSA), have recognised that "Methodologies for integrating (weighing) evidence and assessing uncertainties are of utmost importance to ensure that scientific assessments are transparent, robust and fit for purpose to support decision-makers" [34]. Regarding the latter, Millstone suggest that freedom of information request related to the use of expert advice over GMO decisions, "entailed the disclosure of sufficient information on the science used to support policy to reveal that the science was often profoundly uncertain" [33].

The experience with food safety governance, is clearly different to that of CBW where the evidence base for risk assessment is frequently secreted and sensitive. Nonetheless, the acknowledgement and efforts to "develop a more sophisticated understanding of scientific uncertainty and its treatment" [35], along with experiences, such as the Iraq War (Chilcot) Inquiry and the Butler report in the UK, and WikiLeaks to some extent diminished faith in models of expert advice to policy makers relying exclusively on technocratic or scientific input, unmediated by social, political, ethical, economic and cultural factors. Moreover, it highlights how "risk are routinely predicated on assumptions, which inform the scientific deliberations, but which are not themselves scientific" but rather "hybrid judgements" that draw on scientific as well "normative considerations" [33] (Fig. 2).

4 From Technocratic to 'Decisionist'

As a result of the limitations in technocratic models of risk assessment in other issues areas, the provision of scientific advice shift from a technocratic to a two stage "decisionist model" in which scientific risk assessment was followed by non-scientific process of risk management.

Similar two-phase approaches have been popularised in the US through work of the National Research Council, such as on *Risk Assessment in the Federal Government: Managing the Process.* This model proved highly influential and has been applied by a number of organisations. As Millstone notes "deliberate decisions have been taken to create separate pairs of institutions, with one of the pair labelled as responsible for 'risk assessment', having a scientific mandate, and the other labelled as having responsibility for 'risk management' policy decisions" [33].

More recently an advanced version of this form of risk assessment has been undertaken for Gain of Function research using models draw from the nuclear sector [36] but populated by data on lab incidents and epidemiological data that were passed to NSABB to inform decisions [37] in a manner consistent with the decisionist or "red book" model.

One of the problem with both the decisionist (or "red book") and technocratic models is portrayal of "scientific representations of risk" as if they were entirely free from all social, economic or policy influence, when in fact it is widely now understood as "representations of risk are inevitably hybrid judgements, dependent on both scientific and normative considerations". Several scholars have demonstrated this, illustrating how expert can reach starkly different conclusions from the same body of data because of the different framing assumptions, or as Millstone notes "often because they are asking and answering different questions" [33]. Looking beyond the Gryphon report, the more recent controversy over Gain of Function study also perhaps illustrates how different framings can lead to different conclusions with security and scientific communities – whilst not monolithic – tending towards different conclusion. Indeed, Ron Fouchier, a virologist at the centre of the Gain of Function controversy has remarked "Even if they could be quantified, the weighing of risks and benefits will be a personal (subjective) issue" [37].

5 Towards a Co-evolutionary Model

Millstones' remedy for the limitations of the technocratic and decisionist models, is the notion of a "co-evolutionary model" of science in policy making. The model does not exclude scientific consideration in the form of expert risk assessment, but seeks to contextualise this, by preceding such expert risk assessment with a more explicit process of outlining a risk assessment policy in which socio-economic consideration are used to inform the framing assumptions that feed into expert

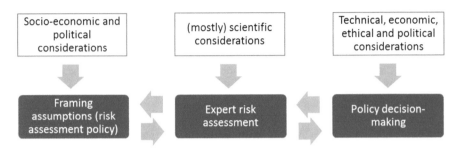

Fig. 3 A co-evolutionary model: reciprocal links between science and policy (Adapted from Millstone 2009)

risk assessment. This could include information such the historical incidents of CBW use which informed assumptions and specifics of the questions asked of assessors. Moreover, rather than the scientifically informed expert risk assessment directly determining risk management policies, policy decision making (risk management) is explicitly informed by technical as well as economical, and political considerations. Put otherwise, "scientific risk assessments are… sandwiched between up-stream framing considerations and down-stream interpretative judgements" [33] (Fig. 3).

6   …And CBW Risk Assessment?

There are significant differences between appraisal of risks related to climate change or GMOs on the one hand, and the risks of chemical and biological weapons on the other. Whilst all these topics are clearly political sensitive, with Climate change and GMOs the body of evidence can largely be made available for external actors to scrutinise. In the case of CBW, much of the information is likely to be security sensitive. This makes it much more difficult to open up CBW risk assessment to form a more discursive, participatory process as has been proposed for other issue areas.

Nonetheless this does not discount the possibility of alternative approaches to addressing CBW related risk that acknowledge that the process of risk identification and characterization are, in part, socially constructed; a notion advanced variously, on risk in general, by Beck, Giddens and Douglas. Under such as model quantitative and qualitative approaches could co-exist combining a more rational phase of risk characterization with a more 'subjective' process of risk evaluation within a ("co-evolutionary") risk assessment process.

In this model the weighting of rational and subjective elements could be determined by the levels of uncertainty and the nature of the risks addressed. In circumstances of increasing uncertainty the employment of qualitative measures could be weighted more heavily; in cases where risks were better characterised quantitative approaches could be given greater weight and the "representations of risks are portrayed as hybrid judgements constructed out of both scientific and non-scientific considerations" [33]. In both cases, reasonably foreseeable risks could be somehow ranked, as studies by Marris et al. indicate, the integration of constructivist or even relativist considerations into objective or quantitative risk assessments doesn't preclude that risks can be categorized and ranked according to a number of factors [38].

However such a process needs to acknowledge "the prospect of radical surprise" or "unknown unknowns" which evade risk identification process – and which obviously cannot be ranked – to avoid generating a false sense of security. Moreover, in seeking to rank risks, assessors would need to acknowledge that "probabilistic reasoning under uncertainty cannot yield a single objectively aggregate value" and open up the framings that inform the process of risk assessment in circumstances where the consequences of an adverse event may be problematic and ambiguous [39]. The latter requires articulating the specific details of the questions that were asked of risk assessors. For example, articulating who (the EU, the Member State, military forces, citizens) or what (human health, welfare, economy, livestock) is at risk; from what (non-state actors, terrorist, states, criminals). Such a step would open-up the process and subject previously hidden assumptions, values and beliefs to reasonable scrutiny rather than black boxing the framings employed. Whilst governments can legitimately withhold details of the data used to inform risk assessment process, there are less grounds for withholding details of the *questions* put to risk assessors and some details of the information that informs assumptions of risk assessors *can* be synthesised and made public.

From the perspective of observers of the risk assessment process and results, it will important to recognise that difficult decisions need to be taken over which risk mitigations measures to pursue to deal with CBW risks in the absence of complete information pertaining to the initial likelihood or consequences of an event. Put otherwise, 'paralysis by analysis' is not an option for policy makers that need to respond – and be seen to respond – to risks, particular for what Slovic refers to as 'dread' risks. It is also needs to be acknowledged that certain CBW mitigations measures can and have lower likelihood and/or even impact in the absence of complete information. Indeed, even when risk assessment is cynically seen as an instrumental tool to justify certain mitigation measures, it does not necessarily mean that the measures themselves are not helpful in mitigating the given risk, also when the exact initial baseline is not known or reliable [40, 41].

7 Conclusions

Risk assessment has become increasingly important in relation to chemical and biological weapons in the twenty-first century with greater emphasis placed on efforts to objectively calculate the likelihood and consequences of adverse CBW events. However, whilst discussion around CBW related risk is important, it needs to be recognised that risks – including CBW risks – are not always objectively analysable, let alone quantifiable and that risk assessment is not always 'rational', but frequently a combination of rationally comparable analysis *and* socially-mediated activity in which risks are socially constructed, and their "importance" subjectively evaluated or constructed.

This is not an argument for giving up on CBW risks assessment. However it does suggest that decision-makers need to be aware of the limits of quantitative (only) risk assessment; and those involved in risk assessment need to be more forthcoming in the uncertainty of probabilistic reasoning and acknowledge that the consequences of adverse CBW-related event may be problematic and ambiguous. It also needs to be recognised that there is a need for caution in claims that all risks have been assessed, a notion that can leave us ever more vulnerable to surprise from unknown unknowns.

References

1. Edmunds, T.: British civil-military relations and the problem of risk. Int. Aff. 88, 265–282 (2012)
2. Slovic, P., Weber, E.U.: Perception of Risk Posed by Extreme Events. (2002)
3. Kaplan, S., Garrick, B.J.: On The Quantitative Definition of Risk. Risk Anal. 1, (1981)
4. Kates, R.W., Kasperson, J.X.: Comparative risk analysis of technological hazards (A Review). Proc. - Natl. Acad. Sci. USA. 80, 7027–7038 (1983)
5. Hohenemser, C., Kates, R.W., Slovic, P.: The Nature of Technological Hazard. Science (80-.). 220, 378–384 (1983)
6. Starr, C.: Social Benefit versus Technological Risk. Science (80-.). 165, (1969)1232-8.
7. Williams, M.J.: NATO, Security and Risk Management: From Kosovo to Khandahar. Taylor & Francis (2008)
8. Leach, M., Scoones, I., Stirling, A.: Pathways to Sustainability: an overview of the STEPS Centre approach. (2007)
9. Christopher Daase, Kessler, O.: Knowns and Unknowns in the `War on Terror': Uncertainty and the Political Construction of Danger. Secur. Dialogue. 38, (2007)
10. Beck, U.: World at Risk. Polity (2013)
11. Council of the European Union: European Security Strategy., Brussels (2003)
12. European Commission: The European Agenda on Security., Strasbourg (2015)
13. Kaplan, S.: The Words of Risk Analysis. Risk Anlaysis. 17, (1997)
14. Garrick, B.J.: Quantifying and Controlling Catastrophic Risks. Elsevier Inc. (2009)
15. George E. Apostolakis: How Useful Is Quantitative Risk Assessment? Risk Anal. 24, (2004)
16. World Health Organization: Laboratory biosafety manual Third edition. (2004)
17. Gormley, Á., Pollard, S., Rocks, S., Black, E.: Guidelines for Environmental Risk Assessment and Management - Green Leaves III. (2011)
18. Caskey, S., Gaudioso, J., Salerno, R., Wagener, S., Risi, G., Kozlovac, J., Halkjær-knudsen, V., Prat, E.: Biosafety Risk Assessment Methodology. (2010).
19. Isukapalli, S.S., Lioy, P.J., Georgopoulos, P.G.: Mechanistic Modeling of Emergency Events : Assessing the Impact of Hypothetical Releases of Anthrax. Risk Anal. 28, (2008)
20. Brown, G.G., Cox, L.A.: How probabilistic risk assessment can mislead terrorism risk analysts. Risk Anal. 31, (2011)
21. Depoy, J., Phelan, J., Sholander, P., Smith, B., Varnado, G.B., Wyss, G.: Risk assessment for physical and cyber attacks on critical infrastructures. MILCOM 2005 - 2005 IEEE Mil. Commun. Conf. 3, 1961–1969 (2005)
22. Keeney, G.L., Winterfeldt, D. Von: Identifying and Structuring the Objectives of Terrorists. Risk Anal. 30, 1803–1816 (2010)

23. Hansard (UK): Strategic Defence and Security Review. In: Oral Answers to Questions, Tuesday 19 October. Hansard, London (2010)
24. Ritchie, N.: Rethinking security : a critical analysis of the Strategic Defence and Security Review. Int. Aff. 87, 355–376 (2011)
25. Stirling, A.: Chapter 2: Risk, uncertainty and precaution: some instrumental implications from the social sciences. In: Berkhout, F., Leach, M., and Scoones, I. (eds.) Negotiating Environmental Change New Perspectives from Social Science. Elgar
26. Yeh, J.-Y., Park, J.-Y., Cho, Y.S., Cho, I.-S.: Animal Biowarfare Research: Historical Perspective and Potential Future Attacks. Zoonoses Public Health. 59, 1–9 (2012)
27. Carus, W.S.: Bioterrorism and Biocrimes - The Illicit Use of Biological Agents Since 1900. (February 2001 revision)., Washington, D.C. (2001)
28. Robinson, J.P.P.: Alleged Use of Chemical Weapons in Syria, HSPOP 4. (2013)
29. Kirby, R.D.: The Sergeant: A Biological Missile. Eximdyne (2014)
30. Epstein, J., Cummings, D., Chakravarty, S.: Toward a containment strategy for smallpox bioterror: an individual-based computational approach. Gener. Soc. Sci. Stud. agent-based Comput. Model. (2006)
31. Carus, W.S.: Chemical weapons in the Middle East. Policy Focus. 15 pp (1988)
32. SIPRI: The Prevention of CBW. The Problem of Chemical and Biological Warfare: Volume V. SIPRI, in association with Oxford University Press (1971).
33. Millstone, E.: Science, risk and governance: Radical rhetorics and the realities of reform in food safety governance. Res. Policy. (2009)
34. Dorne, J.L.C.M., Bottex, B., Merten, C., Germini, A., Georgiadis, N., Aiassa, E., Martino, L., Rhomberg, L., Clewell, H.J., Greiner, M., Suter, G.W., Whelan, M., Hart, A.D.M., Knight, D., Agarwal, P., Younes, M., Alexander, J., Hardy, A.R.: Weighing evidence and assessing uncertainties. Eur. Food Saf. Auth. J. 14, 1–13 (2016)
35. Parliamentary Office of Science and Technology: Handling Uncertainty in Scientific Evidence. (2004)
36. Gryphon Scientific: Risk and Benefit Analysis of Gain of Function Research: Final Report—April 2016. (2015)
37. Reardon, S.: US plan to assess risky disease research takes shape. Nature. October, (2015)
38. Marris, C., Langford, I.H., Riordanz, T.O.: A Quantitative Test of the Cultural Theory of Risk Perceptions: Comparison with the Psychometric Paradigm. Risk Anal. 18, (1998)
39. Stirling, A.: Opening Up the Politics of Knowledge and Power in Bioscience. PLoS Biol. 10, (2012)
40. Comité Européen De Normalisation (CEN): CEN Workshop Agreement - Laboratory biorisk management, Brussels (2011)
41. Caskey, S. and Sevilla-Reys, E.E. IN: Caskey, S., Gaudioso, J., Salerno, R., Wagener, S., Risi, G., Kozlovac, J., Halkjær-knudsen, V., Prat, E.: Biosafety Risk Assessment Methodology. (2010)

Combining Theoretical Education and Realistic, Practical Training: The Right Approach to Minimize CBRNe Risk

Dieter Rothbacher

Abstract CBRNe defense and its effectiveness depend on up-to-date education and realistic training programs. Such programs should allow confirmation through verification; therefore, they also require standards. CBRNe risks have been changing significantly over the last years, and such changes should be reflected in education and training programs. The combination of theoretical education and practical training is the best approach to minimize risks stemming from CBRNe materials and their malicious use. This article will analyze CBRNe education and training; more specifically, focus will be placed on both the theoretical education of First Responders and their realistic, practical training, including live agent training. How realistic does an education and training program have to be to counter current CBRNe risks? Is the use of CBR substances for training an essential element? And if so, can those substances be used safely for individual and/or collective training? According to the Author, realistic training can only be delivered when CBR materials are used; the benefits of combining theoretical education and realistic live agent training with CBR materials outweigh the related occupational risks, thus being essential parts of any effective CBRNE training program. Data from live agent trainings and their analysis will support this theory.

Keywords CBRNe • Training • Didactic

D. Rothbacher (✉)
CBRN Protection GmbH, Vienna, Austria
e-mail: dieter.rothbacher@cbrn-protection.com

© Springer International Publishing AG 2017
M. Martellini, A. Malizia (eds.), *Cyber and Chemical, Biological, Radiological, Nuclear, Explosives Challenges*, Terrorism, Security, and Computation,
DOI 10.1007/978-3-319-62108-1_16

1 Introduction

1.1 Elements of CBRNe Defense

CBRNe events have had extensive media coverage, especially in the last months and years and, consequently, increased the public perception of vulnerability against the use of CBRNe materials.

The various phases of a response to CBRNe events are conceptually largely driven by the definition of this expression; traditionally, civilian and military organizations interpret it differently. Definitions are major factors for stakeholders and their organizations when preparing for all phases of a response to a CBRNe event.

Managing any CBRNe event implies dealing with the following phases: threat, preparedness, response and remediation.

The threat and all its features and components is the main factor influencing and characterizing all other CBRNe defense phases; it is therefore crucial for the stakeholders to identify key areas that require an innovative approach to develop organizational structures, training programs, and CBRNe defense research. The threat sets the context for innovative protective methods in preparedness, response and remediation. How can the threat and its perception drive the development of innovative methods? Or are we still stuck in old ways of thinking, and that is driving the development of CBRNe protective technologies?

From 1990 to 2011, 1431 CBRN related incidents (attacks) have been recorded [1].

Recent, new approaches to analyze the CBRNe threat and incidents have shown that the related industry plays a crucial role in determining which material (C, B, R or N) will be used for unconventional attacks. A study from 2013 shows that high national wealth increases the likelihood of CBRN terrorism; the prevalence of CBRNe materials in society for use by industry, in scientific research and medical diagnostics creates a significant risk of diversion or exploitation by terrorists or criminals.

Preparedness is largely driven by extensive training programs, based upon operational procedures and incident awareness and management. Due to the merging threat posed by chemical, biological, radiological, nuclear and explosive (CBRNE) materials, there has been an increased focus on CBRNE training for first responder and military communities worldwide.

Not only training, but also other methods enhance preparedness; the risk of CBRN attacks and their detrimental consequences in an urban setting has been recognized for some time, but very little has been done to quantitatively assess those risks. Developing defense strategies, identifying safe zones or havens where people can seek shelter, installing CBRN protective equipment are just some of the measures that can be taken to overcome the risk of such events.

Response to a CBRNe event is largely dependent on activities and resources. The response to a CBRNe event may traditionally be deemed a stage at which new technical developments (i.e. detection equipment) are deployed. Examples of recent equipment and developments are robotic platforms, new decontamination technologies (non-aqueous) and respiratory and skin protection. But what about interagency coordination?

The NATO Consequence Management handbook states the following:

> Response to CBRN incident is unlikely to be conducted in isolation. No single civil or military capability, agency or military unit possesses the capacity and expertise to act unilaterally on many complex issues that may arise in response to CBRN incidents. [2]

Barriers to coordination and the lack of understanding about agency roles are fundamental problems to interagency work. Legal and structural barriers, like lack of clarity on chain of command, no protocols on interagency information sharing and exchange are just a few that are worth mentioning at this point. Stakeholders usually complain about the lack of national direction in promoting interagency coordination.

Remediation is the stepchild of all aforementioned phases, but usually the most costly one. Remediation is quite seen as extensive decontamination operations, to render affected areas safe. The question "what is clean"? Is it that there is absolutely no more contaminant or does it mean that it cannot be detected anymore, or even that it is not effective on unprotected people? Can military techniques and equipment be used in a civilian environment? Or are new technologies needed? Recent studies have shown that there is no universal decontamination technology for a clean-up of a CWA in a civilian setting; we need to identify those materials that are easily decontaminated by existing technologies and procedures.

2　CBRNe Standards

2.1　Definitions

What is CBRN? At this point we should look at commonly used definitions of CBRN; CBRNe was only introduced recently.

European Union (EU) There are no commonly accepted definitions of CBRN materials, threats or incidents – for example earlier EU policy documents in this domain merely refer to CBRN incidents without defining what these incidents could be. Other terminology related to CBRN materials refers to terrorist attacks using unconventional means – as opposed to the more conventional means of explosives and arms. In the military context, the terminology mainly refers to the use of non-conventional weapons, or WMD.

> For the purpose of this communication, however, it is most useful to use a rather broad description of the terrorist threat concerning CBRN materials: all uses of chemical, biological, radiological or nuclear substances and materials for terrorist purposes. An approach which looks at all possible ways in which terrorists can use these materials is the only one acceptable from a point of view of prevention and detection, since all possible risks concerning these materials should be covered.
>
> However, when considering preparedness and response in this context, it is unavoidable to start from an all-hazards approach, since no matter whether a CBRN incident is accidental or intentional, man-made or not, the response in terms of civil protection and health is likely to be similar. The CBRN policy package is therefore broadly based on an all-hazards-approach, but with a strong emphasis on countering the terrorist threat, in particular with regard to preventive actions. [3, p. 2]

The European Union CBRN Action Plan defines CBRN materials as follows:

There are no commonly accepted definitions of CBRN materials, threats or incidents – for example earlier EU policy documents in this domain merely refer to CBRN incidents without defining what these incidents could be. Other terminology related to CBRN materials refers to terrorist attacks using unconventional means – as opposed to the more conventional means of explosives and arms. In the military context, the terminology mainly refers to the use of non-conventional weapons, or WMD.

It is most useful to use a rather broad description of the terrorist threat concerning CBRN materials: all uses of chemical, biological, radiological or nuclear substances and materials for terrorist purposes. An approach which looks at all possible ways in which terrorists can use these materials is the only one acceptable from a point of view of prevention and detection, since all possible risks concerning these materials should be covered. [4, p. 3]

NATO NATO distinguishes between Weapons of Mass Destruction and Toxic Industrial Materials (Toxic Industrial Chemicals, Toxic Industrial Biological material, Toxic Industrial Radiological material).

Chemical, Biological, Radiological or Nuclear Weapon A fully engineered assembly designed for employment by the armed forces of a nation state to cause the release of a chemical or biological agent or radiological material onto a chosen target or to generate a nuclear detonation [5].

Chemical, Biological, Radiological or Nuclear Device An improvised assembly or process intended to cause the release of a chemical or biological agent or substance or radiological material into the environment or to result in a nuclear detonation [6].

Chemical, Biological, Radiological and Nuclear Defense Plans and activities intended to mitigate or neutralize adverse effects on operations and personnel resulting from: the use or threatened use of chemical, biological, radiological or nuclear weapons and devices; the emergence of secondary hazards arising from counter-force targeting; or the release, or risk of release, of toxic industrial materials into the environment. Synonym: nuclear, biological and chemical defense [7]. Chemical agent: A chemical substance which is intended for use in military operations to kill, seriously injure or incapacitate man through its physiological effects. The term excludes riot control agents when used for law enforcement purposes, herbicides, smokes and flames [8].

Biological Weapon An item of material, which projects, disperses, or disseminates a biological agent including arthropod vectors can be defined a biological weapon [9].

Radiological Weapon A radiological weapon is any device, including a weapon or equipment other than a nuclear explosive device, specifically designed to employ radioactive material by disseminating it to cause destruction, damage or injury by means of the radiation produced by the decay of such material. Sometimes also described as a radiation weapon or radiation dispersal device [10].

Toxic Industrial Material (TIM) TIM is a generic term for toxic chemical, biological and radioactive substances in solid, liquid, aerosolized, or gaseous form created for industrial, commercial, medical, or domestic purposes. Normally, such materials are retained within their planned manufacturing, storage, and transport facilities. However, their intentional or accidental release may pose a significant hazard. Forces may be exposed to TIMs as a consequence of friendly action, adversary action, or accidents.

Toxic Industrial Chemical (TIC) Industrial chemicals can pose significant toxic hazards and can damage the human body and equipment. Many industrial chemicals are corrosive, flammable, explosive, or react violently with air or water; these hazards may pose greater short-term challenges than the immediate toxic effects. Most but not all TICs will be released as vapor or highly volatile liquid and can have both short-term and long-term health effects. The most important action in case of a massive industrial chemical release is immediate evacuation outside the hazard path. The greatest risk from a large-scale toxic chemical release occurs when personnel are unable to escape the immediate area and are overcome by vapors or blast effects. Military respirators and personal protective equipment may provide limited protection against TICs. Shelter in- place is an alternative action if facilities are available and evacuation is impractical. Useful guidance on the hazards resulting from the release of a wide range of TICs and appropriate initial response can be found in the Emergency Response Guide (ERG) and Agreement on Dangerous Goods by Roads (Europe) – ADR. Allied Tactical Publication (ATP)-3.8.1 Volume I, provides a list of TICs of concern to commanders for the conduct of military operations.

Toxic Industrial Biological (TIB) TIB has the potential to produce significant environmental damage and result in pollution of water supplies, long-term ecological damage, and present a significant hazard to military operations. Possible sources of TIB include hospitals and other medical installations and research, production, storage or recycling facilities for the pharmaceutical or agricultural industries.

Toxic Industrial Radiological (TIR) Possible sources of TIR material capable of producing radioactive hazards are: civil nuclear production, research, recycling and storage facilities; waste containment sites; industrial and medical sources; materials and sources in transit; stolen or smuggled nuclear weapons grade material. The characteristics of radioactive hazards produced will depend on the type of radiation and the nuclide involved. The geographical area of spread of the hazards can vary dramatically according to the source and manner of release [11].

2.2 CBRNe Training

Preparedness is largely driven by extensive training programs, based upon operational procedures and incident awareness and management, as already mentioned in Chap. 1. Specialized training facilities allow enhancing CBRNe preparedness; due to the merging threat posed by chemical, biological, radiological, nuclear and explosive (CBRNE) materials, there has been an increased focus on CBRNE training for first responder and military communities worldwide.

Few countries have live-agent training facilities and while the size varies from one site to another, all live-agent facilities should be configured with roughly the same characteristics if they are going to properly train CBRNe first responders. The main features of training with chemical warfare agents (CWA) and toxic industrial chemicals (TIC), for example, illustrate the broader CBRNE challenges.

One of the main CBRNe training related standards is the NATO standard ATP-3.8.1 Volume III; it provides standards for education, training and evaluation, on all levels: individual and collective, from basic to modular training (including advanced and on-the-job).

Successful CBRN defense depends on a strong, up-to-date education training and exercise program for individual and collective CBRN defense specialists, commanders, and unit and medical personnel that permits a detailed assessment of standards and proficiencies [12, p. 20].

It also recognizes the importance of Live Agent Training and states that Live Agent Training is an element of specialist CBRN defense training and it should be undertaken and conducted in accordance with national regulations [13].

A proper training in accordance with ATP-3.8.1, Volume III (CBRN Defense Standards for Education, Training and Evaluation) is a prerequisite for military forces, especially for CBRN specialists [14, p. 7].

3 Live Agent Training

3.1 Overview

Confidence in equipment, doctrine and technical and tactical procedures can only be achieved through realistic training, and these realistic conditions are extremely important to operational readiness and success.

It is difficult for responders to CBRNe events to gain experience with CBRNe materials, due to the very infrequent occurrence of these incidents.

Using real CBRNe materials for trainings – the so-called live agent training – is the only way to overcome that problem.

The main features of training with chemical warfare agents (CWA) and toxic industrial materials (TIM) – the Chemical Live Agent Training – illustrate the broader CBRNe challenges.

This live agent training does not only include protection and detection, it also highlights – inter alia – the difficulty in decontaminating equipment and infrastructure.

This article is an assessment of the impact of live agent training on operational readiness in general, whilst looking at some aspects of health and safety.

3.2 Occupational Health and Safety

Chemical Live Agent Training is an essential part of CBRNe training, but there is also an absolute duty to ensure that Live Agent Training is conducted in a safe environment and that all hazards are identified and risks controlled.

Prevention Versus Response The highest priority has to be applied to prevention. Safety and risk management systems should also include processes to prepare for, respond to, and recover from any incident including non-agent (medical) incidents should they occur.

Chemical Agent Safety The safety and control systems employed should be designed to reduce the potential for exposure to a trainee to be as low as practicably achievable, whilst maintaining the ability to conduct worthwhile training.

Accepted Occupational Exposure Limits (OEL's) could provide the basis of a safety system. This should be applied in such a way that at 1 m from the contaminated surface the concentration of CW agent (airborne) is less than a Short Term Exposure Level (STEL) for the most volatile agent (GB), a level at which an unprotected worker could work for 15 min. One meter is used as a reference as it is essentially an arm's length from the surface or the working distance, which is the breathing zone of the trainee. This, in combination with highly effective protective clothing and respirators (military standard) provides a large safety margin. Actual vapor levels have to be confirmed through air sampling and analysis for all exercises and scenarios during all courses. Air sampling onto tenax tubes followed by GC/MS analysis should be conducted also during courses to establish vapor levels near contaminated working surfaces and within the breathing zone of trainees and instructors. Through these analyses, one can establish a comprehensive, quantitative understanding of the hazards associated with agents, seasons of the year and agent quantities.

All breathing zone analyses are to be compared to permissible occupational exposure limits.

It also has to be recognized that there will be differences in airborne agent concentrations between summer and winter training and therefore the amount of agent used should be adjusted accordingly.

Environmental Safety The safety of the environment during Live Agent Training is just as important as the safety associated with the agents themselves. All activities are conducted within areas certified and designed for Live Agent Training. Agents are used only within double bund trays to prevent cross contamination from contaminated surfaces to personnel (see Sect.3.3.). At the end of each training activity, all surfaces that have had agent placed on them have to be decontaminated and then disposed of as hazardous waste. Other outer protective surfaces are completely decontaminated. All decontamination liquids need to be disposed of in accordance with site and national requirements.

Personnel Safety and Medical Support While experienced instructors can control the hazards associated with the CW agents used during training, they also require personnel to be certified as fit by their employer to perform the training activities in protective clothing. One should conduct pre-entry and post-entry medical monitoring for all live entries; and for all activities involving the use of nerve agents, a baseline activity of acetyl cholinesterase has to be established for each trainee before live entries commence. All protective equipment and respirators have to be serviced – if need be -and tested on site as part of an equipment maintenance program. When respirators are issued to trainees for use during training they should be expertly fit-tested. Mask fit has to be checked prior to any entry into the hot zone. Where heat stress is an issue in the summer months, the use of meteorological data combined with pre-entry and post-entry medical monitoring along with the use of a Heat Management program based on the fitness of individual trainees should be employed. For all training activities, personnel should carry individual decontamination pouches, decontamination stations are established prior to the use of live agents, atropine auto-injectors have to be readily available for entries involving the use of nerve agents and emergency medical support with a qualified CW trained doctor and a fully equipped ambulance has to be available on-site for the duration of all training activities. A risk assessment for all training activities should be available to all trainees prior to the commencement of training.

3.3   Live Agent Training Technical Data

Activities The activities during a CLAT involve the classical pillars of CBRNe defense: protection, detection, and decontamination. The set up can be a basic one – as described below – and go as far as realistic scenarios, involving inter alia CBRNe reconnaissance, sampling, invasive technology, non-destructive evaluation, CBRNE IED/EOD scenarios, large scale decontamination, leak seal packaging, transport of samples, use of on site identification devices and much more.

Agents Used The agents used during a Chemical Live Agent Training are usually from these two groups:

- Nerve agents;
- Blister agents.

Among lethal CW agents, the nerve agents have had an entirely dominant role since the Second World War. Nerve agents acquired their name because they affect the transmission of nerve impulses in the nervous system. All nerve agents belong chemically to the group of organo-phosphorus compounds. They are stable and easily dispersed, highly toxic and have rapid effects both when absorbed through the skin and via respiration [15]. Agents classically used for CLAT include Sarin (GB), and VX.

Blister agents, or vesicants, are one of the most common CW agents. These oily substances act via inhalation and contact with skin. They affect the eyes, respiratory tract, and skin, first as an irritant and then as a cell poison. As the name suggests, blister agents cause large and often life-threatening skin blisters which resemble severe burns [16]. Examples used for training include sulfur mustard (H, HD, Yperite) and Lewisite (L).

Quantities of Agents Used The quantities of chemical warfare agents used in one week of CLAT are in the range of grams; practical experience has shown that this type of training – in its basic form, as described below – can be run with up to 10 grams of agent per week.

Exercises: Basic Chemical Live Agent Training Stations In Tables 1, 2, 3, and 4, you will find some descriptions of four CLAT Training Stations, to give you an idea on what can be done during basic CLAT. These descriptions are intended only as a guide; its execution will depend on rules and resources of the CLAT Facility.

Table 1 Working station 1 CWA physical properties and use of detection paper

Prerequisite	CBRN trained IAW national and site requirements
Learning objective:	Understand the detection of chemical agents
Sequence:	The focus of this workstation exercise is to gain confidence and employ a detection strategy The exercise is conducted as individuals as follows: I. Visual inspection of petri dish to observe color and viscosity of agent II. Use of glass rod to further investigate viscosity by spreading the agent III. Glass rod made safe, discussion on findings so far and expected results IV. Agent applied on the detection paper by glass rod V. Observation of color change Detection papers to be changed for each student (hot run only)
Equipment:	Work station as per diagram for each agent used: I. Primary trays II. Secondary trays III. Glass petri dish IV. Glass beaker V. Glass stirring rods VI. Detector paper on metal plate VII. White backing paper VIII. Kill bucket mixed to facility requirements Note the detection equipment used should preferably be familiar and in use operationally by students
Chemical agent:	Agents used in this exercise should mirror those used throughout the training. Where possible any available additional agents should also be used but only within the confines of facility safety regulations The amounts of agent to view should consider safety requirements therefore it is recommended no more than 100 microliter of each agent to be used for this exercise
Surface:	One petri dish each
Time:	No time limit
Pass/Fail:	N/A
Workplace set up:	Primary tray to be placed in a larger secondary tray in which two primary trays can be used if space allows (trays may not be necessary in a fume hood as they act as a contamination control measure) White paper to be placed under petri dish to aid observation Detector Paper to be fixed to small metal plates for easy disposal Kill bucket

3.4 Prerequisite for Live Agent Training

Combining theoretical and – at times simulant based training – is a common approach to CBRNe training.

But it cannot be underlined enough – confidence in equipment, doctrine and technical and tactical procedures can only be achieved through realistic training, and these realistic conditions are extremely important to operational readiness and success.

Table 2 Working station 2: use of hand held detectors (vapor)

Prerequisite	CBRN trained IAW national and site requirements
Learning objective:	Understand the detection of chemical agents
Sequence:	The focus of this workstation exercise is to gain confidence and employ a detection strategy The exercise is conducted as individuals as follows: I. Stand- off, observe wind and slowly approach work station (instructor to open petri dish) II. Use a continuous flow vapor detector to establish volatility at varying distances III. Confirm with second type of technology IV. Interpretation of results Exercise end
Equipment:	Work station as per diagram for each agent used: I. Primary trays II. Secondary trays III. Glass petri dish IV. Continuous flow vapor detector V. Wind direction indicator VI. Complimentary detector (different technology) VII. Kill bucket mixed to facility requirements Note the detection equipment used should preferably be familiar to and in use by students
Chemical agent:	Agents used in this exercise should mirror those used in workstation 1 The amounts of agent made available should consider safety requirements therefore it is recommended no more than 100 microliter of each agent to be used for this exercise
Surface:	One petri dish each
Time:	No time limit
Pass/Fail:	N/A
Workplace set up:	Primary tray to be placed in a larger secondary tray in which two primary trays can be used if space allows (trays may not be necessary in a fume hood as they act as a contamination control measure) Detectors made available (should be operating) Placement of petri dish to consider wind, agent volatility and thus vapor obscuration Kill bucket

I am of the firm opinion that using real CBRNe materials for trainings – the so-called live agent training – is the only realistic CBRNe training. But how much theoretical and practical training is required to undergo live agent training effectively and safely?

The group of military responders that was subjected to the survey (see Chap. 4) has to undergo a minimum of 240 h (or 6 weeks) of CBRN related training, prior attending the basic live agent training; but not all of that time is devoted to chemical training only.

Table 3 Working station 3: effects of agents on different surfaces. This station can be repeated in many variations, it is suggested that initial exercises involve the same agent on different surfaces then different agents on the same surface. Final variation should lead to a combination of agents and surfaces where a 'blank' such as motor oil could be considered

Prerequisite	CBRN trained IAW national and site requirements
Learning objective:	Understand the detection of chemical agents
Sequence:	The focus of this workstation exercise is to gain confidence and employ a detection strategy I. The focus is to establish which chemical agent is on a surface likely to be encountered in the operational environment II. With agent on the selected surfaces, trainees should be informed what the surface is and how long agent has been applied III. Instructors should discuss the effects of surfaces to include absorption, physical staining, surface dispersion, cohesion and viscosity IV. With this information trainees should apply a detection strategy and inform the instructor of agent type Exercise end
Equipment:	Work station as per diagram for each agent used: I. Primary trays II. Secondary trays III. Chosen surface IV. Detectors V. Detector paper VI. Kill bucket mixed to facility requirements
Chemical agent:	Agents used in this exercise should mirror those used in workstation 1 and 2 The amounts of agent made available should consider safety requirements therefore it is recommended no more than 100 microliters of each agent to be used for this exercise
Surface:	Concrete, wood, soil, metal, vegetation, sand, rocks
Time:	No time limit
Pass/Fail:	N/A
Workplace set up:	Primary tray to be placed in a larger secondary tray in which two primary trays can be used if space allows Chosen surface contaminated accordingly Detectors made available (should be operating) Kill bucket

Some training providers ask for a minimum of 40 h worth of chemical related CBRN training to be able to participate in a basic CLAT.

At the end it all boils down to national rules and regulations, as even stipulated by NATO in their relevant procedures: Live agent training is an element of specialist CBRN defense training and should be undertaken and conducted in accordance with national regulations [17].

Table 4 Working station 4: decontamination of skin

Prerequisite	CBRN trained IAW national and site requirements
Learning objective:	Understand the execution of individual (immediate/operational) decontamination
Sequence:	The focus of this workstation exercise is to gain confidence in the use of equipment intended for immediate individual decontamination I. The station has two goals: Establishment of decontaminant interference with detection systems II. Use of decontamination systems on contaminated 'skin' Part one: Using pig skin in a tray already decontaminated (no agent to be used); students should utilize detectors to possibly get a response on residues of decontaminant applied Part two: Students should carry out decontamination IAW equipment procedures Confirmatory check if required (vapors may still be present) End of exercise
Equipment:	Work station as per diagram I. Primary trays II. Secondary trays III. 10 cm by 10 cm square of pig skin [2] IV. Decontamination equipment V. Detector paper VI. Detectors VII. Kill bucket mixed to facility requirements
Chemical agent:	30–50 µl of HD. A more miscible, less viscous agent maybe preferred dependent on decontaminant
Surface:	Pig skin
Time:	No time limit
Pass/Fail:	N/A
Workplace set up:	I. Primary trays to be placed in a larger secondary tray, after each student the 'contaminated student sample' is to be renewed II. Decontamination equipment made available III. Detectors made available (should be operating) IV. Kill bucket: Positioned at contaminated student tray V. Decontamination equipment next to contaminated student tray VI. Detectors next to detector response tray VII. Pre-contaminated pig skin should be held downwind from the workstation

4   Training Survey of Participants

The survey's objective was to establish the importance and relevance of chemical live agent training (CLAT) for CBRN(e) responders.

This survey was conducted amongst 200 participants to Chemical Live Agent Training, in one calendar year. All trainings (in total four) had the same structure, course content and objectives, as well as timetable; most workstations fell in the category of basic CLAT, so no scenario type training was part of that 1 week's course. It has to be noted that the participants do not fall in the category of decision makers, they were all part of CBRN Defense units, as enlisted personnel.

The equipment used was identical in all four courses: protective gear, detection and decontamination equipment. The trainings also took place at the same chemical live agent training facility. The questions were asked before and after their first live agent training with chemical warfare agents. The participants – they all came from the same European country and organizational, military CBRN Defense units – had a minimum of 6 weeks (240 h) of CBRN related training prior attending CLAT. The questionnaires were handed out after their preparation for live agent training, but before the first day of CLAT, and after the last day of their CLAT.

All participants to the survey had never gone through CLAT before, so this training was their first time that the participants had the possibility to work with chemical warfare agents.

The CLAT preparatory course – that took place in the week prior CLAT – also involved the use of chemical warfare agent simulants.

The following questions (Q) have been analyzed in detail:

1. Q1: Is CLAT an essential part of your CBRN training?
2. Q2: Is this LAT contributing essentially to your preparedness for C(BRN) related incidents?
3. Q3: Is a 1 week chemical live agent training sufficient?
4. Q4: Could CLAT be replaced with simulant training and achieve the same results in terms of preparedness for C(BRN) related incidents?

4.1   Analysis of Q1

Is CLAT an essential part of your CBRN training? The results of Q1 (Fig. 1) clearly show that on average 8 out of 10 participants believe that CLAT constitutes an essential part of CBRN training, and therefore this element is indispensable. The differences between the two results (before/after) are marginal and can be neglected; not more than 6% of participants believe that this type of training is not an absolutely necessary element of their training, whilst a maximum of 15% could not take a clear stand on this issue.

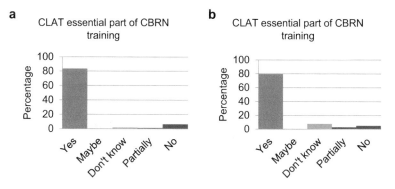

Fig. 1 (**a**) Result 1 – before CLAT (**b**) Result 2 – After CLAT

4.2 Analysis of Q2

Is this LAT contributing essentially to your preparedness for C(BRN) related incidents? The second set of questions shows – like Q1 – very similar results (Fig. 2).

Seventy-five percent of participants (dark and light green) are of the opinion that CLAT contributes essentially to their mission preparedness; this percentage was almost identical after the training, with 74% answering yes (categories 1–3).

CLAT does not contribute indispensably to mission preparedness – less than 5% were in favor of that (categories 8–10, in light and dark red).

The last category (orange) represents a group that was indecisive; they were not sure if this CLAT plays an important role in C(BRN) related mission preparedness.

Again looking at the results of Q1 – the importance of CLAT in CBRN training, and the role in mission preparedness – this Q2 – one can say that the results were almost identical, with a slight difference of 5%.

A minimum of 75% of CLAT participants were of the opinion that this type of training contributes essentially to mission preparedness and constitutes an absolutely necessary part of their CBRN training.

4.3 Analysis of Q3

One week chemical live agent training – is that sufficient?

Five days for this type of training – this represents a quite common approach; also NATO (through the Joint CBRN Center of Excellence) started to conduct basic CLAT, with a duration of 5 days. This course is called CBRN First Responders Live Agent Training Course. This particular group (CBRN Defense units) undergoes CLAT in 5 days, with usually 3 days of actual training with chemical warfare agents.

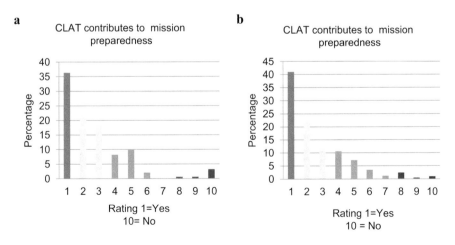

Fig. 2 (a) Result 3 – before CLAT (b) Result 4 – after CLAT

The majority (60%) was of the opinion (before and after) that a 5 days course for a basic CLAT is enough. It has to be noted though that the group of those participants that are of the opinion that 5 days are insufficient increased significantly (before /after), namely by 60% (Fig. 3).

4.4 Analysis of Q4

Could CLAT be replaced with simulant training and achieve the same results in terms of preparedness for C(BRN) related incidents (Fig. 4)? It has to be noted at this point that the participants to this survey underwent simulant based training, the week prior their CLAT; these simulant workstations mirrored the training

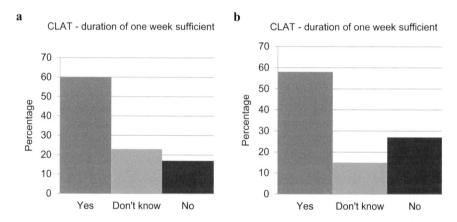

Fig. 3 (**a**) Result 5 – before CLAT (**b**) Result 6 – after CLAT

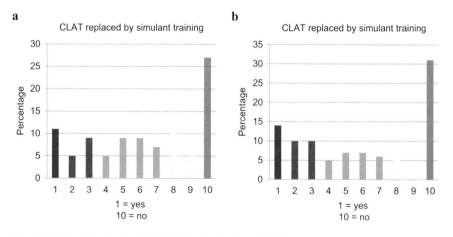

Fig. 4 (**a**) Result 7 – before CLAT (**b**) Result 8 – after CLAT

with chemical warfare agents in the following week. This means that the trainees were given all the necessary tools to compare simulant based and chemical warfare agents (CWA) training. It has to be stressed that these units are using a specific mix of chemicals (purchased from a Dutch based supplier) as simulants; these simulants are very realistic in terms of their physical properties when comparing them to actual CWA. I am of the opinion that this realism of the simulants used had an impact on the result of this survey. The biggest group under this heading is the one that clearly says no, CLAT cannot be replaced by simulant based training (dark and light green/over 40%); however up to 34% of the trainees answered positively (dark and light red), simulant based training can replace CLAT.

This result is in contrast to the analysis of Q1 and Q2, where more than 75% consider CLAT to be essential; this underlines my hypothesis, that the realism of simulants used played an important role in answering with 'yes'. However this has not been verified through the analysis of this survey.

5   Conclusions

The benefits of combining theoretical education and realistic live agent training with CBR materials outweigh the related occupational risks, thus being essential parts of any effective CBRNE training program.

The following conclusions can be drawn:

- Live agent training is an absolutely necessary component of CBRNe training.
- Live agent training prepares enhances mission preparedness of responders.
- Live agent training with CBRNe material builds confidence more effectively than training with simulants.
- A soldier/first responder who has trained with toxic materials is a more confident operator than the one who has trained with simulants alone.
- Live agent training can be conducted safely.

Military/First Responders should therefore attend live agent training; the availability of these types of training is however limited to certain specialized training facilities only.

These facilities allow enhancing CBRNe preparedness – a necessity due to the merging threat posed by chemical, biological, radiological, nuclear and explosive (CBRNE) materials.

Few countries have live-agent training facilities and while the size varies from one site to another, all live-agent facilities should be configured with roughly the same characteristics if they are going to safely and properly train CBRNe first responders.

When I know that our soldiers have seen those drops of deadly nerve agent and been able to detect, identify, and decontaminate them safely, then I am able to certify them as proficient Chemical Corps soldiers. [18]

It is almost universally agreed that countering a CBRNe threat is becoming more difficult and complex. Changes in strategies and scale are forcing both military and civilian organizations to look for and provide solutions for changing needs and threats.

While tactics, techniques and procedures can be developed to meet this demand, only training and exercises ensure that new systems work in the environment they are designed for.

The use of live chemical warfare agents during training creates scepticism and a perception that this type of training can never be safe. However, it is well established that working with live agents provides clear advantages over simulants alone.

References

1. CGJSC/RCESSC Vol. 2, Iss/Num. 2, Fall/Automne C, B, R, or N: The Influence of Related Industry on Terrorists' Choice in Unconventional Weapons Nicole Tishler Carleton University, Norman Paterson School of International Affairs (2013).
2. NATO CBRN Trg WG Consequence Management Handbook Study Draft 3 (2011).
3. Communication from the Commission to the European Parliament and the Council on Strengthening Chemical, Biological, Radiological and Nuclear Security in the European Union – an EU CBRN Action Plan (2009).
4. Communication from the Commission to the European Parliament and the Council on Strengthening Chemical, Biological, Radiological and Nuclear Security in the European Union – an EU CBRN Action Plan (2009).
5. NATO AAP-21 (B) NATO Glossary of Chemical, Biological, Radiological and Nuclear Terms and Definitions, pp. I–10 (July 2006).
6. NATO AAP-21 (B) NATO Glossary of Chemical, Biological, Radiological and Nuclear Terms and Definitions, pp. I–9 (July 2006).
7. NATO AAP-21 (B) NATO Glossary of Chemical, Biological, Radiological and Nuclear Terms and Definitions, pp. I–9 (July 2006).
8. NATO AAP-21 (B) NATO Glossary of Chemical, Biological, Radiological and Nuclear Terms and Definitions, pp. I–8 (July 2006).
9. NATO AAP-21 (B) NATO Glossary of Chemical, Biological, Radiological and Nuclear Terms and Definitions, pp. I–7 (July 2006).
10. NATO AAP-21 (B) NATO Glossary of Chemical, Biological, Radiological and Nuclear Terms and Definitions, pp. I–36 (July 2006).
11. NATO AJP-3.8.1. Allied Joint Doctrine For Chemical, Biological, Radiological, And Nuclear Defense, pp. 2–6, 2–7 (March 2012).
12. NATO ATP-3.8.1, vol. III (April 2011).
13. NATO ATP-3.8.1, vol. III, pp. 2–1 (April 2011).
14. NATO ATP-3.8.1, vol. II. Allied Joint Doctrine For Chemical, Biological, Radiological, And Nuclear Defense (May 2014).
15. Organisation for the Prohibition of Chemical Weapons (OPCW), https://www.opcw.org/about-chemical-weapons/types-of-chemical-agent/nerve-agents/
16. Organisation for the Prohibition of Chemical Weapons (OPCW), https://www.opcw.org/about-chemical-weapons/types-of-chemical-agent/blister-agents/
17. CBRN Defense Standards For Education, Training And Evaluation. ATP-3.8.1, vol. III (April 2011).
18. ROBERT D. ORTON Brigadier General, USA Commandant Department of the Army US Army Chemical School Fort McClellan, Alabama (March 1992).

Chemical Security Culture in an Insecure World: The Experience and Understanding of the Chemical Industry

Timothy J. Scott and Carola Argiolas

Abstract "Chemical security" is not a new topic, and in fact has been for some time an integral requirement in the daily operations in the chemical industry. Chemical security has been discussed and addressed for many years, but became an international issue after key events like Seveso and 9/11. These two events were followed by regulations that focused on security and in the case of 9/11 the creation of a federal agency in the United States with a primary focus on the security of critical infrastructure including the chemical industry. After years of being an industry-driven effort of less international significance than biological, radiological and nuclear events, the chemical industry and "chemical security" in general came to the forefront of government and public concern.

Keywords Chemical security • Chemical industry

1 Introduction

The first reaction around the world was the concern of a direct terrorist attack against a chemical manufacturing facility. The general thought was that such an attack would cause innumerable injuries and fatalities to a company's employee population, but of higher concern was the potential impact on the communities surrounding the chemical facilities.

Regulations were created – some were already in place – with the focus on securing chemical manufacturing facilities from a direct terrorist attack. Little thought was given to the size of the community, the proximity to the chemical facility or the

T.J. Scott (✉)
The Dow Chemical Company, Midland, MI, USA
e-mail: TJScott@dow.com

C. Argiolas
Insubria Center on International Security, The University of Insubria, Como, Italy
e-mail: Carola.argiolas@uninsubria.it

© Springer International Publishing AG 2017
M. Martellini, A. Malizia (eds.), *Cyber and Chemical, Biological, Radiological, Nuclear, Explosives Challenges*, Terrorism, Security, and Computation,
DOI 10.1007/978-3-319-62108-1_17

chemicals and processes that were in place at the facility. The idea of a risk-based approach to the issue was lost in the urgency to take action. Industry responded to the call and either voluntarily or to meet the new regulations spent hundreds of millions of dollars per company – billions when the final effort was complete – to minimize the impact of a direct attack on a chemical facility. This rather chaotic response was impressive but misdirected and didn't truly improve the security of the industry, surrounding communities or the nation. Out of this chaos however came a partnership among industry, government, and communities; and with everyone around the table focused on a common goal progress was made. While the initial focus was off target the effort was worthwhile in many ways. The increased awareness within the industry as well as the focus by government, regulators, and communities and general public was worthwhile. Partnership became the key word and process to success [1].

Discussion around regulations became a top priority and action was expected immediately. The concept of a "risk-based" approach to chemical security regulations was embraced by both the government regulators and the chemical manufacturers. With this common goal in place the "partnership" was launched and reasonable regulations and initiatives were established [2].

Initially the focus remained on a direct terrorist attack at a chemical manufacturing facility. Most regulations focused on the consequence of a successful attack, the likelihood of such an attack and the capability of an adversary successfully planning and conducting such an attack.

Regulatory requirements were established in the following areas:

- Restricting the area perimeter
- Securing site assets
- Screening and controlling access
- Deterring, detecting and delaying an attack
- Shipping and storage of chemicals
- Sabotage
- Theft and/or diversion
- Cyber security
- Response to security/emergency situations
- Monitoring of the manufacturing facilities
- Training of the security and emergency response personnel
- Personnel surety (background checks)
- Elevated threats/security procedures
- Specific threats and vulnerabilities
- Reporting to the regulatory authorities

These requirements are critical to an integrated security plan for any site (Fig. 1).

Over the years the chemical industry in general has consistently demonstrated its resiliency – recovering from hurricanes on the Gulf Coast; tragic fires, explosions, chemical releases and various kinds of operational emergencies; financial and economic collapse at the national level; and expanding environmental regulations. As an industry we've made significant improvements in each of these scenarios, but "security" – while important – was not a top priority.

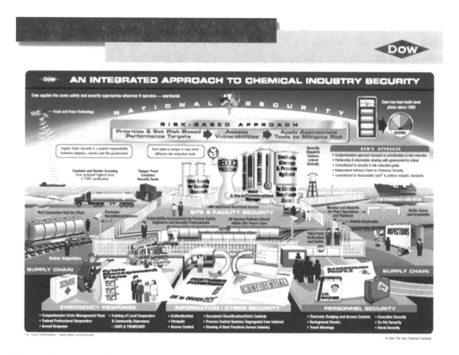

Fig. 1 An integrated approach to chemical industry security

During this same timeframe the industry strengthened its internal emergency response capabilities and organized mutual aid response plans and teams, created in 1985 the Community Awareness and Emergency Response or CAER program and in 1986 added the TransCAER program. In the late 1980s we implemented multi-level security contingency plans that allowed us to increase our security profile to meet potential security risks resulting from Desert Storm. But success was based on the actions of each company's individual effort. There were some loosely organized coalitions but in general each company developed its security and emergency response capabilities and operating discipline on its own and it was primarily a US effort and focused on the physical protection of our manufacturing facilities [3–5].

The focus was on the threat of a terrorist attack on our facilities – based on the government's priorities. DHS was new and trying to get organized and develop regulations, and industry and government came together in a partnership to develop risk-based regulations that made sense and have made significant progress together. As the threat changed the industry and government focus changed from the concern of a direct attack to the potential for theft and diversion of chemicals that could be weaponized.

The industry secured its manufacturing operations around the world and regulations remain in place to ensure the safe operations of our facilities. The industry is still under attack – not as a specific target – but because the enemy has changed from organized attacks on specific targets to random targets attacked by a lone wolf or small independent terrorist groups or radicalized youths in our own back yard – and many of them are using chemicals as their weapon of choice.

Additional regulations have been added to close these gaps in security and address these new scenarios, but not everyone is playing under the same rules and many of these chemical weapons precursors can be purchased or stolen from various sources.

Industry and government have made substantial progress but more need was and still is needed to address the changing threat. Regulatory action is being taken toward not only the chemical industry, but small shops and stores, local suppliers and others who have never been on the radar screen but now are identified as potential targets for theft and diversion by these new terrorists. In addition the list of potential targets is growing with a focus on "soft targets". These targets range from shopping malls and transportation hubs – so-called soft targets – to the men, women and children of Syria targeted by their own government and its allies.

These potential targets for theft and diversion are typically smaller non-regulated facilities or those who are not members of any industry association – orphan facilities – retail stores and those like the fertilizer distribution site in West, Texas. The "chemical industry" now includes everything from the largest manufacturers to places like beauty parlors, WalMart or Home Depot.

The larger members of the "chemical industry" need to share their expertise in partnership with the small unassociated companies throughout the US and around the world. We are responsible for the chemical sector's success, but industry large and small, the government, industry associations and retailers can't address the challenge individually. But together – as partners – they can make a significant difference.

A united global partnership is needed – with all the players at the table. DHS and similar regulating authorities around the world, the US State Department and its global counterparts and the Organization for the Prohibition of Chemical Weapons are establishing a path forward on a global basis. We need to challenge our US and global industry associations to renew their emphasis on Responsible Care implementation with a focus on security. And we need the individual companies, the regulators and elected government officials globally to join the partnership. We need to be more visible and active as an industry on a national and international level (Fig. 2).

The industry has achieved much in improving the security of its companies, the industry at large, and the communities in which they operate. The progress and partnership have earned the respect of many, but that respect can be lost in an instant. Industry has a voice and needs to use that voice to challenge the status quo, to bring clarity to the confusion and order to the chaos – to bring those who use chemical weapons under the microscope of humanity and the rule of international law to put a stop to the use of chemical weapons.

The chemical industry is a resilient industry – and it has proven that it can respond, recover and resume operations after any crisis – natural or man-made.

Our partnerships with others – industry, government, regulators and communities give us the ability to bring to justice those using chemical weapons and eliminate the access to chemical weapons precursors in order to reduce the risk and avoid a crisis before it occurs.

Personnel surety is essential to a safe and secure environment in the chemical industry. Thorough and repetitive background screening on all employees – company

Partnership = Success

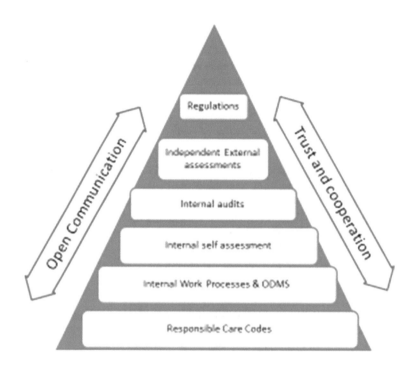

Fig. 2 The importance of partnership

and contractor – is the foundation of any security program. This screening should include the obvious criminal checks, but also checks of specific importance to the chemical industry and government.

Working partnerships and open communication are critical with government agencies at the higher levels in order to access more detailed and specific information of concern. In the US the Department of Homeland Security will require companies hiring new employees in the chemical sector – and specifically those working in high risk facilities – to submit personal data in order to check these individuals against the terrorist database. In Germany employees working in specific process areas of concern or higher risk are required to submit additional personal data in order to obtain an Authorized Operator credential. Industry in general and the chemical industry specifically have made significant progress in their safety performance over the last few years. Much of this success is due to improved manufacturing processes and personnel training, but the bulk of this success comes with open and honest two-way communication and the total involvement of the employees and contractors. The people working in the facilities take great ownership of their areas and take personal responsibility for their performance and personal safety. They most often see "safety" as their personal responsibility as much as that of the company. We need to raise this same level of awareness and personal ownership in security.

In April of 2007 I was quoted on the front page of USA Today stating my opinion that the insider threat was the highest risk scenario for the chemical industry. Today in 2014 the insider threat continues to be the highest threat to the chemical industry. We've not seen direct attacks – or even a verified threat of attack – on a chemical manufacturing facility, but we have seen what chemicals in the wrong hands can do – whether it's individuals, terrorist cells or governments. The focus of the chemical industry has shifted to address this theft and diversion scenario. This is the most difficult security process to implement effectively. The insider threat is real and will never go away, but there are ways to minimize the risk:

1. Personnel screening – thorough and specific to the chemical industry, to include criminal history of any kind, including performance history within the local community and industry. It must be repetitive and consistent. It must also include a partnership with the government for a screening against the terrorist database or similar screenings. And government must but open and honest about the screening.
2. Additional physical security around the chemicals or processes of interest.
3. Personal commitment of the employees – they must take responsibility for "security" just as they have for "safety".
4. Total involvement and commitment to those throughout the supply chain – customers, distributors, carriers – they all need to have the same robust personnel surety programs in place as you have.

2 Community Awareness and Emergency Response (CAER)

Developing a chemical safety and security culture is best accomplished in the early planning stages for the creation or expansion of the chemical industry and supporting infrastructure.

The "culture" needs to be one of open communication, understanding and cooperation, i.e. "partnerships" among all interested parties and for the benefit of all. Critical partners include not only those in the chemical industry, but just as importantly the local government officials, regulatory agencies, community leaders, local emergency responders, local health care officials, local businesses and the public at large – especially those identified as either public or industrial "nearby neighbors". The understanding, acceptance and support of these partnerships are essential to the success of the industry and the community. All have a vested interest in the success of the partnership, but all have different perspectives. A foundation of knowledge and trust about what industry is doing, and just as importantly understanding the value of why they are doing it is essential. Sharing information is critical – information about the operations of the facilities; the company's safety programs, processes and history; the nature of the chemicals being used; and the plans for emergency situations – both inside the facility and in the surrounding community. Information answers questions, replaces fear and builds a common platform for progress and success. This communication, awareness and training effort needs to be open and continuous; and needs to include the communities near chemical manufacturing operations, but also along the transportation routes used to distribute these chemicals [3].

The "culture" inside the company must also be inclusive of all aspects of safety and security throughout the life cycle of the products, must include all aspects of safety and security, and must take a risk-based approach to solutions. Within the company a safety and security culture must be in place from the development of a product, through the manufacturing process, to the customer ordering process, through transport and distribution to the end-user and safe use by that end user, to the recycle or safe disposal of any remaining product. Security throughout this life cycle is just as important and just as complex – encompassing physical site and product security, but also cyber and process control security, security of proprietary information, supply chain security, and personnel security. In addition, the company must have established and tested emergency response, crisis management, business resilience and business continuity plans in place in order to respond effectively should an incident occur. Included in these emergency plans and drills are the local responders and communities near the manufacturing operations or along distribution routes.

Developing a chemical safety and security culture is not an easy task, but it can be accomplished if all the groups and individuals who have a vested interest in the task are involved from the beginning. The chemical industry is the foundation – a building block – for many industries and products essential to improving life around the world. Communication, knowledge and partnership are proven keys to success.

As the partnership and process matured the "chemicals of interest" were identified which focused industry and government on the most important areas of concern.

A key asset to the successful partnership was the many relevant voluntary programs and initiatives that were already in place with many members of the chemical industry. These voluntary programs served as a foundation for additional risk-based based initiatives that addressed similar goals.

In the mid-1980s many chemical companies adopted voluntary initiatives to minimize operational upsets that could potentially impact employees and the surrounding communities. In addition many companies implement voluntary initiatives to communicate, train and prepare employees and the general public in the surrounding communities for potential on-site and off-site emergencies. In the United States the initiative was called Community Awareness & Emergency Response (CAER). This initiative was implemented globally by many companies and CAER or similar initiatives remain in place today on an international scale. While the goal of every company was to operate safely and minimize operational upsets or emergencies there were and still are significant internal plans and procedures for responding to these events and protecting the safety of employees. The CAER initiative this safety net to the surrounding communities and the public at large. CAER later became the foundations of the internationally recognized Responsible Care Code adopted and implemented by most chemical industry associations around the world [4] (Fig. 3).

Responsible Care® Security Code

Evaluation Process (Check)

- Internal Auditing
- Third-party verification of site security

Review (Act)

- Management of Change
- Continuous Improvement Commitment

Code has **13** specific elements which

companies implement

Fig. 3 Evaluation process check

3 Security Plans vs Emergency Plans: Not All Plans Are for Everyone

One learning from the initial regulatory efforts and public participation and communication initiatives was that security plans and emergency response address very different goals and have very different audiences. Security plans should be unique and specific to an individual chemical manufacturing facility and should only be shared and coordinated with the appropriate local authorities – law enforcement, fire department, and national regulatory agencies. Security plans should be closely guarded, strictly secured and shared on with those who have a "need to know". This poses a challenge for the sites and their voluntary programs focused on openly communicating with the community. The inappropriate sharing of security plans weakens the overall security of the facility.

On the other hand everyone needs to be aware and engaged in the emergency response plans of the facility in order to effectively respond to the emergency. Emergency planning is critical to the safety of the people within the chemical facility as well as the surrounding community. Emergency response plans need to be specific to the many different scenarios that could occur, and every person in the facility or in the surrounding community has a personal need to know how they will be notified of the emergency and trained to know how they should respond to any and all potential emergency scenarios. Planning for such emergencies should take place on a regular basis inside the facility, but should also include participation and communication with the community on a regular basis [6].

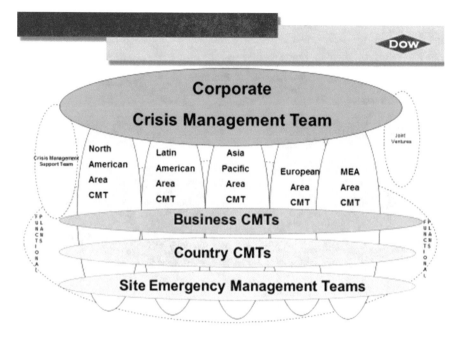

Fig. 4 Evaluation process check

Integrated Crisis Management Plans Every manufacturing site, community and government should have an emergency response plan and crisis management plan for all scenarios that should be considered. These scenarios should be included in emergency response and crisis management plans at each level of the industry as well as the surrounding community and governments. The response to an "emergency" should be handled at the lowest level. If the situation escalates a crisis management plan should be implemented to minimize the impact and avoid escalation of the event (Fig. 4).

Plans, work processes and operating discipline are the first step – the foundation – for the successful operation of chemical facilities during normal operations, in the process of preparing and responding to emergencies, or in the implementation of securing the operations from attack, theft or diversion or products or information. Each step needs to be reviewed on a regular basis to insure the processes are being implemented and maintained appropriately; and that adequate training and testing is applied to achieve expectations of all parties concerned – internal and external. Multi-layered compliance plans are required to achieve the expected level of security and emergency response readiness. The compliance plan includes all the partners: employees, communities, regulators and stockholders – anyone who could be impacted by a security breach at the company or during the transportation of the chemicals. Security is everyone's business and a partnership ensures success [2] (Fig. 5).

Inherently Safer Technology (IST) is often discussed and used as a foundational tool for chemical safety and security. IST is not a one-size fits all tool and is best

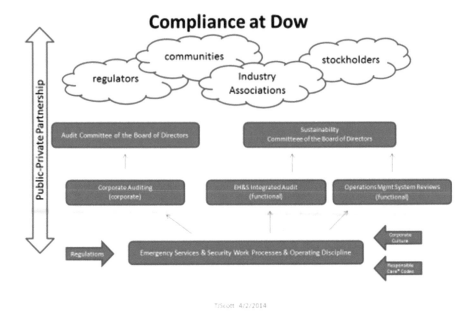

Fig. 5 Compliance at DOW

implemented during the early construction planning of a facility. The chemical industry in general embraces the implementation of IST.

1. Intensification (Minimize) – using smaller quantities of hazardous substances within a process. For example:
2. Substitution – a strategy to replace one material with a less hazardous substance if the chemistry of the process will allow. For example:
3. Attenuation (Moderate) – using less hazardous conditions, such as lower temperature or pressure or a less hazardous form of material, if the chemistry allows. For example:
4. Limitation of effects – designing a plant or process to minimize the impact of a release of material or energy. For example:
5. Simplification and error tolerance is designing to eliminate or tolerate operating errors by making the plant more user-friendly and reliable.

Other Safe Operations and Security Strategies Inherently safer technology, when properly applied, generally can become a competitive advantage. Having said that, inherent safety is not the solution to every challenge, but rather is an evolutionary process that takes several strategies into account. With inherently safer operations as a first step, from there Dow employs:

• Passive safeguards, which involve highly reliable systems that have a small likelihood of failure.
• Active controls and mitigation, including state-of-the-art Safety Instrumented Systems (SIS) designed and tested in accordance with internationally recognized standards (e.g., IEC-61511) to ensure high reliability.

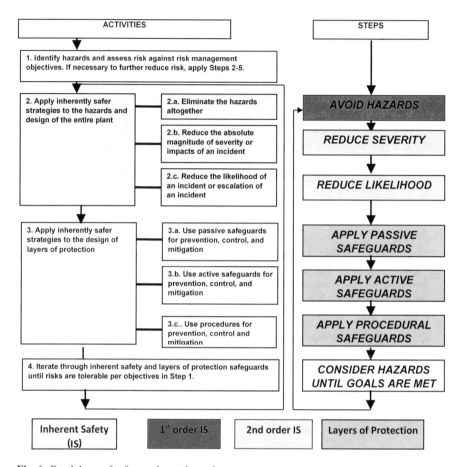

Fig. 6 Breakdown of safety and security action

- Procedures to assure that operating and maintenance practices are being carried out in a consistently safe fashion.
- Completion of detailed safety and security risk assessments, and taking direct actions to reduce risk based upon the outcome. This may even include the closure of plants and relocation to locations with a lower surrounding populations or additional of secondary containment systems to reduce the risk to the public.
- State-of-the-art Security detection, deterrent, and delay mechanisms.
- A formal Operating Discipline Management System to ensure that all layers of protection necessary to improve chemical process safety are carried out for the life of our plants.

Implemented alone, these strategies may not achieve the desired result. Together, they make a notable impression on process safety (Fig. 6).

4 Conclusion

On a final personal note, it is not enough to regulate the chemical industry or the distribution and routine use of chemicals in our global society. It is not enough that industry meets and often exceeds the regulations, or that emergency response plans and procedures are in place globally. We – as individuals, as nations, as an industry and as the global conscience must strongly respond to the intentional misuse of chemicals as weapons of mass destruction against civilian populations and humanity in general. Such acts should be addressed as war crimes and crimes against humanity and those convicted should receive the harshest punishment.

References

1. Chemical Facilities Anti-Terrorism Standards (CFATS) – US Department of Homeland Security https://www.dhs.gov/chemical-facility-anti-terrorism-standards
2. An Integrated Approach to Chemical Industry Security – copyright of The Dow Chemical Company
3. Responsible Care Community Awareness & Emergency Response initiative – American Chemistry Council https://responsiblecare.americanchemistry.com/Management-System-and-Certification/
4. TransCAER® – created in partnership by The Dow Chemical Company and Union Pacific Railroad in 1986. https://www.transcaer.com/
5. Responsible Care® – International Council of Chemical Associations https://www.icca-chem.org/responsible-care/
6. "Guidelines for Analyzing and Managing the Security Vulnerabilities of Fixed Chemical Sites" – Center for Chemical Process Safety and the Occupational Health & Safety Administration – US Department of Labor (2003)

Living with Chemicals: How to Prevent Their Use for Hostile Purposes and Mitigate Chemical Risks

Ralf Trapp

Abstract Chemicals are part of daily life. They are essential, amongst others, to fight disease, produce and preserve food, provide clean water, manufacture goods, and provide energy. But chemicals also can be exploited for hostile purposes. Certain precursor chemicals, for example, can be converted into chemical weapons, prohibited under the Chemical Weapons Convention. Other industrial chemicals can be used directly as improvised weapon given their toxicity or explosiveness. Access to, use and transfers of chemicals therefore are subject to strict controls. Balancing chemical security objectives with ensuring that chemical products can legitimately be use is complicated and requires a multi-stake-holder governance approach. It requires legislation and administrative measures, but also voluntary compliance assurance by the manufacturers and users of chemicals. This paper provides an overview on the chemical risks embedded in today's society, explains how trends in science, technology and commerce may affect this threat environment, and discusses good practices to prevent misuse and mitigate risks.

Keywords Chemical safety • Chemical security • Chemical weapons convention

1 Introduction

When the OPCW and the UN Security Council decided in September 2013 to eliminate Syria's chemical weapons programme [1, 2], many observers assumed that this would end the use of poison gas weapons in that ever-worsening conflict. Syria's chemical weapons were put under international control, removed from the country and destroyed. Its chemical weapons production, mixing and filling

R. Trapp (✉)
Independent Disarmament Consultant, Chessenaz, France
e-mail: ralf.trapp@gmail.com

© Springer International Publishing AG 2017
M. Martellini, A. Malizia (eds.), *Cyber and Chemical, Biological, Radiological,*
Nuclear, Explosives Challenges, Terrorism, Security, and Computation,
DOI 10.1007/978-3-319-62108-1_18

facilities were functionally disabled under the watchful eyes of OPCW inspectors and subsequently destroyed. Despite these successes, however, it took less than 6 months before new reports emerged about the use against civilian quarters of non-lethal chemical agents and barrel bombs releasing chlorine gas. Subsequently, investigations confirmed [3] that both the Syrian armed forces and Daesh had used toxic agents.

Some commentators were quick to question the effectiveness of the approach adopted by the international community to eliminate the Syrian chemical weapons programme, and raised doubts about whether all chemical weapons had in fact been put under international control and destroyed. Others wondered whether there was a way of denying access to toxic chemicals such as chlorine that were being used in improvised devices. Whilst both ways of looking at the problem have merit – there have been persistent claims that Syria may have retained part of its CW stockpile, and there is indeed a need to control access to and prevent the misuse of hazardous chemicals for hostile purposes – they fail to appreciate the role that chemical products play in human society, and their inherent dual (or rather: multiple) use potential.

Chemical products have been used in society ever since humans understood the properties of natural materials and learned how to employ them for their purposes. From the use of poisoned arrows to using minerals and extracts from plants and animals as medicines, or the use of Greek fires to attack enemy ships with incendiary weapons that would continue to burn under water, the use of chemicals for good or bad has been a constant feature in history. Over the centuries, experimentation and advances in technology have made it possible to extract metals from minerals, to make paints from natural pigments, to make gunpowder for use in rifles and artillery, to manufacture pottery and porcelain to prepare and store food and drink, to increase food production and provide safe drinking water – the list is endless.

The expansion of scientific knowledge, together with the technological and industrial revolution of the nineteenth century and the emergence of a chemical industry at the beginning of the twentieth century, fundamentally changed the scale at which human society was able to use chemical products and processes. Chemistry became one of the foundations of modern society; it enables the delivery of many goods and services. But equally, chemistry has been used for hostile purposes including as a weapon of war, or by criminals and terrorists to cause harm and deaths.

This paper first looks at the use of chemicals as weapons of war, and then at the way chemicals are used in society for beneficial purposes. It gives an overview of the risks associated with chemicals, from accidents to malevolent uses, and then discusses what is done to strengthen governance frameworks as well as practical measures to increase resilience against and preparedness for incidents (natural, accidental, hostile) involving hazardous chemicals.

2   Chemicals as Weapons of War

2.1   Overview

Different properties of chemicals have been exploited in the design of weapons, including amongst others toxicity, explosiveness, flammability and combustibility, corrosiveness, and stickiness. The intended targets of such chemical means of warfare can be enemy troops, their equipment and vehicles, their supplies including food and water, farm animals and crops, or aspects of their environment (for example, vegetation coverage). The ICRC in its study of customary rules of International Humanitarian Law (IHL) specifically addresses chemical means of warfare in rules 21 (poisons and poisoned weapons), 24 (chemical weapons), and 30 (incendiary weapons) [4]. Other rules that also are relevant include rules 1–6 dealing with the principles of distinction (between military and civilian targets or objectives), and the rules under chapter 44 (war crimes).

In addition, chemical products are used for a range of other military purposes, including as fuels, propellants, lubricants, paints, protective coatings, for cleaning and decontamination purposes, and for other purposes.

2.2   Explosive Weapons

High Explosives (HE)   Weapons using (conventional) high explosives employ the physical force of a detonation (heat, blast, fragmentation). The explosive is usually encased in shells, bombs or warheads, and fired at the enemy by delivery systems such as artillery, aircraft or missiles. Alternative, they are used in the form of mines or other devices such as manually placed munitions and devices. HE weapons are detonated on impact/contact or whilst in the final approach to the target. HE weapons that fail to explode are designated as unexploded ordnance (UXO).

High explosives are chemical compounds that store large amounts of energy in their chemical bonds, and when detonated form stable, strongly bonded molecules such as carbon monoxide and dioxide, or nitrogen. Usually they are nitrated organic compounds (e.g., nitro-glycerine, TNT, HMX, PETN, RDX, nitrocellulose) or inorganic salts (e.g., ammonium nitrate), and tactical mixtures may also contain other components such as powdered metals, or plasticizing oils or waxes to modulate stability and performance.

Thermobaric Weapons   These are weapons that utilize ambient oxygen rather that an oxidiser; they disperse a cloud of combustible fine particles (droplets or dust of certain metals or organic fuels) in air, usually by using a small dispersion charge. The fuel cloud suspended in air is detonated with a secondary ignition charge (typical for fuel air explosives), or the weapons design itself results in a superheated fuel that auto-ignites upon dispersion. As the cloud detonates, it generates secondary combustion effects, consuming the oxygen in the surrounding air and creating a heat and shock wave; the burning plasma cloud penetrates structures and confined spaces unless they are hermetically sealed. Once the fuel is consumed, the resulting vacuum creates a powerful back-blast.

Thermobaric weapons can be used to clear minefields and aircraft landing zones, and to destroy vegetation coverage. They have been used against living targets, including in confined spaces such as tunnel systems. If used in such circumstances, the weapons kill by blast and the subsequent rarefaction (vacuum) effect, which ruptures the lungs. Brain damage, also, has been reported [5–7].

Legal Issues The indiscriminate use of explosive weapons against civilians and non-combatants is subject to the general rules and prohibitions of humanitarian international law. But there is no consensus as to what constitutes a legitimate use of explosives causing incidental harm to non-belligerents (proportionality; discrimination between combatants and non-combatants). Furthermore, under the Amended Protocol II (dealing with mines, booby-traps and other devices) and Protocol V (explosive remnants of war) of the UN Convention on Certain Conventional Weapons (CCW), States have a responsibility to record and retain information on their use or abandonment of explosive ordnance (including locations, types and quantities of weapons used), to provide such information to parties in control of the territory that may be affected by UXO, and to assist with the removal of this threat [8].

Recent conflicts have again underscored the vulnerability of populations and civilian infrastructure when explosive weapons are used in populated areas [9, 10]. In asymmetrical conflicts, these risks can be aggravated by fighters "blending in" with local populations, or when control over populations itself has become a strategic objective [11]. The UN Secretary-General reports since 1999 periodically to the UN Security Council on the protection of civilians in armed conflict, and since 2009 addresses specifically the use of explosive weapons as so-called "area weapons", pointing to their indiscriminate and severe humanitarian impact when used in densely populated environments [12]. The Anti-personnel Landmines Convention [13] (opened for signature in Ottawa in 1997 and in force since 1 March 1999) and the Convention on Cluster Munitions [14] (opened for signature in Oslo in 2008, in force since 1 August 2010) are examples for steps taken to strengthen the norms against HE weapons that cause unacceptable harm to civilians. Nevertheless, much remains to be done to reduce civilian casualties not only during wars but also after hostilities have ended, when remnants of HE weapons (abandoned, lost, UOX) continue to pose high risks for people that live in former conflict zones.

2.3 Toxic Chemicals

Toxicity has been used as a weapons principle in several ways: for poisoning targeted individuals, for killing or injuring large numbers of enemy troops in battle or denying them access to terrain, for harassing enemy troops and degrading their fighting ability, for killing animals such as horses used to pull military equipment, for destroying vegetation cover to expose enemy supply and transport lines, or for destroying crops.

There is broad consensus that the 1925 Geneva Protocol prohibits all these forms of using toxic chemicals as weapon of armed conflict. The provisions of the 1997 Chemical Weapons Convention [15], on the other hand, are more narrowly focussed on toxic chemicals that affect humans or animals.

Chemical Weapons The term "chemical weapons" applies equally to the toxic chemicals used as weapons, the precursor chemicals used to manufacture them, and to specialised munitions, equipment and devices used to disseminate and employ them.

Chemical agents can be categorised in several ways, but usually they are distinguished into nerve agents (V agents, G agents such as Sarin or Soman, and next-generation nerve agents – "novichoks" and Intermediate-Volatility Agents (IVA)), blister agents such as sulfur and nitrogen mustards as well as Lewisite, pulmonary agents such as phosgene and chlorine, blood agents such as hydrogen cyanide, and incapacitating agents such as BZ. Certain toxins, too, have been weaponised in past military programmes (ricin, saxitoxin).

Most of the chemicals used in military programmes have little or no legitimate applications, but small amounts are often used in research, for chemical defence purposes, and some agents are used in small amounts as active ingredients of medicines (for example, sulfur mustard for the treatment of certain types of cancers). Older chemical agents such as phosgene or chlorine, on the other hand, are produced at industrial scale and used as intermediates in the manufacturing of a range of other chemical products.

Riot Control Agents Riot control agents (RCAs) are chemicals that cause temporary sensory irritation – the symptoms usually disappear shortly after exposure ends. They form a separate category of toxic chemicals under the Chemical Weapons Convention: their use in warfare is prohibited but they can legitimately be used for law enforcement purposes.

In modern history, RCAs were the first types of toxic chemicals to be used on the battlefield – during World War I. Although their battlefield effectiveness is limited, they can act as a force multiplier and their use carries a high risk of escalation towards more lethal chemical weapons. Because of their use in law enforcement, many States stock RCAs. Amongst the agents used today are CN, CS, CR and capsaicin.

Herbicides Herbicides are plant poisons; they have been used in several armed conflicts to destroy crops and vegetation, most extensively during the Viet Nam war in the 1960s and early 1970s (agent orange). The chemicals used as defoliants in Viet Nam included 2,4-D and 2,4,5-T. Significant environmental and toxic damage was attributed in subsequent years not only to these two herbicides, but also to an impurity that was present in comparatively high concentrations – 2,3,7,8,-TCDD or "dioxin".[1]

---

[1] The same material was released in 1976 in Seveso, Italy, as the result of an industrial accident leading to large scale contamination of the environment, thousands of dead animals and many people living near the chemical plant showing a range of symptoms. A 2001 study confirmed that TCDD is carcinogenic to humans and associated with cardiovascular- and endocrine-related effects [16]. In 2009, an update found an increase in lymphatic and hematopoietic tissue neoplasms as well as breast cancer [17].

Herbicides continue to be used in large amounts in agriculture and urban areas (including by private users) to control weed growth. Some herbicides are toxic to humans and require respiratory protection when used.

Legal Issues Chemical weapons are banned under the Chemical Weapons Convention, which entered into force on 29 April 1997 and has today 192 States Parties [18]. The CWC also prohibits the use in armed conflict of RCAs. With regard to the use of herbicides as a means of warfare, the CWC contains reference to a general prohibition of their use as a method of warfare.

The CWC, which today is almost universally adhered to,[2] sets stringent rules with respect to the destruction of all CW stockpiles, the elimination or conversion for permitted purposes of the facilities for their production, for international verification at military and chemical industry facilities as well as for the clarification of any non-compliance concerns including by on-site challenge inspections, and for national implementation. It also aims at facilitating international cooperation among States Parties including the strengthening of their capacity to respond to chemical attacks.

2.4 Anti-Material Weapons

Certain properties of chemical materials can be used for destroying or interfering with enemy military equipment. Examples discussed in the literature include anti-traction materials (lubricating polymers sprayed on the ground or surfaces to deny access by people or vehicles to certain areas or structures), obscurants (smoke), rigid or sticky foam barriers, and chemicals that interfere with the functioning of equipment or affect structures (combustion modifiers, fuel contaminants, super-corrosives, embrittling agents, super-adhesives, depolymerisation agents) [19]. There also have been suggestions for using biological agents as anti-material weapons (e.g., bacteria that degrade plastics or metals).

Little more is published about the integration of such weapons into contemporary military doctrine, but developments should be closely monitored to assess any particular risks they may pose to existing arms control regimes or humanitarian laws.

2.5 Incendiaries

Incendiaries are chemical products that set fire to an object or cause burn injuries to people. Incendiary weapons include, for example, flamethrowers, fougasses, or weapons (shells, rockets, bombs, mines, etc.) that are specifically designed to employ incendiary substances. Examples for chemicals that can be used as

---

[2] Only North Korea, Egypt and South Sudan have not taken any action on the treaty; Israel has signed the CWC but has yet to ratify it.

incendiaries include magnesium, thermite (a mixture of powdered aluminium and ferric oxide), napalm, white phosphorous and triethylaluminium [20]. Weapons that have an incendiary effect which is incidental rather than the design effect, or that have combined effects but are not specifically designed to employ the incendiary effect, are not covered under the term.

Napalm is a mixture of a gelling agent and either gasoline (petrol) or a similar fuel that sticks to surfaces and burns at temperatures well above 800 °C. It was first used at the end of the Second World War, and subsequently in several other wars (Korean War, Greek Civil War, Viet Nam war, the French wars in Indochina and Algeria, the Yugoslav war, and others). The effects to humans include severe burning, asphyxiation and death.

Legal Issues Incendiaries are covered under the 3rd Protocol of the CCW. The Protocol prohibits any deliberate use of incendiary weapons against the civilian population, or to use air delivered incendiaries against military targets located within a concentration of civilians. Care must be taken to clearly separate the military objectives from the concentration of civilians. Forests and other vegetation cover may not be attacked except when they are used to cover, conceal or camouflage combatants or other military objectives. The ICRC goes further and states that under customary international humanitarian law, the anti-personnel use of incendiary weapons is prohibited, unless it is not feasible to use a less harmful weapon to render a person *hors de combat* [21]. Despite these restrictions, many States retain incendiary weapons in their military arsenals.

3 Chemicals in Society

3.1 Overview

Modern society could not function without access to a wide range of chemical products with a multitude of different properties. Today's global challenges – feeding a growing world population, lifting people out of poverty and promoting economic development, ensuring adequate health care and combating disease, managing the consequences of global warming, providing appropriate transportation and housing – have been embedded in a new global agenda for sustainable development [22]. To achieve these targets and objectives will, amongst others, require the (appropriate) use of chemical materials and processes.

But chemicals also pose risks to society. The very properties that we use for beneficial purposes can also cause injury or other harm to humans or the environment. Accidents involving chemical materials can occur throughout their life cycle, from manufacturing through storage, transport, and use to final disposal. These risks may involve any of the hazards already discussed in the previous section, and not a few of the chemicals discussed there as weapons of war have legitimate applications, too.

The ILO estimates that 2.34 million people die each year from work-related accidents and diseases. The annual global number of cases of non-fatal work-related diseases is estimated at 160 million. A significant part for these occupational diseases and deaths is caused by exposure to chemicals [23]. The International Centre for Chemical Safety and Security maintains a Timeline of major chemical accidents and disasters [24], starting with the 1794 Grenelle Gunpowder Factory explosion in France (1000 deaths) and ending (for the moment) with the 2014 methyl mercaptan release at a DuPont facility in Houston, Texas, that killed four workers and injured a fifth.

A recent study by the US Centre for Disease Control and Prevention that covered nine US States and the period from 1999 to 2008 identified a total of 58,000 chemical incidents; almost ¾ had occurred at fixed facilities (plants, storage facilities) and about 1/3 were transport related. Most accidents happened in the chemical industry itself, albeit with a significant downward trend over time. Incidents in the educational sector and in agriculture, on the other hand, had increased with time. The incidents involved 15,506 injured people including 354 deaths. 49% of the injured people were employees of the party responsible for the incident, 31% were members of the general public, 11% students exposed at school, and 9% first responders. The death rate was highest among affected members of the public (54%), followed by employees (44%) and responders (3%). The chemicals that caused the largest number of injured persons were carbon monoxide, ammonia, chlorine, hydrochloric acid and sulphuric acid [25–28].

These data are consistent with the experience of other countries. The accident potential is often largest at the manufacturing stage when chemicals undergo reactive conversions with chemical bonds being broken and new bonds being formed. Chemical reactions as well as the physical processes used for product separation and purification often involve significant energy transfers. Some of these processes have a fairly narrow margin of safe operation with regard to such parameters as temperature or pressure, and can get out of control when operators make errors or critical equipment fails.

A classical example was the Seveso disaster: on 10 July 1976, an explosion occurred at the ICMESA chemical plant near Milan, Italy. The plant was manufacturing 2,4,5-trichlorphenol, and the cloud accidentally release into the atmosphere contained 2,3,7,8-TCDD ("dioxin"), contaminating a densely populated area of approximately 6 by 1 km. Eleven communities were directly affected by the release, others became subject to post-incident restrictions. There were a range of initial clinical signs but by 1984, an international steering committee concluded that the only ascertained health effect was chloracne, mostly in children. Subsequent long-term studies, however, confirmed excesses of lymphatic and hematopoietic tissue neoplasms in the zones with higher contamination. These zones also showed increased mortality from circulatory diseases in the first years after the accident, from chronic obstructive pulmonary disease, and from diabetes mellitus among females [29]. Extensive clean-up and restoration measures were implemented, and in regulatory terms, the disaster led to the adoption of the Seveso directive [30, 31].

Another well-know example was the Bhopal disaster of 1984, when a Union Carbide pesticide plant leaked more than 40 t of methyl isocyanate, caused by a combination of operator error and faulty or inactivated process and safety equipment. Some 3800 people died shortly after the release (mostly dwellers in a slum area next to the plant); eventually the casualty numbers reached 25,000 deaths and 500,000 exposed persons [32].

Partly as a result of lessons learned from such past accidents, chemical plants today operate within a rigid regulatory framework and implement a range of measures to prevent accidents and disasters, control and contain the damage should they nevertheless occur, and respond to incidents. They have trained staff, emergency protocols and dedicated response systems in place (for details in the European context see [31]).

Much of today's regulatory framework has evolved in response to growing environmental and health concerns, as well as lessons learned from industrial accidents. The focus of these measures is on prevention and safety. Security concerns existed all along (historically associated mostly with the use of improvised explosive devices), but were seen to be primarily governmental responsibility. After the Sarin attack on the Tokyo subway in 1995, and more so after the attacks on the United States on 11 September 2001, the international regulatory framework to prevent, deter and counter the use of hazardous chemicals by terrorists and criminal organisations has been further extended, including by voluntary measures in industry.

Given the overlap between safety and security systems and preparedness measures, the following discussion will address safety and security issues alongside. In the future, it will also be important to pay increasing attention to hybrid threats – for example threats posed by radicalisation and cyber attacks combined with the risk potential inherent in chemical infrastructure. Such hybrid threats are a reflection of the dramatically changing security and technology environment. The concept of hybrid threats "aims to capture the mixture of coercive and subversive activity, conventional and unconventional methods (i.e. diplomatic, military, economic, technological), which can be used in a coordinated manner by state or non-state actors to achieve specific objectives while remaining below the threshold of formally declared warfare. There is usually an emphasis on exploiting the vulnerabilities of the target and on generating ambiguity to hinder decision-making processes. Massive disinformation campaigns, using social media to control the political narrative or to radicalise, recruit and direct proxy actors can be vehicles for hybrid threats" [33, p. 2]. The use of toxic chemicals in the Syria conflict is a pertinent example.

3.2   Chemicals in Industry and Agriculture

Many industries use chemical products and processes in their manufacturing processes. From a security perspective, of particular interest are materials that are (or can be turned into) explosives or toxic chemicals.

Explosives and Combustible Materials The most common civilian use of explosive materials is in mining and quarrying. Both high explosives (chemicals that detonate) and low explosives (chemicals that deflagrate) are used. The mining industry tends to use nitrate-based inorganics such as ammonium nitrate as the source of oxygen, mixed with combustible fuel sources. An example is ANFO, a mixture of ammonium nitrate and fuel oil such as diesel or kerosene. Ammonium nitrate, of course, is also used in agriculture as a fertilizer.

Explosives are also used for tunnelling, road building and other construction works, for the demolition of old buildings and structures, as propellants, to bond metals, in the production of industrial diamonds, in fireworks, in firearms ammunitions and in other applications.

Safety hazards associated with the use of explosives are often associated with either accidents in storage and transport, or use. A recent example for the destructive potential inherent in these materials was the explosion at the fireworks market outside Mexico City on 21 December 2016, killing at least 31 people, injuring 70 more and leaving massive physical damage [34]. This was the 3rd explosion at this location since 2005, signifying that fireworks are an important part of the local economy.

Another example was the devastating explosion in Tanjin, China, on 12 August 2015: 50 people died and 700 were injured in a series of explosions in a warehouse that stored high explosives; another nearby warehouse that stored sodium cyanide (a highly toxic material) was also affected; the explosion resulted in massive structural damage and a high deaths toll among the first responders [35].

When addressing explosive risks associated with industrial or agricultural activities, one also needs to consider risks associated with chemicals that are not generally considered as explosive materials. An example is powder explosions, which can occur with any sort of combustible material. Such explosions are associated with certain types of activities that create fine dust (such as milling), or with conditions of transport or storage vessels such as fuel tanks, that can result in the formation of explosive mixtures of air and combustible material.

Refineries and fuel storage facilities also pose significant safety and security risks. Large amounts of combustible materials are stored and used in industry or private households as fuels for transport and as heating materials or for energy production. This hazard potential needs to be addressed in particular when facilities are located near densely populated areas (or when transports move through such areas).

Toxic Materials In addition to explosive and combustible materials, industry and agriculture use a wide range of chemicals that are toxic to humans, animals and/or the environment. Many of the intermediates produced and used in the chemical industry fall into this category. The examples of the Seveso and Bhopal disasters have already been highlighted, as well as the regulatory advances that have been made to increase the safety of chemical operations. The chemical industry has made significant efforts to increase the safety of its operations, including the use of safer processes and more reliable equipment, avoidance where possible of the use of highly toxic intermediates or products, staff training, and measures to prevent accidents and to respond to them should they nevertheless happen. As a result, the general trend over recent decades in the chemical industry has been towards fewer incidents, fewer injured or killed people and less physical damage. Nevertheless, there remains a significant hazard potential in the chemical industry and its associated industries.

But toxic chemicals are also used outside the chemical industry, for a variety of purposes: pesticides in agriculture, chlorine in water treatment, phosphine in semiconductor manufacturing, acids and caustic materials in surface treatment, and the active ingredients used by the pharmaceutical industry are examples for the use of toxic chemicals outside the chemical industry. This is why regulators as well as the industry have taken a life-cycle approach towards the management of chemical risks, addressing chemical risks from cradle to grave.

Precursor Chemicals In addition to toxic chemicals, industry and agriculture also use chemical products that could be used as precursors for the production of toxic chemicals. The traditional precursors of chemical weapons have been included in the Schedules of the Chemical Weapons Convention, and are thereby subject to a range of national implementation requirements (export and import controls, export restrictions to States not party of the CWC, declarations of exports, imports, production, in case of Schedule 2 chemicals also processing and consumption). Facilities that produce (in case of Schedule 2 chemicals also process or consume) these chemicals above certain annual amounts are declared to the OPCW, and above certain production (processing, consumption) thresholds become liable to on-site inspection by OPCW inspectors to check that these facilities are not used for the illicit production of chemical weapons.

These non-proliferation systems have been designed based on the experience with past State chemical weapons programmes. An example is the inclusion of thiodiglycol in Schedule 2 – a chemical that has applications as a solvent in the production of elastomers, lubricants, stabilizers, antioxidants, inks, dyes, photographic and copying processes, antistatic agents, epoxides, coatings, metal plating and in the textile industry. But it also is a precursor for the production of the chemical warfare agent sulfur mustard [36].

In today's security environment, States increasingly need to take account of terrorist and criminal threats. The criteria these actors use for the selection of precursors and chemical agents differ from those used in State programmes. Whilst State programmes attempt to optimise a range of parameters (including toxicity, cost, suitability for long-term storage under a wide range of conditions, effectiveness of dissemination, environmental stability, and availability of protection for one's own troops), criminal and terrorist organisations are more likely to look at expected effect, accessibility, and how to overcome technical challenges to manufacture an effective improvised dissemination device. It will be important therefore to take measures also to control access to 'unorthodox' toxic and precursor chemicals available in industry, agriculture, or elsewhere (including in relatively small amounts). That includes the use of toxic chemicals that from a military point of view are obsolete given their limited battlefield utility but that are widely available (such as chlorine), or of ways of releasing toxic chemicals that would not be considered efficient for military use but that can nevertheless be very effective in different scenarios (such as barrel bombs that release chlorine).

New Materials and Technologies Advances in science and technology, too, may create new risks. One area of research that has received particular attention in recent years is the intersection of chemistry and biology ("convergence"), including certain applications of synthetic biology, bio-manufacturing of chemical products, genome editing, and nanotechnology. In fact, the United States in 2016 classified genome editing as a new global security threat [37, 38]. Convergence and advances in bio-nanotechnology are profoundly changing the landscape of the life sciences, and some argue there are indicators of the onset of a biological revolution. As these transformational changes manifest themselves in the large-scale use of new products, technologies and solutions in industry and society, they may also change the (bio)chemical risk landscape. A recent workshop concluded that these developments "offer promising benefits for society, but their exploitation for harmful purposes by various actors in the context of their dual use potential cannot be disregarded. [But] scientists and their work are part of the solution and not part of the problem. A conversation about security threats should be multidisciplinary, while at the same time recognizing that the same thing may be understood differently depending on the community or the context" [39, p. 10].

3.3   Chemicals in Daily Life

The use of hazardous chemicals is not limited to industry and agriculture. Private consumers, too, use chemicals on a daily basis for a wide range of purposes. By comparison to industrial uses, the amounts are smaller, but on an individual basis the risks potential cannot be neglected.

The WHO estimates that 35% of ischaemic heart disease and 42% of stroke, the two largest contributors to global mortality, could be prevented by reducing or removing exposure to chemicals such as from ambient air pollution, household air pollution, second-hand smoke and lead [40]. Chemicals such as heavy metals, pesticides, solvents, paints, detergents, kerosene, carbon monoxide and drugs lead to unintentional (and largely preventable) poisonings at home and in the workplace, estimated to cause 193,000 deaths annually. The annual global disease burden in 2012 (*not* including deaths caused by air pollution) has been estimated at 1.3 million (2.3% of the total death burden for the 7 leading diseases).[3] Amongst the sectors WHO considers as targets for effective interventions are agriculture (toxic pesticides), industry and commerce (air and water pollution and industrial chemicals), transport (air pollution, lead in petrol), housing/communities (solid fuel smoke, home safety with respect to, for example, medicines and cleaning products), and clean water.

Gas explosions in households are an example for the omnipresence of hazardous chemicals in daily life. In the USA, from 2007 to 2011, an estimated 51,600 fires per year started with ignition of a flammable gas and another 160,910 fires with ignition of a flammable or combustible liquid, resulting in a combined 622 civilian deaths, 4393 civilian injuries, and $2.2 billion in direct property damage per year [41]. Spontaneous combustion and chemical reactions accounted for an estimated 14,070 fires per year between 2005 and 2009 (23% structural fires, 8% vehicle fires, 37% outside non-trash and unclassified fires, and 32% outside trash or rubbish fires). The structural fires included residential buildings (50%), storage places (12%), mercantile or business structures (9%), and manufacturing or processing structures (9%) [42].

Another example for explosives present in society is fireworks. In 2013, fireworks reportedly caused some 15,600 fires in the United States alone [43]. In Europe, the fireworks market is worth €1.4 billion, 90% of which are imported from China. In France, for example, over 15,000 firework shows are held every year, nearly half of which are professional displays [44]. In a survey of 316 accidents in France and abroad, nearly 65% of accidents were caused by human or organisational breakdown (handling errors, loss of vigilance, overzealousness, inaccurate guidelines or lacking protocols, inadequate ergonomics, lack of verification).

Another area of risk to human health from exposure to chemicals in daily use is poisoning. In the United States in 2014, 2.2 million cases of poisoning of humans were recorded (the majority in the age group 0–6 years; 79% accidental, 14% intentional). In children younger than 6 years, cosmetics and personal care products were in the lead (14%), followed by cleaning substances (11%) and analgesics (9.3%). In adults, analgesics were the most common cause of poisoning (12%), followed by sedatives, hypnotics and antipsychotics (10.4%), antidepressants (6.7%), cardiovascular drugs (6.1%), household cleaning products (5.7%), anticonvulsants (3.7%), and pesticides (3.5%). In 8% of all cases, the effect was moderate or major (including death) [45].

These examples illustrate a wider point: hazardous chemicals are ubiquitous in human life. Risk management strategies, whilst aiming at reducing the abundance of toxic, explosive or inflammable chemicals in society, must also recognize that chemical products and processes cannot simply be taken out of society altogether.

---

[3] Poisoning, leukemia, lung cancer, ischaemic heart disease, stroke, intellectual disability, chronic obstructive pulmonary disease.

4   Chemicals, Organised Crime and Terrorism

4.1   Chemical Terrorism[4]

The use of hazardous chemicals as weapons by non-State actors is not a new phenomenon. Arson and poisoning have been features of criminal acts for centuries. Improvised bombs and explosive devices (or explosives stolen from available sources) have been used throughout the centuries by plotters who wanted to topple governments or remove rulers, including parties of civil unrest. Improvised explosive devices have been made from fertilizer and other legitimate products, acquired via open or black markets, or diverted from sources such as military or law enforcement arsenals.

What has changed in recent years is the destructive force that non-State groups are able and willing to command, and the sophistication of the weapons design they use. Some of the explosive devices used in the Syrian conflict are a close match to military weapons in terms of destructive power and battlefield utility. Also, there is an increasing willingness and capability to use un-conventional weapons. The terrorist attacks on the United States on 9/11 were a first example for how aspects of the transport system could be turned into weapons or become the subject of devastating attacks. Subsequent attacks on trains (Madrid), the local public transport system (London), the use of transport vehicles as weapons in crowed places (Nice, Berlin), and attempted attacks on aeroplanes (the "shoe bomber") have underscored the willingness of certain terrorist groups and associated individuals to use non-conventional attack approaches to cause mass carnage.

The Anthrax letter attacks that followed 9/11, even though not linked to terrorism, underscored the potential inherent in bioterrorism in terms of human lives at risk, and economic costs as well as social disruption. Fears of bioterrorism have increased; so have concerns about the risks associated with improvised radiological devices ("dirty bombs") and makeshift nuclear explosive devices. There are also fears that terrorists may be developing thermobaric weapons for use in confined spaces such as rail tunnels or subway systems, and some evidence exists that terrorist groups have been working on the design of fuel air explosives [46].

The use of toxic chemicals by non-State groups has always been a security concern, given the relative easy with which toxic chemicals or precursor materials can be acquired from black markets or legitimate sources in amounts sufficient to cause significant damage yet small enough to make detection by national and international control systems difficult. The use of Sarin in the Tokyo subway system in 1995 was a first indication that terrorists can acquire the technical capability to produce sophisticated chemical weapons such as nerve agents in amounts that pose significant threats to populations. In subsequent years, improvised explosive devices were

---

[4]This paper uses the term "terrorism" in a generic sense, relating to any form of terrorism, in whichever manifestation, and recognising that there is no international consensus of the definition and scope of the term "terrorism".

combined with toxic chemicals such as chlorine gas cylinders and used in insurgencies in Iraq. Whilst these early attempts of chemical terrorism were relatively inefficient given the challenges inherent in effective design of chemical dissemination devices, subsequent improvised chemical weapons used in Syria and Iraq have become much more lethal. The chemical weapons used in Ghouta in August 2013 – in spite of who the perpetrators were – were highly effective and according to some estimates caused as many as 1400 death [47] (other estimates ranged in the hundreds), and the barrel bombs used to disseminate chlorine gas, too, are highly effective when used in built-up populated areas [3].

A key question is how terrorists acquire hazardous chemicals for use in explosive devices or as improvised chemical weapons. The experience of ISIS/Daesh in Syria seems to suggest that acquisition patterns depend primarily on context and accessibility [48]. Acquisition sources include weapons stashes such as unguarded weapons sites of other belligerents (including military forces), black markets, diversion from "legitimate transfers", extraction of explosives for manufacturing IEDs from ammunitions (tank shells, mortars, anti-aircraft) and unexplored ordnance, use of empty gas cylinders (oxygen cylinders, liquefied gas bottles) filled with high explosives such as TNT, mixtures of ammonium nitrate or other fertilizer and TNT, or C4 (obtained from black markets), extraction of explosive materials from fireworks, or diversion of pesticides from agricultural users. TNT and other explosives components (RDX, ANFO, engine oil used as moderator) are also diverted from legitimate producers or traders, or supplied by "supportive countries".

4.2   Chemicals and Organised Crime

Organised crime, too, has long been associated with handling and using chemical products and processes. In addition to the use of explosives and toxic materials for criminal purposes, the most significant aspect to consider is the manufacturing and trafficking of narcotic/psychotropic drugs – one of the major sources of revenue of organised crime. Analyses of Europol and UNODC show that the market for illicit drugs is the most dynamic among the criminal markets. This has led to increased cooperation between Organised Crime groups but also competition between these groups across national, linguistic and ethnic divisions. As a consequence, the drug trafficking phenomenon is highly complex and difficult to address effectively [49, 50].

The health burden of drug abuse is considerable. UNODC estimates that there have been 207,400 drug-related deaths in 2014 worldwide; over 247 million people use drugs, of them 12 million people inject drugs (of which 14% are living with HIV), and approximately 29 million drug users suffer from drug-related disorders. The most widely cultivated drugs are cannabis (129 countries reported its production over the period from 2009 to 2014), followed by opium (49 countries) and coca (7 countries). Among synthetic drugs, amphetamines take the lead – they also are the second most commonly used drug type. The prevalence of drug use has remained relatively stable over recent years – approximately 5.2% of people are reported to use drugs.

An important evolving feature is the marketing of new psychoactive substances (NPS). UNODC reported that between 2008 and 2015, 644 NPS have entered the markets as reported by 102 countries. Most of these NPS are synthetic cannabinoids, but increasingly other types of chemicals are reported including opioids, fentanyl derivatives and sedatives [50].

From a security perspective, one feature of particular concern is the nexus of drugs trafficking and terrorism. Cooperation between terrorism and organised dugs has been reported for many years, in such areas as acquisition of firearms, manufacturing and use of IEDs, and money laundering. As terrorist groups are taking a closer interest in chemical warfare options, the possible transfer from criminal to terrorist actors of technical expertise and equipment used in illegal drugs laboratories for chemical synthesis ought to be considered as a potential emerging threat. Also, should terrorist organisations take an interest in the use of psychoactive compounds (as incapacitating agents), as has been suggested in the literature [51], a closer association of organised crime and terrorism could result in new avenues for terrorists to acquire "exotic" weapons based on psychoactive compounds.

5 Prevention and Mitigation

5.1 General Principles and Concepts

The risks associated with the manufacturing, storage, use and disposal of chemical materials cannot be reduced to zero, but a range of strategies can be implemented to reduce and manage them. These strategies involve, in particular, regulatory and other governance measures enacted by governments as well as voluntarily implemented by the different communities that handle chemical materials, including industry; technical measures to prevent or reduce the likelihood of chemical incidents; measures to harden chemical infrastructure against the impact of external forces (natural events such as flooding, tsunamis or earthquakes; man-made events such as accidents or hostile acts); substitution of hazardous chemicals by less dangerous materials whenever possible; technical and organisational measures to respond to chemical incidents and minimise their impact; and measures to ensure recovery after chemical incidents.

In all these strategies to strengthen resilience, the human factor is critical. Leadership and commitment to reduce chemical risks and improve response capacity are essential for any effective risk mitigation strategy; so is constant improvement through training, exercises, sharing of good practices and learning from past successes and failures.

In such a complex and comprehensive approach, there are no firm demarcation lines between chemical safety (the protection of people and the environment from hazardous properties of chemicals) and chemical security (the protection of chemical materials, equipment, technologies and knowledge from misuse for hostile purposes). Instead, chemical security measures need to be integrated with chemical safety measures.

5.2 Laws, Regulations and Other Governance Measures

There is a vast range of regulatory measures that apply to chemicals, from manufacturing to storage, transport, use and disposal. Many of them have been developed in recognition of the safety risks to humans, animals, plants, and the environment.[5] Here is not the space to give a detailed overview of this complex regulatory framework, but some of the instruments and regimes that are of particular relevance, at the international level, will be highlighted.

Regulatory Framework for Chemical Safety The Strategic Approach to International Chemicals Management (SAICM) is a key international policy framework for the sound management of chemicals. It was adopted by the International Conference on Chemicals Management in 2006 [52, 53]. SAICM recognises the need for a multistakeholder, multisectoral approach and has the stated aim to ensure that by 2020, chemicals are produced and used in ways that minimize significant adverse impacts on the environment and human health. UNEP and the WHO support the SAICM Secretariat. SAICM's focus is on developing and implementing a comprehensive strategy for chemical safety and involving governments, the private sector and relevant international organisations. It involves risk reduction (prevention, reduction, remediation, minimization and elimination), knowledge and information sharing and awareness raising, governance, capacity building and technical cooperation (including access to financial and other resources needed), and measures to address illegal international traffic in hazardous substances and dangerous goods. SAICM makes an explicit link to advancing public health and human security.

An important programme under SAICM is the Quick start programme, which helps countries develop capacity for the sound management of chemicals. Independent evaluations have underscored the inclusive and participatory nature of the programme. Almost every project had improved awareness of the importance of

---

[5]Amongst others, these include: Chap. 19 of Agenda 21 adopted by the UN Conference on Environment and Development in Rio de Janeiro in 1992, ILO Conventions No. 170 on Safety in the Use of Chemicals at Work and No. 174 on the Prevention of Major Industrial Accidents, Basel Convention on the Control of Transboundary Movements of Hazardous Wastes and Their Disposal, Rotterdam Convention on the Prior Informed Consent Procedure for Certain Hazardous Chemicals and Pesticides in International Trade, Stockholm Convention on Persistent Organic Pollutants, Globally Harmonized System for the Classification and Labelling of Chemicals; within the EU: Seveso III Directive.

the sound management of hazardous chemicals, and had laid the foundations for stakeholder cooperation, with many examples of increased trust between government, civil society and the private sector. Chemicals were being mainstreamed into national legislation, policies and institutions, although gaps remained in some countries as well as with regard to risk assessment and better management of highly hazardous pesticides [54].

The ILO Conventions and Guidelines including the Guidelines on occupational safety and health management systems (ILO–OSH 2001), and the social dialogue in promoting the sound management of chemicals throughout their life cycle are another set of international requirements that aim at increasing chemical safety at the work place [23]. With regard to human health and environmental protection and chemical safety in society, the EU's REACH (Registration, Evaluation, Authorization, and Restriction of Chemicals) regulation (in force since 2007) establishes an integrated approach to risk management in the EU Member States [55]. It places responsibility on industry with regard to risk management, the gathering and registration of safety information, the step-by-step substitution of the most dangerous chemicals by suitable alternatives, and it established the European Chemicals Agency with seat in Helsinki.

Regulatory Framework for Chemical Security With regard to strengthening chemical security, an important framework was established by UN Security Council Resolution 1540 (2004) [56]. In response to the 11 September 2001 attacks, it recognized a growing terrorism threat associated with CBRN materials and weapons of mass destruction (WMD). States were called upon to refrain from any form of support for non-State actors with regard to NBC weapons and related delivery systems. They were to adopt and enforce appropriate effective prohibitions, and to take and enforce effective measures to establish domestic controls to prevent WMD proliferation. The resolution established the 1540 Committee (its mandate has been extended several times, most recently until 2021 by UNSC Resolution 1977 (2011)). States are requested to submit reports of how they intend to implement these requirements, and to adopt voluntary action plans, taking account of advances in science, technology and commerce as they affect proliferation trends/risks. Among the measures that States should implement are enforcement measures; measures to address the financing of WMD proliferation; measures to ensure accounting for and securing or relevant materials and controls of national exports and transhipments; controls of access to intangible transfers of technology and information that could be used for WMD purposes; best practices should be shared and assistance provided (among States, in coordination with international and regional organisations, and the UN Trust Fund for Global and Regional Disarmament Activities) [57]. The 1540 Committee coordinates its activities with the Security Council's Committee concerning ISIL (Da'esh), Al-Qaida and associated individuals and groups, and with the Counter-terrorism Committee established by the Security Council Resolution.

A significant contribution to the regulatory framework of chemical security emanates from the Chemical Weapons Convention. The OPCW in 2003 adopted an action plan on improving national implementation, and reiterated a previous decision on the contribution of the CWC to the fight against terrorism [58]. The initial focus of this action plan was on enacting national laws to domesticate the prohibitions and requirements of the CWC into the laws of all States Parties. This included adaptation of penal laws, and enactment of other laws and regulations to ensure the full implementation of the Convention. A second key aspect was the establishment of competent and appropriately empowered National Authorities in all States Parties. Some progress has been made in both areas, but the Third CWC Review Conference in 2013 had to conclude that only 91 States Parties had fully enacted legislation and/or adopted administrative measures to meet their obligations. Full implementation was yet to be achieved, and the Review Conference called for increased efforts to guard against the possible hostile use of toxic chemicals by non-State actors such as terrorists [59]. States Parties and the Secretariat were called upon to promote a high level of readiness to respond to chemical weapons threats, amongst others by making full use of regional and subregional capacities and expertise.

A feature that distinguishes chemical security from nuclear/radiological security is that there is no single international lead agency. The UN's Counter-terrorism Implementation Task Force (CTITF) has established a working group on preventing and responding to WMD terrorist attacks, which in 2011 prepared a report on inter-agency coordination in the event of a terrorist attack using chemical or biological weapons or materials. This report mapped out the different institutional mandates, capacities and protocols that must be synchronised at the international level to ensure an effective response. In 2014, the Working Group initiated a new project on "Ensuring Effective Inter-Agency Interoperability and Coordinated Communication in Case of Chemical and/or Biological Attacks"; it started in 2015 and is expected to lead into table-top and field exercises in 2017 [60].

In the transport sector, chemical safety and security issues are regulated under the UN's Regulations for the Transportation of Dangerous Goods. For several years now, these regulations include a set of provisions regarding the security of the trans-portation of dangerous goods [61]. They address all parties involved in the transport chain (individuals, carriers, vehicle crews) and deal with a range of requirements: certification, security features at temporary storage terminals, safety inspections, security training, and special provisions for high consequence dangerous goods which have the potential for misuse in a terrorist event and which may, as a result, produce serious consequences (mass casualties, mass destruction, mass socio-economic disruption).

Another important legal instrument which criminalises any use of a civil aircraft as a weapon, including for the discharge of chemical (and biological as well as nuclear) weapons and materials is the 2010 Convention on the Suppression of Unlawful Acts Relating to International Civil Aviation [62].

Security threats associated with explosive materials are covered under several of the 19 counter-terrorism conventions, the custodian of which is the UN Office for Drugs and Crime (UNODC). Of particular relevance are the 1997 International Convention for the Suppression of Terrorist Bombings, the 1991 Convention on the

Marking of Plastic Explosives for the Purpose of Detection, the 1971 Convention for the Suppression of Unlawful Acts against the Safety of Civil Aviation and the related 1988 Protocol for the Suppression of Unlawful Acts of Violence at Airports Serving International Civil Aviation [62].

Despite this extensive system of regulations, there remain concerns about the effectiveness and comprehensiveness of the international legal framework to address threats of chemical terrorism. In March 2016, Russia proposed to negotiate a new International Convention for the Suppression of Acts of Chemical Terrorism. Whilst some of the elements of this proposed new treaty have raised concerns about their compatibility with the Chemical Weapons Convention, the underlying concern is real that there is a need to increase practical cooperation among states in preventing and prosecuting acts of chemical terrorism, including by sharing information and applying a "prosecute or extradite" approach [63].

Voluntary Governance Measures Many of these regulatory measures require active support from a range of non-governmental stakeholders, including industry. The chemical industry has become an active partner in this respect, and has taken on responsibility to ensure internal compliance with chemical safety and security requirements. Other industries, too, have adopted voluntary steps to prevent the misuse of their technologies and products for hostile purposes. For example, certain biotechnology companies have adopted voluntary screening procedures of customers as well as individual orders, with regard to DNA sequences ordered by Internet.

A pertinent example for productive industry involvement in security-driven issues is the relationship that has evolved between the International Council of Chemical Associations (ICCA) and the OPCW. This collaboration has its origin in the chemical industry's involvement in the negotiations of the CWC [64]. Today, the ICCA's flagship programme Responsible Care® requires its member companies to support the CWC, and industry continues to support verification visits and provides peer-to-peer assistance [65]. Responsible Care® measure are being certified as an industry standard (RCMS® or RC14001® if combined with ISO 14001) [66, 67], so that compliance with pertinent legislation and regulations are built into the management systems and culture of chemical companies. The Responsible Care® Security Code guidance and best practice document adopted in 2010 by the European chemical industry association CEFIC sets out leadership commitments for security as a fundamental part of the overall management system, and provides details on security risk analysis; the implementation of security measures along "deter, detect, delay, respond" principles; the training of staff, contractors, suppliers and service providers; as well as awareness raising, information of staff in critical functions, and communication within the company and with external stakeholders. It aims at full integration of security into "management of change" [68].

There are also industry guidelines that integrate security measures into the safety and management systems of companies involved with the transport of highly dangerous goods [69]. These guidelines are voluntary measures and indicative – the companies themselves need to decide how to apply them using their own judgement. The guidelines contain detailed recommendations on management routines and operating practices such as vetting individuals (employees and job seekers), ascertaining the reliability of business partners and contractors, technical options for securing temporary storage areas including access control systems and protections (physical, information technology), security training for staff, and the possible content and processes for adopting and implementing a security plan.

One important aspect to note is that when industry organisations and their associated companies commit to adopting such measures as an integral part of their management systems and culture, these measures are rolled out across entire industries (within companies, in their subsidiaries, with global reach in case of multinational companies). This has a global dissemination and multiplier effect, complementing the efforts by governments to enact and enforce global legal frameworks to fight chemical terrorism.

5.3   Resilience to Chemical Threats

Regulatory and other governance measures create a legal and policy framework for addressing and managing chemical threats. But equally important are the practical preparations that States and other actors are making to increase resilience to chemical incidents. With respect to the threat of malevolent use of chemicals, deterrence is not only the result of strong laws and effective law enforcement, but also of robust protections and response systems that limit the impact of any such hostile uses of chemicals. For any chemical incident – natural, accidental or malevolent – well-functioning protection, response and recovery systems are essential to minimise the loss of life, reduce the impact on society and the environment, and ensure swift recovery.

Efforts to increase resilience must address the entire spectrum of practical measures, from deterrence of hostile acts and prevention to early detection, measures to protect people (including incident responders) and the environment, response measures to contain an incident and mitigate its effects, medical countermeasures to reduce the number of casualties and treat victims (including with regard to long-term care), investigative measures to identify the causes of an incident (in case of malevolent or negligent acts also attribution), measures to clean up and decontaminate people, equipment, structures and buildings, and measures necessary for recovery. A systematic approach is needed that ensures strong leadership, builds up human capacity and ensures that the necessary mix of skills and competencies is maintained through training and exercises, and it is necessary to ensure that there are adequate and tested procedures and protocols as well as sufficient fit-for-purpose and compatible equipment.

The main responsibility for developing and implementing such response systems rests with governments. Other stakeholders, too, have responsibilities and make important contributions. It is important to bear in mind that in the case of chemical incidents, local response capacity is critical (backed up by national response systems).

Chemical incident, as a rule, are acute in nature and require a swift local response. Response times measure in hours rather than days or weeks, as may be the case for other types of threats (such as biological or radiological). This is why it is important to build resilience into the local as well as national institutional structures, to enhance and maintain response capacities where there is a high risk of chemical incidents such as near chemical infrastructure, and to develop and maintain response plans and protocols that ensure the interaction of the different actors involved, at local as well as national levels.

International Cooperation The international community can do a lot to help countries strengthen their resilience to chemical incidents. Several international organisations have developed cooperation programmes in this field – some under the umbrella of SAICM, others as part of their respective institutional mandate.

One indicator for the state of preparedness of countries is the (self-assessed) compliance with the requirements of the International Health Regulation (IHR) 2005. It requires States to put in place (minimum) core capacities to detect, assess and report public health events including chemical accidents and emergencies. Whilst there has been some noticeable progress with regard to countries meeting targets set with regard to, amongst others, legislation, coordination and disease surveillance, core capacity in the field of chemical health risks remains comparatively low. Only 57% of countries have reported that they have minimum core capacity in place [40].

The Inter-Organization Programme for the Sound Management of Chemicals (IOMC) has developed a toolbox to help countries in decision making with regard to chemicals management, starting with a gap analysis through a Country profile or SAICM Implementation plan [70]. The toolbox helps countries to address chemical accidents (major hazards prevention, preparedness and response) and includes an assessment phase, a systematic scheme for identifying all key stakeholders, and tools for implementing policies and practical measures with regard to planning, implementation and coordination regarding chemical accidents. It addresses resource requirements, information needs, administrative and legal structures, leadership and enforcement requirements.

The OECD, too, has made significant contributions to chemical safety and security. It has set up a Working Group on Chemical Accidents (its current mandate covers the period from 2013 to 2016), which works in three areas: development of common principles, procedures and policy guidance; analysis of issues of concern and best practices; and information and experience sharing and communication. It has developed a series of top-level guidance documents for governments, which include Guiding Principles for Chemical Accident Prevention, Preparedness and Response (2003); Guidance on Developing Safety Performance Indicators for Industry, and Guidance on Developing Safety Performance Indicators for Public Authorities and Communities/Public (2008); and Corporate Governance for Process Safety: Guidance for Senior Leaders in High Hazard (2012) [71].

Finally, a number of countries implement international cooperation programmes to assist other countries in their effort to enhance chemical safety and security. In many cases, these assistance and cooperation programmes mirror internal efforts (such as in the US the work of the Department of Homeland Security; in the EU the work under the CBRN Action Plan and the different Framework Programmes including Horizon 2020).

An example is the U.S. Department of State Chemical Security Engagement Program, implemented by the State Department's Office of Cooperative Threat Reduction. The programme seeks to reduce global chemical threats by strengthening the ability of partner nations of the US to disrupt chemical attacks, and to strengthen security with regard to weapons-applicable chemical materials, expertise, and related infrastructure. It offers workshops and training on issues such as chemical security management, security in the chemicals supply chain, detection and chemical forensics, chemical security for industry and academia, and prevention and deterrence of chemical incidents for law enforcement [72].

Another example is the EU's Instrument contributing to Security and Peace (IcSP), which under it's long-term capacity building component includes a range of programmes that are directed at helping partner countries enhance their systems in chemical safety and security, as part of a wider approach towards cooperation and technical assistance in CBRN risk mitigation. This includes the CBRN Centres of Excellence (CoE) initiative which was first set up under the Instrument for Stability in 2010. The initiative involves the establishment of National Focal Points for CBRN risk mitigation and National CBRN Teams by the partner countries; these national entities are linked into regional networks coordinated by Regional Secretariats. There are today 54 partner countries across eight regional settings that participate in the CoE network. International Organisations such as the WHO and the OPCW collaborate with the initiative and advise on programme directions and priorities. The CoE methodology includes voluntary national needs assessments in the CBRN field, and national CBRN action plans. These national action plans and the discussions in the CoE network provide a framework for the planning of regional and trans-regional projects. Whilst the initiative's project portfolio covers the entire CBRN range, taking an all-hazards approach, many of the projects focus on chemical safety and security. Increasingly, the direction of these projects has been shifting from general awareness raising and promoting the concepts of CBRN safety and security, to practical training, exercises and the provision of specialised equipment. Thirty-seven out of sixty-one CoE projects include and specifically address chemical safety/security issues, 23 involve training, and 7 the delivery of equipment [73].

6   Final Thoughts

Chemicals are an essential aspect of modern life. But they can do harm when released into the environment, or involved in accidents, or misused for self-harm or hostile purposes. It is important therefore that governments, industries that handle or manufacture chemicals, and the users of chemical products take measures to ensure the safety of populations and the environment. Enhancing chemical safety and security involves many stakeholders with at times divers interests and perceptions; it builds on a range of laws and regulations, and requires an array of practical measures.

But it also requires a genuine dialogue between the world of technical (chemical, engineering, legal, safety, security) experts and the rest of society. Public perceptions of chemical risks, and of what should be done to minimize them, are often very different from the assessments and recommendations put forward by experts. Such mismatches of perception can undermine the political support for measures necessary to ensure chemical safety and security, but they can also obstruct the use of chemistry for beneficial purposes. It is therefore important to maintain a broad-based conversation about chemical safety and security involving all stakeholders, to share information about benefits and risks associated with chemicals and chemical technologies, and to improve education in the field of chemical safety and security.

References

1. United Nations Security Council Resolution 2118 (2013), http://www.un.org/en/ga/search/view_doc.asp?symbol= S/RES/2118(2013)
2. OPCW Executive Council Decision EC-M/33/Dec.1 (2013), https://www.opcw.org/fileadmin/OPCW/EC/M-33/ecm33dec01_e_.pdf
3. OPCW UN Joint Investigation Mechanism Third Report S/2016/738 (2016), http://www.un.org/ga/search/view_doc.asp?symbol=S/2016/738
4. ICRC Customary International Humanitarian Law Database., https://ihl-databases.icrc.org/customary-ihl/eng/docs/home
5. Huber, D.: ARMOR (November-December 2001), pp. 14–17 (2001)
6. Human Rights Watch (2000), https://www.hrw.org/report/2000/02/01/backgrounder-russian-fuel-air-explosives-vacuum-bombs
7. Defence Update., http://www.defense-update.com/news/6702carpet.htm
8. Convention on Prohibitions or Restrictions on the Use of Certain Conventional Weapons which may be Deemed to be Excessively Injurious or to Have Indiscriminate Effects, as amended on 21 December 2001, and Protocols, https://www.un.org/disarmament/geneva/ccw/
9. Brehm, M., Borrie, J.: "Explosive weapons – framing the problem" Background Paper No. 1 of the Discourse on Explosive Weapons (DEW) UNIDIR Project (2010), http://www.unidir.org/files/publications/pdfs/explosive-weapons-framing-the-problem-354.pdf
10. Borrie, J, Brehm, M,: Int. Rev. Red Cross vol. 93, no. 883, 1–28 (2011)
11. Smith, R.: The Utility of Force: The Art of War in the Modern World, Allen Lane, London (2006), as cited by Borrie and Brehm, op. cit. [10]
12. United Nations Security Council, Report of the Secretary-General on the Protection of Civilians in Armed Conflict, Document S/2009/277 (2009), http://www.un.org/ga/search/view_doc.asp?symbol=S/2009/277
13. Convention on the Prohibition of the Use, Stockpiling, Production and Transfer of Anti-Personnel Mines and on their Destruction, https://www.un.org/disarmament/geneva/aplc/
14. Convention on Cluster Munitions., https://www.un.org/disarmament/geneva/ccm/
15. Convention on the Prohibition of the Development, Production, Stockpiling and Use of Chemical Weapons and on their Destruction, https://www.opcw.org/chemical-weapons-convention/
16. Bertazzi, P.A., Consonni, D., Bachetti, S., Rubagotti, M., Baccarelli, A., Zocchetti, C., Pesatori, A.C.: Amer. J. Epidemiol. vol. 153 no. 11, 1031–1044 (2001)

17. Environmental Health: a global access science source. 8 (1): 39; http://ehjournal.biomedcentral.com/articles/10.1186/1476-069X-8-39
18. OPCW., https://www.opcw.org
19. Lewer, N, Davison, N: Disarmament forum no. 1, 37–51, (2005) https://www.peacepalacelibrary.nl/ebooks/files/UNIDIR_pdf-art2217.pdf
20. Weapons Law Encyclopaedia., http://www.weaponslaw.org/weapons/incendiary-weapons
21. ICRC: Customary International Humanitarian Law, Rule 85., https://ihl-databases.icrc.org/customary-ihl/eng/docs/v1_rul_rule85
22. UN General Assembly Resolution A/Res/70/1 (2015), http://www.un.org/ga/search/view_doc.asp?symbol=A/RES/70/1&Lang=E
23. ILO: Safety and Health in the use of chemicals at work (2013), http://www.ilo.org/wcmsp5/groups/public/@ed_protect/@protrav/@safework/documents/publication/wcms_235085.pdf
24. ICSS., http://www.iccss.eu/media/timeline-for-chemical-incidents/
25. Orr, M.F., Wu, J., Sloop, S.L.: Surveillance Summaries, vol. 64, no. 2, 1–9 (2015)
26. Ruckart, P.Z., Orr, M.F.: Surveillance Summaries, vol. 64, no. 2, 10–17 (2015)
27. Duncan, M.A., Wu, J., Neu, M.C., Orr, M.F.: Surveillance Summaries, vol. 64, no. 2, 18–24 (2015)
28. Anderson, A.R.: Surveillance Summaries, vol. 64, no. 2, 39–46 (2015)
29. Consonni, D., Pesatori, A.C., Zocchetti, C., Sindaco, R., D'Oro, L.C., Rubagotti, M., Bertazzi, P.A.: Amer. J. Epidemiol., vol. 167 no. 7, 847–858 (2008)
30. European Union Directive 82/501/EEC., http://eur-lex.europa.eu/legal-content/EN/TXT/PDF/?uri=CELEX:31982L0501&from=EN
31. European Union Directive 2012/18/EU., http://eur-lex.europa.eu/legal-content/EN/TXT/PDF/?uri=CELEX:32012L0018&from=EN
32. Broughton, E.: Environ Health, 4:6 (2005), http://ehjournal.biomedcentral.com/articles/10.1186/1476-069X-4-6
33. European Commission Joint Communication to the European Parliament and Council "Joint framework on countering hybrid treats – a European response" JOIN(2017)18 final, p. 2
34. The Guardian (21 December 2016), https://www.theguardian.com/world/2016/dec/21/mexico-firework-market-explosion-death-toll
35. The Telegraph (12 August 2015), http://www.telegraph.co.uk/news/worldnews/asia/china/11799581/Huge-explosions-rock-Chinese-city.html
36. Lundin, J. (Ed.): Verification of dual-use chemicals under the Chemical Weapons Convention: the case of Thiodiglycol, SIPRI CBW Studies Series no. 13, OUP Oxford (1991)
37. US Senate Armed Services Committee (2016), https://www.dni.gov/index.php/newsroom/testimonies/217-congressional-testimonies-2016/1313-statement-for-the-record-worldwide-threat-assessment-of-the-u-s-ic-before-the-senate-armed-services-committee-2016
38. Gerstein, D.M.: Bull. Atomic Scientists (2016), http://thebulletin.org/how-genetic-editing-became-national-security-threat9362
39. Spiez Laboratory and CSS ETH Zurich (2016), http://www.css.ethz.ch/content/dam/ethz/special-interest/gess/cis/center-for-securities-studies/pdfs/SpiezConvergenceReport-2016.pdf
40. WHO: The public health impact of chemicals: knows and unknowns (2016), http://apps.who.int/iris/bitstream/10665/206553/1/WHO_FWC_PHE_EPE_16.01_eng.pdf
41. National Fire Protection Association., http://www.nfpa.org/news-and-research/fire-statistics-and-reports/fire-statistics/fire-causes/chemical-and-gases/fires-starting-with-flammable-gas-or-flammable-or-combustible-liquid
42. National Fire Protection Association., http://www.nfpa.org/news-and-research/fire-statistics-and-reports/fire-statistics/fire-causes/chemical-and-gases/spontaneous-combustion-or-chemical-reaction
43. National Fire Protection Association., http://www.nfpa.org/news-and-research/fire-statistics-and-reports/fire-statistics/fire-causes/fireworks
44. Ministry of Ecology, Sustainable Development and Housing of France (2012), http://www.aria.developpement-durable.gouv.fr/wp-content/files_mf/Fireworksaccidentanalysis_jan2012.pdf

45. Poison Control National Capital Poison Center (USA)., http://www.poison.org/poison-statistics-national
46. Defence Update (2006), https://web.archive.org/web/20110607231414/http://www.defense-update.com/commentary/24-1-06.htm
47. US White House Press Office (2013), https://www.whitehouse.gov/the-press-office/2013/08/30/government-assessment-syrian-government-s-use-chemical-weapons-august-21
48. The New Arab: Where do Islamic State and al-Qaeda get their explosives? (3 March 2016), https://www.alaraby.co.uk/english/indepth/2016/3/23/where-do-islamic-state-and-al-qaeda-get-their-explosives
49. Europol Strategic Analysis Reports., https://www.europol.europa.eu/content/page/strategic_analysis_reports
50. UNODC World Drug Report (2016), http://www.unodc.org/doc/wdr2016/WORLD_DRUG_REPORT_2016_web.pdf
51. Wheelis, M.: Nonconsensual manipulation of human physiology using biochemicals, in: Pearson, A.M., Chevrier, M.I., Wheelis, M: Incapacitating Chemical Weapons, pp. 1–17, Lexington Books Lanham MD (2007)
52. SAICM., http://www.saicm.org/
53. UNEP: SAICM texts and resolutions of the International Conference on Chemicals Management (2006), http://www.saicm.org/images/saicm_documents/saicm%20texts/SAICM_publication_ENG.pdf
54. Report of the International Conference on Chemicals Management, Fourth Session, SAICM/ICCM.4/15(2015), http://www.saicm.org/images/saicm_documents/iccm/ICCM4/Re-issued_mtg_report/K1606013_e.pdf
55. Regulation (EC) 1907/2006 (2006), http://eur-lex.europa.eu/legal-content/EN/TXT/PDF/?uri=CELEX:32006R1907&from=EN
56. UN Security Council 1540 Committee., http://www.un.org/en/sc/1540/
57. UN Security Council Resolution 2325 (2016), http://www.un.org/ga/search/view_doc.asp?symbol=S/RES/2325(2016)
58. OPCW Conference of the States Parties Decision C-8/DEC.16 (24 October 2003); https://www.opcw.org/fileadmin/OPCW/CSP/C-8/en/c8dec16_EN.pdf
59. OPCW Report of the Third Review Conference RC-3/3 (19 April 2013), https://www.opcw.org/fileadmin/OPCW/CSP/RC-3/en/rc303__e_.pdf
60. UN CTITF Working Group on Preventing and Responding to WMD Terrorist Attacks (2011), https://www.un.org/counterterrorism/ctitf/en/preventing-and-responding-wmd-terrorist-attacks
61. UN Economic Commission for Europe, ADR 2015, Chapter 1.10, https://www.unece.org/fileadmin/DAM/trans/danger/publi/adr/adr2015/ADR2015e_WEB.pdf
62. UN Office on Drugs and Crime., http://www.un.org/en/counterterrorism/legal-instruments.shtml
63. Meier, O., Trapp, R.: Bull. Atomic Scientist (7 June 2016). http://thebulletin.org/russia%E2%80%99s-chemical-terrorism-proposal-red-herring-or-useful-tool9531
64. McLeish, C., Lak, M.: The Role of Civil Society and Industrial Non-state Actors in Relation to the Chemical Weapons Convention, in: Krutzsch, W., Myjer, E., Trapp, R.: The Chemical Weapons Convention – A Commentary, pp. 37–48, OUP Oxford (2014)
65. ICCA Responsible Care ®., https://www.icca-chem.org/responsible-care/
66. Responsible Care Management System technical Specifications (2013), https://responsiblecare.americanchemistry.com/Responsible-Care-Program-Elements/Management-System-and-Certification/RCMS-Technical-Specifications.pdf
67. Responsible Care ® (American Chemistry Council)., https://responsiblecare.americanchemistry.com/Management-System-and-Certification/
68. CEFIC: Responsible Care Security Code Guidance and Best Practice for the Implementation of the Code (2011), http://www.rcsk.sk/mix/Responsible%20Care%20Security%20Code%20-%20Guidance.pdf
69. CEFIC: Industry guidelines for the security of the transport of dangerous goods by road (2015), http://www.cefic.org/Documents/IndustrySupport/RC%20tools%20for%20SMEs/

Document%20Tool%20Box/Security%20Guidelines%20of%20the%20transport%20of%20dangerous%20goods.pdf

70. IOMC/WHO., http://www.who.int/iomc/en/
71. OECD: 25 Years of chemical accident prevention – history and outlook (2013), https://www.oecd.org/chemicalsafety/chemical-accidents/Chemical-Accidents-25years.pdf
72. United States State Department, Chemical Security Programme, http://www.csp-state.net/
73. Trapp, R.: *The EU's CBRN Centres of Excellence Initiative after Six Years,* EU Non-proliferation Papers No. 55 (2017)

Security, Development, and Governance CBRN and Cyber in Africa

Adriaan van der Meer and Alberto Aspidi

Abstract Nowadays threats are diffuse and hybrid with a pertinent "role" of non-state actors. Criminal networks are capable of quickly exploiting vulnerabilities in the security environment.

The global security environment remains extremely dynamic. New security challenges and threats are increasingly blended and integrated. Many of the governmental structures in the countries with whom the EU is working in the field of external relations will struggle to cope with these challenges.

All these developments require the EU to revisit its response options. This paper proposes a number of steps to simplify the overall architecture of the EU's external financing assistance. It makes a number of recommendations on the focus of cooperation. Based on the experience under the EU CBRN risk mitigation Centres of Excellence initiative and EU policy to promote cybersecurity, it recommends that greater emphasis be placed on addressing governance issues in order to deal more effectively with the new security challenges.

Keywords Security • Governance • Development • CBRN • Cybersecurity

1 Introduction

Our experience of working on issues related to Security, Development and Governance has given us the opportunity to look back and reflect on what has worked, what has not, and what can be done to deliver more efficient and sustainable results in fighting multifaceted global threats. This reflection is pertinent in light of the recent scourge of radicalisation and terrorism in Europe.

A. van der Meer (✉) • A. Aspidi
Directorate-General for International Cooperation and Development of the European Commission, Brussels, Belgium
e-mail: vandermeer09@gmail.com; alberto.aspidi@gmail.com

© Springer International Publishing AG 2017   385
M. Martellini, A. Malizia (eds.), *Cyber and Chemical, Biological, Radiological,*
Nuclear, Explosives Challenges, Terrorism, Security, and Computation,
DOI 10.1007/978-3-319-62108-1_19

2   A Dynamic Security Environment

Today's threats are broad and shifting. They are growing in number and complexity. Interesting indicators in this respect are the Nuclear Security Index of the Nuclear Threat Initiative[1] and the cases of missing nuclear materials as reported to the IAEA under the Incident and Trafficking Database. Since 1995, over 2700 incidents of illicit trafficking and other related unauthorized activities involving nuclear and other radioactive materials have been confirmed, among which 46 involve criminal possession of HEU or plutonium.[2] The increased use of radioactive sources in industrial and medical facilities is leading to security concerns. So-called orphan sources could be used for the preparation of dirty bombs.[3]

Challenges in the field of biosafety and chemical security remain. The recent attack in Kuala Lumpur with VX agent[4] and the use of chemical weapons in Iraq and Syria by Daesh (IS)[5] shows the weaknesses in international control with respect to WMD. Bill Gates warned in February 2017 that rapid advances in genetic engineering have opened the door for small-scale terrorism. A single bioterrorist could kill hundreds of millions of people. He predicted that: "A highly lethal global pandemic will occur in our lifetimes".[6]

The Global Terrorism Index of 2015 indicates that since the beginning of the twenty-first century, there has been more than a ninefold increase in the number of deaths from terrorism, rising from 3329 in 2000 to 32,685 in 2014.[7]

Criminal networks are mobile in terms of the geography and nature of the criminal activities. Their business model is flexible and changes from commodity to commodity. It ranges from illicit trafficking of human beings and drugs, through illegal selling of cultural goods to illicit trade in wildlife.[8] Trafficking in falsified medicines

---

[1] The NTI Nuclear Security Index is a first-of-its-kind public benchmarking project of nuclear materials security conditions on a country-by-country basis. The 2016 NTI index shows that progress to secure weapons-usable nuclear materials has slowed down in comparison to previous years.

[2] The number of cases is in reality probably higher because of the number of unreported cases.

[3] A dirty bomb or radiological dispersal device is a radiological weapon that combines radioactive material with conventional explosives.

[4] See Decision of OPCW Executive Council of 10 March 2017, https://www.opcw.org/news/article/opcw-executive-council-condemns-chemical-weapons-use-in-fatal-incident-in-malaysia/

[5] IHS Conflict Monitor, London 2017.

[6] Speech held at the 53rd Munich Security Conference, 18 February 2017.

[7] The Global Terrorism Index, Institute for Economics and Peace, 2015. Terrorism in the year of reporting remains highly concentrated in Iraq, Nigeria, Afghanistan, Pakistan and Syria. Unofficial information indicates that in 2015 on the African continent 4000 attacks took place killing over 30,000 persons. See also the Global Peace Index 2016 of the same institute. Figures on the economic impact of various terrorist attacks vary considerably depending on the methodology used and time period covered. The OECD calculated that the direct costs related to the 9/11 attacks are above $27 billion. Other sources put the total cost much higher than the OECD estimate.

[8] The total value of the illicit trade in wildlife is estimated to be around 20 billion euro per year.

is a multi-billion illegal business.[9] In the EU alone, more than 5000 international organised crime organisation are currently under investigation, involving more than 180 different nationalities. This is in increase in comparison to 2013: when around 3600 groups were active in the EU.[10]

In security terms, terrorism, hybrid threats and organised crime know no borders.[11] Threats inside the EU often originate outside the EU. For example, this is recognized in the EU Policy towards the Western Balkans. The EU has therefore agreed to strengthen cooperation with the countries of the Western Balkans region to fight against security threats.[12] There are indications that the scale of global cybercrime business is now greater than that of drug trafficking.[13] There is a continued interest in dangerous materials from terrorist groups such as Daesh (IS) which certainly has the capacity to make chemical weapons. The Molenbeek (IS) network showed interest in obtaining nuclear materials. The growing capabilities of terrorist groups with respect to CBRN materials present a serious and growing threat.[14] In this respect, the human factor is an underestimated element in the security discussion: the world relies on scientists to discover and develop new inventions to bring solutions to everyday societal challenges. As research communities increase their knowledge on a day-to-day basis, they become increasingly capable of carrying out cutting-edge research in laboratories worldwide. CBRN know-how is widespread and access to such information is easily obtained. These developments are positive but also bring about new risks. Technical know-how, high-risk materials and technologies can be used for different purposes than originally intended. The G8, in a set of recommendations in July 2009, drew attention to the spread of sensitive know-how worldwide, highlighting the importance of "engaging scientists and raising awareness and responsibility among them to divert their knowledge in legitimate scientific disciplines to unintended malicious purposes".[15]

---

[9] According to some sources, in some West African countries 60% of anti-malarial medicines available on the market are ineffective fakes. According to Al Jazeera (19 March 2017) is nearly 1 out of every 3 drugs in Africa illicit or counterfeited. Ivory Coast destroyed 50 tonnes of fake medicine worth more than one million USD.

[10] EUROPOL Report, European Union Serious and Organised Crime Threat Assessment Crime in the age of technology, March 2017.

[11] A Global strategy for the EU's Foreign and Security Policy, p. 53, June 2016.

[12] Council of the European Union, Council Conclusions on strengthening the EU internal security's external dimension in the Western Balkans including via the Integrative Internal Security Governance (IISG), 9 December 2016.

[13] The retail value of the European drugs market is estimated to be at least two billion euro per month. A recent report (Rand Corporation, Brussels, July 2016) on internet-facilitated drugs trade (cryptomarkets) estimates this market to have a monthly revenue of 12–22 million euro. See also The Globalization of Crime, A Transnational Organised Crime Threat Assessment, UNODC, 2010.

[14] Opening Statement, NATO Deputy General Ambassador Alexander Vershow at the Annual NATO Conference on WMD Arms Control, Disarmament and Non-proliferation, Ljubljana, Slovenia, 9 May 2016.

[15] G8-Global Partnership Working Group (GPWG), recommendations for a coordinated approach in the field of global weapons of mass destruction, knowledge proliferation, and scientists' engagement, 2009.

Newly emerging technologies give birth to new types of threats.[16] For example, from a non-proliferation point of view, new high-technologies such as cloud computing, robotics, autonomous systems (industry 4.0), Unmanned Aerial Vehicles (UAV), additive manufacturing, bio, chemical, and nano technologies, as well as intangible transfers[17] could be used for malicious purposes. More research institutes and private companies are involved in developing such technologies. In addition, the dividing line in each of these technologies between civilian and military purposes is becoming increasingly blurred. Their peaceful use must be ensured.[18] Illicit procurement networks enable proliferation.[19] Massive disinformation campaigns, often combined with misuse of social media, are adding to our security challenges. Interestingly it is in this context that in 2015, the Russian Defence Minister Sergei Shoigu qualified the media as a weapon.[20]

"Hybrid threats" are the new order of the day.[21] While not an entirely new concept, hybrid threats have taken on a far greater profile as they have become increasingly effective in the twenty-first century, making full use of the advances of globalisation and technology. These threats affect our partner countries and a response is therefore needed in order to strengthen the nexus between security and development. In Africa governmental structures are often confronted with cyberattacks, infiltration, and economic pressure. In parts of Eastern Europe governments are trying to cope with misinformation campaigns that distort perceptions. At the same time they are subject to economic pressure, to non-attributable cyberattacks, to infiltration and deliberate attempts to corrupt authorities, let alone to so-called "peacekeeping operations" or "humanitarian interventions".

It is becoming increasingly clear that the various security threats to our societies which manifest themselves in different forms – illicit trafficking of human beings,

---

[16] The Global Risks Report, 2017, 12th Edition, pp. 42–47, World Economic Forum, Davos, Switzerland.

[17] Intangible transfers refer, inter alia, to the transmission of software, DNA sequences and technology by electronic media, new cyber-tools (cyber-surveillance technologies), and new forms of financial transactions such as Bitcoin.

[18] Commission Staff Working Document of 28 September 2016, Impact Assessment, Report on the EU Export Control Policy Review, SWD (2016), page 3.

[19] In order to provide a concrete framework for collective EU commitment to the fight against proliferation, the European Union adopted the so-called EU New Lines for Action. The document, as endorsed by EU Council Conclusions in 2008 as well as in 2010 and 2013, includes actions such as the review and strengthening of export controls on dual-use items. Concerns are expressed and an increase of vigilance is recommended with respect to "…protecting the access to proliferation-sensitive knowledge" and "…further strengthening protection of our scientific and technical assets against unintended transfers of sensitive technology and know-how, including dual-use items"; http://www.consilium.europa.eu/uedocs/cms_data/docs/pressdata/EN/foraff/139067.pdf

[20] Interfax, May 2015.

[21] The joint Framework on Countering Hybrid Threats, 6 April 2016, JOIN (2016)18 final defines hybrid threats as a mixture of coercive and subversive activity, conventional and unconventional methods (i.e., diplomatic, military, economic, technological) which can be used in a coordinated manner by state and non-state actors to achieve specific objectives while remaining below the threshold of formally declared warfare.

drugs, arms, cybercrime, misuse of hazardous CBRN materials, wildlife crime, as well as terrorism, and attacks to critical infrastructures – are becoming more and more intertwined.

The presence of transnational organised crime can be "a symptom of state weakness but once it has created its own dynamics this intensifies the causes of state failure and organised crime can be then best conceptualised as a proximate cause".[22] Even more worryingly, the connection between transnational organised crime and terrorism is growing.

For example, with regard to the terrorist attacks in Paris, Brussels and London, the linkages between criminal activities and terrorism have come closer. The perpetrators of such attacks use experience gained from their criminal activities and their criminal networks in order to obtain arms and falsified identity papers and create safe houses as they "graduate" to the next step in the planning and implementation of terrorist attacks. We now speak of gangster-jihadists. Equally, the attacker at a Paris airport and at the British Houses of Parliament had a strong criminal record which illustrates again that there is a crime- terror nexus.[23] Another example is the ever closer connection between cybersecurity and nuclear safety. The operation of nuclear power plants will be at risk if insufficient action is taken by governments and operators to protect the installations from cyberattacks. This is a real concern. But, from a substantive point of view, is this a cybersecurity issue i.e., protection of critical infrastructure, or a nuclear safety one, or both?

Other examples of interlinkages are those between the use of CBRN materials and the fight against terrorism or measures to prevent radicalisation in the cyberspace (fight against cyber-jihadism).

In conclusion, the threats are borderless, fluid and of important magnitude. There is an increasing blurred delineation between different types of security threats. Moreover, our opponents are as mobile and flexible as multi-nationals shifting activities from country to country or from region to region.

This begs a number of questions. Should work on addressing global security threats be done in a more synergised manner, rather than a stove-pipe approach? Should a different approach be adopted? Does the international community works with a too limited threat perception, and do the sub-categories lead to limited results? But how can we achieve this?

---

[22] West, Jessica, The Political Economy of Organised Crime and State Failure: The Nexus of Greed, Need and Grievance, Norman Paterson School of International Affairs, Carleton University, Ottawa 2006, page 11.

[23] Attacks on 18 March 2017 and on 22 March 2017, See: Criminal Pasts, Terrorist Futures: European Jihadists and the New Crime - Terror Nexus, The International Centre for the Study of Radicalisation and Practical Violence (ICSR) Authors: Raman Basra, Peter Neumann, Claudia Brunner, London, UK, 2016.

3   EU Tools

The EU has various tools at its disposal to deal with the root causes of and the fight
against external threats. They are notably defined under the EU's Comprehensive
Approach to external conflict and crises.[24] Programmes and projects are being
developed under the EU's financial instruments to promote peace and security especially under the Instrument contributing to Stability and Peace, the Instrument for
Nuclear Safety Cooperation, the CSDP budget lines and other financial
instruments.[25]

The EU reckons that 30% of development assistance is spent on peace, security
and governance related actions.[26]

The renewed Neighbourhood policy now places more emphasis on security
related work. The new focus on security opens up a wide range of new areas of
cooperation under the European Neighbourhood Policy. Cooperation could include
security sector reform, border protection, tackling terrorism and radicalisation, and
crisis management.[27] The same applies to the European Union Emergency Trust
Fund for Africa.[28] It supports improvements in overall governance, in particular by
promoting conflict prevention and enforcing the rule of law through capacity-
building in support of security and development as well as law enforcement, including border management and migration related aspects. Actions could also contribute
to preventing and countering radicalisation and extremism. The proposed renewed
partnership with the countries of Africa, the Caribbean and the Pacific is explicitly
aimed at strengthening the joint commitment to combat the proliferation of WMD,
including control of dual-use items, and to limit the uncontrolled spread of small
arms and light weapons. In addition, it identifies, as a specific objective for Africa,
the fight against organised and transnational crime, terrorism and radicalisation, and
illicit trafficking in human beings, wildlife, drugs and hazardous materials and
related illicit financial flows.[29]

In the fight against proliferation of Weapons of Mass Destruction, the EU actively
participates in export control regimes, strengthens common rules governing Member

---

[24] Joint communication from the European Commission and the High Representative of the
European Union for Foreign Affairs and Security Policy, JOIN (2013) 30 final, Brussels 11
December 2013.

[25] For a comprehensive overview of EU policies and initiatives in the world, see: The EU and the
World: Player and Policies, post-Lisbon, A Handbook, Edited by Antonio Missiroli, EU Institute
for Security Studies, 2016, Paris, France.

[26] Speech Commissioner Mimica at the event of the Overseas Development Institute, London,
Europe in the World: Promoting Peace and Security, 26 January 2016.

[27] Joint Communication by the European Commission and the High Representative of the Union
for Foreign Affairs and Security Policy, Brussels, 18 November 2015, JOIN (2015) 50 final, page
12–15.

[28] A European agenda on migration, 2015, Valletta summit on migration.

[29] Joint Communication by the European Commission and the High Representative of the Union
for Foreign Affairs and Security Policy, Strasbourg, 22 November 2016, JOIN (2016) 52 final,
page 15.

States' export policies of military – including dual-use – equipment and technologies, and supports export control authorities in third countries and technical bodies that sustain arms control regimes.[30] The European Commission has proposed a recast of the existing EU regulation on dual-use export controls as proposed on 28 September 2016. The purpose of the recast is to strike a balance between ensuring a high level of security and adequate transparency, and maintaining the competitiveness of European companies and legitimate trade in dual-use items.[31] The new piece of legislation will be part of the EU outreach programme to partner countries on export control in dual-use items.

Legal provisions providing for a stronger EU as actor in external relations are included in the Lisbon Treaty. The Treaty explicitly mentions, for the first time, that one of the objectives of the EU's external action is: "to preserve peace, prevent conflicts and strengthen international security [...]".[32] This is an overarching objective that shall be pursued by all external policies, instruments and tools. This would also relate to the EU's activities in multilateral frameworks. However, progress is slow, for example, in the framework of the reviews under the various international conventions related to the fight against Weapons of Mass Destruction (nuclear, chemical, and biological) or related to UNSCR 1540 on the prevention of proliferation. During a recent review the so-called 1540 Committee "took note of the increasing risks of proliferation in relation to non-State actors arising from developments in terrorism and in relation to the potential for misuse arising from the rapid advances in science, technology and international commerce".[33] Subsequently, the UNSC in a resolution of December 2016 re-emphasised the importance of all states to implement fully and effectively.[34] It stressed the need for States to pay constant attention to these developments to ensure effective implementation of the 1540 resolution.[35]

---

[30] A Global Strategy for the EU's Foreign and Security Policy, June 2016, page 42.

[31] Proposal for a Regulation of the European Parliament and of the Council setting up a Union regime for the control of exports, transfer, brokering, technical assistance, and transit of dual-use items (recast) of 28.09.2016, COM (2016) 616 (final).

[32] Article 21(2) TEU.

[33] Final document on the 2016 comprehensive review, S/2016/1038 of the status of implementation of resolution 1540 (2004), New York, 9 December 2016.

[34] UNSCR 2325 (2016) of 15 December 2016.

[35] For a more personal account of the review, see message of the Chair: Two Years before the Mast, by H.E. Mr. Roman Marchesi, 1540 Compass, Winter 2015, issue 11, pages 3–10.

4 Chemical, Biological, Radiological, and Nuclear Risk Management in Africa: Emergence of All Hazards Approach

Chemical, Biological, Radiological and Nuclear (CBRN) risks – natural, accidental or criminal – represent a key threat to the security and health of human beings, the environment and critical infrastructures. Promoting a multi-hazards culture of safety and security in this area, from prevention to consequence management, is fundamental to development and stability. Disease surveillance, waste management, emergency planning, civil protection and cross-border trafficking of CBRN materials are areas of concern to the EU and its partner countries.

An innovative approach has been developed under the EU chemical, biological, radiological and nuclear (CBRN) Centres of Excellence initiative, with a budget of more than 200 million euro for the 2010–2020 period. The essence of the work is a voluntary demand driven (bottom up) approach that promotes ownership of the work in the partner countries. National Focal Points, local and international experts meet on a voluntary basis to identify and discuss needs and solutions. To that extent, regional Round Tables are organised twice a year to identify regional priorities, cross-border CBRN cooperation (including table top and real time field exercises) and follow-up of regional activities.

This initiative, driven and implemented by the European Commission (DG International Cooperation and Development-DEVCO), establishes regional platforms to tackle all aspects of chemical, biological, radiological and nuclear risks arising from natural disasters, accidental catastrophes and criminal behaviour, by involving all the key stakeholders at a very early stage, thereby fostering the development of expertise in the countries concerned.

Each partner country – of which there are 56 in total and 28 from the African continent – appoints its national focal point (NFP) to coordinate the work to be done in his country. National CBRN inter-ministerial teams, gathering representatives from all relevant agencies dealing with CBRN issues (diplomats, civil protection, police, first responders, judges, etc.) are established. They total around 1000 officials worldwide. The partner countries choose the national focal points and national teams themselves.

This methodology is well tested, but has taken longer than expected to set up structures in partner countries. This was due to political circumstances, the varying extent of high level national support, the heterogeneous nature of the partner countries, the amount of previous experience in the area of CBRN risk mitigation, and the extent to which the relevant structures (national teams for instance) had already been established.[36]

---

[36] European Court of Auditors, Special Report 2014/17, Can the EU's Centres of Excellence initiative contribute effectively to mitigating chemical, biological, radiological and nuclear risks from outside the EU?, page 14.

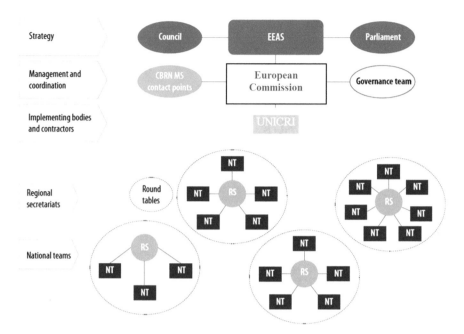

Fig. 1 EU risk mitigation centres of excellence structure

Partner countries define their needs on a voluntary basis with the help of a needs assessment toolkit (gap-analysis) and develop national action plans in which they prioritise their needs. As the Figs 2 and 3 show, 20 African countries are involved in the needs assessment process and 11 actions plans have been completed.[37]

In total 32 projects have been funded in Africa, covering a broad range of subjects mostly relating to biological, chemical and border management issues. In some cases awareness raising on CBRN matters was supported as well as the establishment of relevant regulatory frameworks. A flagship project was the training of staff and early mobilization of mobile laboratories in 2014 in Western Africa to fight against Ebola.

Partner countries cooperate through eight regional secretariats of which there are three in Africa, namely North and Sahel (Algiers), Atlantic Façade (Rabat) and East-Central (Nairobi). They ensure the regional aspect of the work and house a regional coordinator as well as the expert providing direct on-site assistance. This all-hazard CBRN security governance initiative is steered by the European Commission and the European External Action Service (EEAS). Main implementing partners are the United Nations Interregional Crime and Justice Research Institute (UNICRI) and the EU Joint Research Centre (JRC), European Commission's in-house scientific body. In addition, two facilities have been established: a governance team advising on

---

[37] A comparison shows that Senegal is the only African country that fully participates in both the EU's CBRN imitative and in executing UNSCR 1540.

	Planned NAQ (2017)	Algeria, Benin, Cameroon, Mauritania, Burkina Faso, Mali, Ghana
	Completed NAQ	Cote D'Ivoire, Gabon, Liberia, Senegal, Togo, Niger, Burundi, DRC, Kenya, Uganda
	Completed (Self)	Morocco, Tunisia, Seychelles
	No Information available	Sierra Leone, Libya, Ethiopia, Malawi, Rwanda, Zambia

Fig. 2 CBRN needs assessments questionnaires in African CoE partner countries (May 2017)

governance issues and a team of technical experts providing on-site assistance. Projects are mostly implemented by consortia pooling expertise available in the EU Member-States.

The need in Africa to address the full scope of CBRN threats and risks is substantial. However, the EU initiative is not implemented in the southern part of Africa due to resource limitations.[38]

---

[38] For an overview of CBRN assistance and capacity-building programmes for African states, see report of the Institute for Security Studies, Pretoria, South Africa, 2016.

	Planned NAP	Algeria, Benin, Cameroon, Mauritania, Burkina Faso, Mali, Ghana
	Completed NAP	Cote D'Ivoire, Gabon, Liberia, Senegal, Togo, Niger, Burundi, DRC, Kenya, Uganda, Seychelles
	No Information available	Sierra Leone, Libya, Ethiopia, Malawi, Rwanda, Zambia

Fig. 3 CBRN national action plans in African CoE partner countries (May 2017)

5 Cybersecurity in Africa[39]

The exponential spread of digitalisation in all facets of life has brought profound changes in societies. Impressively, two-thirds of Internet users live in the developing world where access to the Internet is growing almost four times faster than in developed countries. Countries in Africa are the new cyber frontier in terms of mobile internet penetration. But at the same time, the digital system is full of vulnerabilities. Cybercrime and cyberattacks are rampant and the benefits of ICT cannot be enjoyed without an open, safe, and secure underlying digital environment. In

---

[39] This section is based on an oral contribution by the EC to a high level meeting on cybersecurity held in Amsterdam on 13 May 2016 but it has been updated to take recent developments into account.

this context, any effort to promote digital technologies and services in Africa, and elsewhere, needs to be coupled with comprehensive measures tailored to strengthen the cybersecurity of partner countries and increase their capacity to protect their strategic assets (e.g. energy, telecommunications, financial, transportation, water) as well as their ability to effectively cooperate in addressing cybercrime and cyberattacks.

With this in mind, the EU has taken a number of initiatives to seek cooperation with various African countries.

In general, the EU's external support to foster cybersecurity focuses on three areas:

1. Developing or reforming appropriate regulatory and legal frameworks in compliance with international standards. This includes the promotion of the Budapest Convention on Cybercrime;
2. Strengthening the capacities of criminal justice authorities, such as law enforcement, prosecutors, and judges, in order to enable them to effectively investigate, prosecute, and adjudicate cases of cybercrime and other offences involving electronic evidence;
3. Supporting the development of organisational and technical mechanisms increasing cyber resilience and preparedness. These efforts are aimed at facilitating the development of national cybersecurity strategies, promoting effective inter-institutional, inter-agency and international cooperation as well as public-private exchanges, and the setting up of functional national Computer Emergency Response Teams.

Particular attention is required in ensuring that the implementation of such activities is done in a policy coherent manner, through the promotion of a multi-stakeholder model that also ensures compliance with human rights values and principles. Considering that the technological developments in the cyber domain are rapidly increasing, and with the law enforcement and judiaciry authorities trying to catch up, a key element of the EU's cyber capacity building efforts is the incorporation of the rights-based approach, including safeguards against potential flow-on risks, such as the misuse of the assistance provided for censorship, surveillance, unlawful interception. A manual has been developed to ensure that these principles are adhered to in every step of the project preparation and implementation – it is available on the EC's website and is called "Operational Human Rights Guidance for EU external cooperation actions addressing Terrorism, Organised Crime and Cybersecurity: Integrating the Rights-Based Approach (RBA)".[40] In 2013 the EU launched the "Global Action on Cybercrime (GLACY)" jointly with the Council of Europe. This was the first programme of its kind with a focus on providing to African and Asian partners comprehensive capacity building measures, encompassing institutional, legal and operational perspectives that also create the enabling environment for these countries to further serve as regional hubs of South-South

---

[40] https://ec.europa.eu/europeaid/operational-human-rights-guidance-eu-external-cooperationactions-addressing-terrorism-organised_en

cooperation and share experiences with their neighbours. Financed under the EU's Instrument contributing to Stability and Peace, GLACY was a project aiming to enable criminal justice authorities in priority countries to investigate, prosecute and adjudicate cases involving cybercrimes and electronic evidence and to effectively engage in international cooperation. With a budget of 3.35 million euro and a 3 year duration, GLACY provided institutional and capacity-building support to seven priority countries (Mauritius, Morocco, the Philippines, Senegal, South Africa, Sri Lanka, and Tonga) which were selected based on their political commitment to implement the common standards of the Budapest Convention, in that they had signed or requested accession to this Convention.[41]

It was decided to continue this programme as GLACY+ with a financial allocation of 10 million euro.[42] The implementing period for this programme is 2016–2020 and the most recent additions to priority countries have been Ghana and the Dominican Republic, while this programme serves also as a facility to help any country requesting support with developing or revising its cybercrime and electronic evidence legislation. In this work on fighting cybercrime the EU promotes a holistic approach with the creation of national coordination teams in priority countries. This approach involves authorities across the criminal justice chain, as well as partnership with national judicial and law enforcement academies.

Given the vast needs of developing countries in building up their cybersecurity and to effectively responding to cybercrime and cybersecurity challenges, in 2014 it was decided to make an additional 22 million euro available under the EU's Instrument contributing to Stability and Peace during the period 2014–2017 for actions in these two fields. Projects take into account the experience gained in similar activities in the Western Balkans and in Eastern Europe.

Complementary to the actions on cybercrime, are efforts that tackle the strengthening of organisational frameworks of cybersecurity, as well as the technical incident response capabilities of developing countries. A new programme on the broader cyber resilience framework is planned to start in 2017. It will focus on increasing the security and resilience of critical information infrastructure and networks supporting critical services of countries mainly in Africa, and some key countries in Asia. With a budget of 11 million euro, this action envisions to support the adoption and implementation of a comprehensive set of policy, organisational, and technical measures including the development of national cybersecurity strategies, the establishment of Computer Emergency Response Teams (CERTS), as well as the increased capacities for regional and international cooperation on cybersecurity.[43]

All these programmes are in line with the European Cybersecurity Strategy (2013)[44] and the political guidance provided by EU Member States in the Council

---

[41] Budapest Convention on Cybercrime, signed in Budapest on 23/11/2001, Council of Europe.

[42] Officially announced on 23 January 2017.

[43] Commission Implementing Decision of 27 July 2016 on the Annual Action Programme 2016 for Article 5 of the Instrument contributing to Stability and Peace.

[44] Joint Communication from the European Commission and the High Representative of the European Union for Foreign Affairs and Security Policy on Cybersecurity of the European Union, JOIN (2013) 1 final, 07.02.2013.

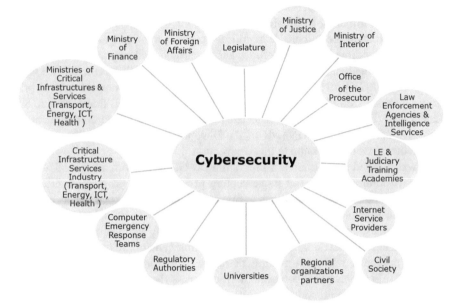

Fig. 4 Minimum number of governmental agencies needed to achieve an effective cybersecurity policy

Conclusions on Cyber Diplomacy (2015).[45] Moreover, they contribute to the implementation of the 2030 Agenda for Sustainable Development and the EU's policy to counter hybrid threats.[46]

The cyber needs of developing countries especially with regard to institutional capacities (law enforcement, judiciary, incident response agencies) are so high that effective and consistent cooperation in capacity building is a crucial factor in coordinating limited resources and avoiding fragmentation. Cyber capacity building is a complicated matter as can be seen from Fig. 4.

Therefore, the creation of hubs of local experts in different regions is needed to scale up cybersecurity capacity building programmes that require long-term expert commitment. In future, the established CBRN network could be used to exchange views on cybersecurity matters and to develop cyber policies at a regional level.

At the national level different agencies working on cyber issues need to cooperate more closely. The policy, technical, business/industry and civil society communities rarely cooperate. In order to overcome this situation, the facilitation of a functional multi-stakeholder and multi-dimensional engagement is fundamental.

---

[45] See Council Conclusions on Cyber Diplomacy adopted on 11 February 2015.

[46] Joint Communication from the European Commission and the High Representative of the European Union for Foreign Affairs and Security Policy on a Joint Framework on Countering Hybrid Threats, JOIN (2016) 18, 06.04.2016.

A policy can only be established successfully if a holistic approach is pursued. An effective and efficient national cyber policy involves numerous actors and multi-layer cooperation. Important administrative reforms often need to overcome vested interests. For these reasons, there is a need to promote working in a broad state context rather than operating in thematic and sectorial cooperation silos.

The quality of administration directly impacts on governments' ability to provide public services – including effective law enforcement, and to prevent and fight against cybercrime. It is important to engage first or at least at the same time, on broader administrative reforms in developing countries before assistance is given in other related areas in the fight against cybercrime such as the purchase of equipment. Governance issues need to be tackled upfront. Critical to this process will be the successful mainstreaming of cyber as a crosscutting issue across policies and practices both in developed and developing countries.

6 Responses

The frequently changing and shifting security challenges, including the interconnections between global threats and the blurring of division lines between them, necessitate a rethinking of the EU approach to its financial cooperation programmes. In this context, the EU's agenda on a Multi-Annual Financial Framework (2014–2020) (MFF) and beyond provides an opportunity to respond to the new security landscape. This Multi-Annual Framework lays down the maximum annual amounts which the EU may spend in different political fields over a certain period. Within the MFF, Heading 4 – Global Europe, covers all external action of the EU and is mainly composed of the External Financing Instruments which provide support to third countries. According to Article 17 of the Common Implementation Regulation, a mid-term review of the instruments covered by this Regulation has to be presented to the European Parliament and the Council by the end of 2017.[47] To assist the Commission in this task, independent evaluations have been carried out per instrument. In order to gain the views of the wider public, an open public consultation on the preliminary results of the evaluations was launched on 7 February 2017 with the deadline of 3 May 2017. The results of the evaluations and consultations will inform the Commission both for the mid-term review report and for the work on future instruments.[48]

---

[47] OJ L 77,15.03.2014, p. 95.

[48] https://ec.europa.eu/europeaid/public-consultation-external-financing-instruments-european-union_en

6.1   An Integrated Approach

An integrated approach should be followed i.e., to mainstream security aspects into development cooperation programmes as a way of addressing the root causes of poverty, fragility and vulnerability. The above-mentioned Renewed Partnership document with the countries of Africa, the Caribbean and the Pacific and the Renewed European Neighbourhood policy document recognise this approach.

6.2   Simplification

The overarching structure of the EU's external financing instruments is under a review, which could lead to could lead to changes . As Fig. 5 shows there is a multiple set of instruments that deal with security related issues. Some funding comes from inside the EU budget and some from the outside. For example, the financing of the African Peace Facility, which is a major source of funding for peace operations in Africa, and the costs relating to EU military peace operations are funded by mechanisms falling outside the EU budget.[49]

With respect to more specifically security related activities there are various options.

One option would be to fully integrate the financing of security related activities – including a quick response capacity in case of conflicts or as part of conflict prevention – within the various geographical financing instruments such as for the pre-accession countries, those falling under the EU's Neighbourhood Policy or for Africa. This would be in addition to bilateral security related actions and some regional activities, for example those related to border management and maritime security. Such an option would entail the creation of a thematic financing facility inside each of the respective geographical programmes. The advantage to this option is that this would probably lead to a more integrated approach per region. It could also lead to savings in human and financial resources. The disadvantage is that this could lead to fragmentation and hence an inadequate response to the various (increasing) global and hybrid threats as described. Moreover, expertise would be split up in regional silos, therefore running the risk that essential overall interconnections are not made in the fight against global security threats.

Another option relates to improved economies of scale by integrating a number of external financing instruments dealing with security issues on a global basis. For example, by integrating the Instrument contributing to Stability and Peace, the Instrument for Nuclear Safety Cooperation, and the African Peace Facility (despite being outside the EU budget) while at the same time updating their objectives.

---

[49] The African Peace Facility is funded by the European Development Fund and EU military operations under the EU's Common Security and Defence Policy (CSDP) through the so-called external Athena Mechanism by 27 Member-States with an opt-out for Denmark. Missions are taking place for example in the Horn of Africa, Mali, and Somalia.

Main EU Instruments for External Cooperation

Horizontal instruments with a specific legal basis

Geographical instruments with a specific legal basis

Horizontal instruments included in DCI legal basis

Fig. 5 Main instruments for EU external cooperation

Moreover as far as legally possible, current relevant activities under the Common Security and Defence Policy (CSDP) could be incorporated into this new overarching security instrument, including the objectives and tasks of the CSDP non-military missions. Exploring this option is recommended in view of the new EU Multi-Annual Financial Framework 2020–2027. A reflection group could be established to examine ways to reinforce the EU's internal and external dimension in dealing with security matters as mentioned in the European Agenda for Security.[50]

In this context, further attention could be paid to the updated and modernised Overseas Development Assistance (ODA) reporting directives on peace and security expenditures that clarify the eligibility of activities involving the military and police as well as activities related to the prevention of violent extremism. They also set outs the boundaries for development related training of military personnel.[51]

---

[50] COM (2015) 185 final, Strasbourg, 28.04.2015. It identifies, inter alia, three priorities, namely the fight against terrorism, the fight against organised crime and the fight against cybercrime.

[51] Revised Reporting directives on ODA in the field of Peace and Security as agreed on 18–19 February 2016 by the OECD Development Assistance Committee. A casebook on peace and security related activities in accordance with the updated ODA reporting directives, is currently being prepared.

Highly relevant for the discussion of these matters are the 2016 proposals related to (i) capacity building in the security sector, which covers security sector reform[52] and (ii) the legislative proposal to extend the Union's assistance under exceptional circumstances. The latter details the possibility to use military actors in partner countries to contribute to sustainable development and in the development of peaceful and inclusive societies.[53]

Clearly, there is not necessarily need for any new external financing instrument because this could lead to further complexity in an already complex situation. In addition, allocation of already limited human and financial resources to another instrument is not warranted under the present circumstances.

6.3   Intermediate Steps

A number of intermediate steps could be proposed, keeping in mind the Mid-Term Review 2014–2017. A pragmatic approach is suggested given that the majority of the instruments are in place, including the repartition of the budgets per instrument. Some changes in priorities could be made inside the various instruments. For instance, with regard to the Instrument contributing to Peace and Stability the number of priorities could be reduced in view of the fact that other instruments have refocussed to deal with security issues. The priorities for the period 2018–2020 could be aligned with those of the priorities of the European Agenda for Security, i.e., countering terrorism, promoting cybersecurity and CBRN Risk Mitigation.[54] In terms of delivery mechanisms, the eight Regional Secretariats under the CBRN initiative could be gradually transformed into regional security platforms. Other security related activities during the upcoming financing period could only be financed if there is a clear link with one of the three above-mentioned priorities.

A clearer division of work should be ensured between the so-called geographical financing instruments (such as DCI, EDF, ENP, Pre-accession Instrument, and including the recently established EU Emergency Trust Fund for Africa) and the thematic instruments (IcSP, INSC, EIDHR as well as the Partnership Instrument and the EU's Security Fund). This would result in stronger synergies and complementarity between the various external financing instruments as was identified by the President's Juncker priority 9, Europe as a Stronger Global Actor, which made

---

[52] Joint Communication by the European Commission and the High Representative of the Union for Foreign Affairs and Security Policy, Strasbourg, 5 July 2016, JOIN (2016) 31 final. It defines, inter alia, possible EU support for the security sector in partner countries.

[53] Proposal of the European Commission amending Regulation (EU) No. 230/2014 of the European Parliament and of the Council of 11 March 2014 establishing an instrument contributing to stability and peace, Strasbourg, 5 July 2016, CON(2016) 447 final.

[54] See also Professor Beatrice de Graaf, Terrorism as a Historical Phenomenon, NPO 1 broadcast), 12 March 2016.

the call: "to bring together the tools of the EU's external action in order to be more effective".[55]

7   Immediate Steps

A number of immediate steps could be foreseen to gain efficiency and impact. These could be introduced within the existing frameworks without the need for specific legal actions. A number of adjustments could be made to programming and project implementation under relevant EU external financing instruments, taking into account the innovative approaches under the CBRN programme and in promoting cyber security. Greater focus should be placed on establishing, building, and maintaining administrative structures in partner countries so that they are able to effectively prevent and respond to security threats and risks.

Under the current MFF (2014–2017) period, nine national and eight regional or thematic programmes are aimed at supporting conflict prevention and resolution, peace, and security-related activities. In addition, in 45 countries programmes are being developed with a broader focus on governance and the rule of law, including support for them to make the financing transition from missions and operations under CSDP to other instruments. This support has allowed various countries to embark upon reforms such as strengthening transparency and providing basic state services to their citizens including access to justice.

Other initiatives have a similar objective. For instance, the State Building Contracts under the EU's[56] budget support programme was conceived as part of the EU's contribution to improve governance in fragile states. Another example is the African Governance Architecture (AGA),[57] which was designed to assist countries to achieve democratic governance and enhance respect for human rights and the rule of law. The EU has also provided such support in the framework of the recently established Emergency Trust Fund for Africa.

External cooperation programmes might be reinforced by focusing more on governance issues, including administrative reforms and public administration reforms.

As an example, the importance of good governance was recently underlined in a study on the situation in Central Asia.[58] The study criticized western assistance for not focusing enough on institutional reform to post-Soviet states. It makes clear that as a result, reforms have been piecemeal and incomplete due to the fact that strong

---

[55] The Juncker priority 9, Europe as a Stronger Global Actor, A new start for Europe, Jean-Claude Juncker, Strasbourg, 15 July 2014.

[56] See for example state building contracts with Mali and South Sudan.

[57] This facility is the overall political and institutional framework for the promotion of democracy, governance, and human rights in Africa. It was established in 2011 in the Headquarters of the African Union in Addis Ababa, Ethiopia.

[58] Johan Engvall, the State as an investment Market: Kyrgyzstan in Comparative Perspective, Pittsburgh: University of Pittsburgh Press, 2016.

vested interests continue to oppose reform in many sectors and a Soviet-like mindset continues to permeate state institutions.

Of course, it is essential to take a comprehensive approach which includes building close links between humanitarian, stabilisation and development policies. The recent initiatives for capacity building in support of security and development including new initiatives in the area of Security Sector Reform – as announced in early July 2016 – will certainly build up the security capacity in partner countries. But, this is not good enough and focus should be shifted.

The work on the nexus between security and development needs to focus more on a broad state context putting emphasis on the delivery of effective and efficient public services in the area of security.

The quality of administration directly impacts on a government's ability to provide public services, including fair distribution of scarce public resources. The 2015 EU flagship project: "collect more – and spend better"[59] is hereby crucial, as is effective law enforcement, for example in the fight against and prevention of corruption and organised crime. There is a need to engage in broader administrative reforms before the provision of further assistance. A paradigm shift from "extractive" to "inclusive" institutions is a prerequisite for a well-functioning public administration that is resilient to complex security threats and that is able to properly utilise the targeted, technical assistance provided to address organised crime, illicit trafficking, and terrorism.

In order to deal effectively with security issues, governmental administrations need to take a multi-agency/multi-layer approach and to be highly coordinated, i.e., to ensure that all agencies are working well together. The Fig. 6 below shows the minimum administrative requirements to effectively execute a CBRN policy at national level.

The availability of a special EU funded Governance assistance team advising and training participants in interagency cooperation and collaboration under the CBRN programme as well as the principle: "administrative reforms first" that is applied in improving cybersecurity are good practices to be followed.

But, before acting, there should be a deep analysis of the actors, including senior management, and their vested interests, motivations, and interests in the administrative reform process. Such a mapping exercise, including professionalization of the public administration, could be carried out at the inception stage of each programme or project. During the implementation stage, a "culture" of trust and collaboration should be promoted allowing, for example, for the smooth exchange of information. In this respect much can be learnt from the methodology applied by the IAEA and relevant national bodies in promoting a nuclear safety, respectively nuclear security,

---

[59] The aim of this initiative is to improve fiscal outcomes and address related implementation challenges in order to implement the post 2015 development agenda, European Commission (2014), Commission Implementation Decision of 23.07.2014 adopting a Multiannual Indicative Programme for the Thematic Programme "Global Public Goods and Challenges" for the period 2014–2020, C (2014) 5072 final.

Fig. 6 Minimum number of governmental agencies needed to achieve an effective CBRN risk mitigation policy

culture.[60] The Science Centres in Ukraine and in Kazakhstan could assist in these matters given their past record on dealing with these issues.

The "more for more" principle and stronger conditionality from the EU side could be applied across the board. More binding contractual arrangements with the partner countries could be considered to achieve the necessary administrative reforms before the remainder of the assistance is delivered. Political will and leadership in institutional (administrative) reform should be rewarded, i.e., delivering credible track-records in the area of public administration reform with the ultimate aim of more effective and accountable governance delivering basic services. For the Western Balkans an Integrative Internal Security Governance concept has been developed, which is composed of Representatives of Agencies active in the region, representatives of the EU and of other international partners. The aim is to set up a number of coordinated activities in the fight against global threats. Such a concept could also be considered for other priority regions.[61] Equally, under the EU's cooperation programmes various governance facilities – drawn up together with other

---

[60] See for example the IAEA Implementing guide on Nuclear Security, IAEA Nuclear Security Studies, IAEA, 2008.

[61] Council of the European Union, Council Conclusions on strengthening the EU internal security's external dimension in the Western Balkans including via the Integrative Internal Security Governance (IISG), 9 December 2016, p. 6

partners – exist. Further use of such facilities in other parts of the world should be considered.[62]

Certainly, this requires long-term engagement and often the EU's efforts are limited due to time-sensitive programming cycles. Moreover, this is a resource intensive exercise.

Nevertheless, the following direct, hands-on support measures are suggested:

- embedding twining-type assistance and introducing the concept of peer reviews to all geographical areas;
- screening exercises should promote the role of various agencies in terms of competence, functioning, resources and mutual cooperation;
- providing long-term on-site mentoring and advice;
- creating special governance assistance teams that foster genuine and efficient multi-agency cooperation through trainings, table-top exercises, and handling of specific cases to build trust and establish regional networks of cooperation (security platforms);
- utilising new technologies in creating an enabling environment at all levels (e-government), including regulatory and governance frameworks, as is recognized in the Addis Ababa Action Agenda[63] as well as the promotion of merit based civil servant systems in various partner countries (civil service reforms).[64]

In general, the effectiveness of cooperation in the security area is hampered by high turn-over of staff in the partner countries' administrations. Bold steps to stop the turnover of trainees under the EU's external assistance programmes could be contemplated. Should more frequent use be made of the Philippines model, i.e., obliging those taking part in the programmes to sign a declaration that they will stay in the same post for at least 2–3 years??

8 Agenda 2030: UNSDG16

SDG 16 does reflect on the inter-linkages between the different threats mentioned above and also recognises that effective, accountable, resilient and inclusive institutions are essential for the creation and sustenance of secure and peaceful societies, as well as promoting the rule of law and eradicating poverty.[65] The achieved consensus on this point provides for the necessary framework to rethink EU actions, and

---

[62] See for example the Governance Facility for Uganda, created by eight of Uganda's International Development Partners among others the EU.

[63] Addis Ababa Action Agenda of the Third International Conference on Financing for Development, Addis Ababa, Ethiopia, 13.07.2015.

[64] On 30 May 2016, President Nazarbayev of Kazakhstan instructed the Government to introduce innovations and best international practices into the national civil service of Kazakhstan.

[65] Transforming our World: The 2030 Agenda for Sustainable Development, Resolution adopted by the United Nations General Assembly on 25 September, A/RES/70/1.

grasp the opportunity of incorporating the lessons from the past into future work. SDG 16.a. specifically mentions the need to strengthen relevant national institutions in developing countries to prevent violence and combat terrorism and crime. This objective alone is an appeal for the EU to increase its emphasis on administrative and security governance reforms in its external cooperation programmes.

Disclaimer The authors write in their personal capacity. The views set out in this article are those of the authors and they do not represent in any way the official point of view of the European Commission or any other EU institution. Neither the European Union institutions and bodies nor any person acting on their behalf may be held responsible for the use which may be made of the information contained therein. Any mistakes or omissions are solely those of the authors.

Printed in the United States
By Bookmasters